Roger Lee and Haeng-Kon Kim (Eds.)

Computer and Information Science

Studies in Computational Intelligence, Volume 131

Editor-in-Chief
Prof. Janusz Kacprzyk
Systems Research Institute
Polish Academy of Sciences
ul. Newelska 6
01-447 Warsaw
Poland
E-mail: kacprzyk@ibspan.waw.pl

Further volumes of this series can be found on our homepage: springer.com

Vol. 111. David Elmakias (Ed.)
New Computational Methods in Power System Reliability, 2008
ISBN 978-3-540-77810-3

Vol. 112. Edgar N. Sanchez, Alma Y. Alanís and Alexander G. Loukianov
Discrete-Time High Order Neural Control: Trained with Kalman Filtering, 2008
ISBN 978-3-540-78288-9

Vol. 113. Gemma Bel-Enguix, M. Dolores Jiménez-López and Carlos Martín-Vide (Eds.)
New Developments in Formal Languages and Applications, 2008
ISBN 978-3-540-78290-2

Vol. 114. Christian Blum, Maria José Blesa Aguilera, Andrea Roli and Michael Sampels (Eds.)
Hybrid Metaheuristics, 2008
ISBN 978-3-540-78294-0

Vol. 115. John Fulcher and Lakhmi C. Jain (Eds.)
Computational Intelligence: A Compendium, 2008
ISBN 978-3-540-78292-6

Vol. 116. Ying Liu, Aixin Sun, Han Tong Loh, Wen Feng Lu and Ee-Peng Lim (Eds.)
Advances of Computational Intelligence in Industrial Systems, 2008
ISBN 978-3-540-78296-4

Vol. 117. Da Ruan, Frank Hardeman and Klaas van der Meer (Eds.)
Intelligent Decision and Policy Making Support Systems, 2008
ISBN 978-3-540-78306-0

Vol. 118. Tsau Young Lin, Ying Xie, Anita Wasilewska and Churn-Jung Liau (Eds.)
Data Mining: Foundations and Practice, 2008
ISBN 978-3-540-78487-6

Vol. 119. Slawomir Wiak, Andrzej Krawczyk and Ivo Dolezel (Eds.)
Intelligent Computer Techniques in Applied Electromagnetics, 2008
ISBN 978-3-540-78489-0

Vol. 120. George A. Tsihrintzis and Lakhmi C. Jain (Eds.)
Multimedia Interactive Services in Intelligent Environments, 2008
ISBN 978-3-540-78491-3

Vol. 121. Nadia Nedjah, Leandro dos Santos Coelho and Luiza de Macedo Mourelle (Eds.)
Quantum Inspired Intelligent Systems, 2008
ISBN 978-3-540-78531-6

Vol. 122. Tomasz G. Smolinski, Mariofanna G. Milanova and Aboul-Ella Hassanien (Eds.)
Applications of Computational Intelligence in Biology, 2008
ISBN 978-3-540-78533-0

Vol. 123. Shuichi Iwata, Yukio Ohsawa, Shusaku Tsumoto, Ning Zhong, Yong Shi and Lorenzo Magnani (Eds.)
Communications and Discoveries from Multidisciplinary Data, 2008
ISBN 978-3-540-78732-7

Vol. 124. Ricardo Zavala Yoe
Modelling and Control of Dynamical Systems: Numerical Implementation in a Behavioral Framework, 2008
ISBN 978-3-540-78734-1

Vol. 125. Larry Bull, Bernadó-Mansilla Ester and John Holmes (Eds.)
Learning Classifier Systems in Data Mining, 2008
ISBN 978-3-540-78978-9

Vol. 126. Oleg Okun and Giorgio Valentini (Eds.)
Supervised and Unsupervised Ensemble Methods and their Applications, 2008
ISBN 978-3-540-78980-2

Vol. 127. Régie Gras, Einoshin Suzuki, Fabrice Guillet and Filippo Spagnolo (Eds.)
Statistical Implicative Analysis, 2008
ISBN 978-3-540-78982-6

Vol. 128. Fatos Xhafa and Ajith Abraham (Eds.)
Metaheuristics for Scheduling in Industrial and Manufacturing Applications, 2008
ISBN 978-3-540-78984-0

Vol. 129. Natalio Krasnogor, Giuseppe Nicosia, Mario Pavone and David Pelta (Eds.)
Nature Inspired Cooperative Strategies for Optimization (NICSO 2007), 2008
ISBN 978-3-540-78986-4

Vol. 130. Richi Nayak, Nikhil Ichalkaranje and Lakhmi C. Jain (Eds.)
Evolution of WEB in Artificial Intelligence Environments, 2008
ISBN 978-3-540-79140-9

Vol. 131. Roger Lee and Haeng-Kon Kim (Eds.)
Computer and Information Science, 2008
ISBN 978-3-540-79186-7

Roger Lee
Haeng-Kon Kim
(Eds.)

Computer and Information Science

With 132 Figures and 54 Tables

Professor Roger Lee
Computer Science Department
Central Michigan University
Pearce Hall 413
Mt. Pleasant, MI 48859
USA
lee1ry@cmich.edu

Professor Haeng-Kon Kim
Department of Computer Engineering
Catholic University of DaeGu
GyeonSan, GyeongBuk 712-702
Rep. of Korea
hangkon@cu.ac.kr

ISBN 978-3-642-09807-9 e-ISBN 978-3-540-79187-4

DOI 10.1007/978-3-540-79187-4

Studies in Computational Intelligence ISSN 1860-949X

Cover Design: Deblik, Berlin, Germany.

Printed on acid-free paper

9 8 7 6 5 4 3 2 1

springer.com

Preface

The purpose of the 7th IEEE/ACIS International Conference on Computer and Information Science (ICIS 2008) and the 2nd IEEE/ACIS International Workshop on e-Activity (IWEA 2008) to be held on May 14–16, 2008 in Portland, Oregon, U.S.A. is to bring together scientists, engineers, computer users, and students to share their experiences and exchange new ideas and research results about all aspects (theory, applications and tools) of computer and information science; and to discuss the practical challenges encountered along the way and the solutions adopted to solve them.

In January, 2008 one of editors of this book approached in house editor Dr. Thomas Ditzinger about preparing a volume containing extended and improved versions of some of the papers selected for presentation at the conference and workshop. Upon receiving Dr. Ditzinger's approval, conference organizers selected 23 outstanding papers from ICIS/IWEA 2008, all of which you will find in this volume of Springer's Studies in Computational Intelligence.

In chapter 1, Fabio Perez Marzullo et al. describe a model driven architecture (MDA) approach for assessing database performance. The authors present a profiling technique that offers a way to assess performance and identify flaws, while performing software construction activities.

In chapter 2, authors Huy Nguyen Anh Pham and Evangelos Triantaphyllou offer a new approach for testing classification algorithms, and present this approach through reanalysis of the Pima Indian diabetes dataset, one of the most well-known datasets used for this purpose. The new method put forth by the authors is dubbed the Homogeneity-Based Algorithm (HBA), and it aims to optimally control the overfitting and overgeneralization behaviors that have proved problematic for previous classification algorithms on this dataset.

In chapter 3, Marcus Figueredo et al. give us a new solution for the compression of electrocardiogram (ECG) transmissions in hopes of preventing the leakage of network bandwidth, which is caused by the generation of excessive data on the part of the ECG. The approach offered by the authors splits the ECG transmission signal into two parts, labeled "plain" blocks and "complex" blocks, and applies separate compression methods to each, resulting in an overall approximate compression ratio of 28:1.

In chapter 4, Akiyoshi Wakatani presents a modified version of the Pairwise Nearest Neighbor (PNN) algorithm, which he calls the "multi-step parallel" PNN algorithm. Wakatani seeks to reduce the potentially large communication costs associated with a parallel system that uses distributed memory by restructuring the PNN compression algorithm for use with these systems.

In chapter 5, Eun-Ju Park et al. suggest the use of an SIR (SPIC Integration Repository) prototyping system for the effective management and support of heterogeneous assets and tools of SPIC (Software Process Improvement Center in Korea) in order to acquire CMMI certification. The authors provide the main design and modeling issues for the organizing and managing of SIR for software process improvement. The use of SIR, they explain, will provide an API that is suitable for the manipulation of the SPIC's meta-assets, related tools and models in order to help small and medium sized enterprises acquire CMMI certification.

In chapter 6, Antonio Ruiz Martínez et al. present a way to include payment information in Web Services Description Language (WSDL), WS-Policy and Universal Description, Discovery and Integration (UDDI) for those web services that require payment in order to be used.

In chapter 7, Tokuro Matsuo and Takayuki Fujimoto propose a new software engineering education model based on analogical thinking with respect to students' histories and experiences.

In chapter 8, Thomas Seemann and Harald Hungenberg present us with an empirical comparison of the performance differences between a prediction market and more traditional forecasting methods.

In chapter 9, Pablo Cabezas et al. propose an agent-based novel architecture with a semantic in-memory OSGi service registry based on OWL. They suggest that this architecture enhances the potential of OSGi with semantic data extracted from services deployed in the framework, using software agents in conjunction with Java Annotations and a Java Reflection API to dynamically obtain and invoke all required information.

In chapter 10, authors Seung Woo Shin and Haeng-Kon Kim put forth a new framework for Service Oriented Architecture (SOA) based on the use of Xplus with Agile methodologies for use by small and medium sized enterprises. They explain that the use of this system with will provide these organizations with successful results for a wide variety of cases.

In chapter 11, Hamid Khosravi et al. offer a new a new method for text summarization using fuzzy logic. They compare this method with other methods of text summarization in their study.

In chapter 12, Hee Won Lee et al. propose a new proactive reuse method to increase reusability in developing network management systems using the KT platform by use of a flexible network management topology architecture.

In chapter 13, Sussy Bayona Luz et al. present a case study on the deployment process for five software development and maintenance sites within one organization, and discuss the results of that study. They focus on the deployment process elements, critical success factors and impact on the process. They also explain the levels of acceptance and use of these software processes.

In chapter 14, Gongzhu Hu and Xiaohui Huang introduce a method for medical image retrieval and classification using low-level image features. The described method is based on the selection of prominent features in the high dimension feature space and the parameter of the k-NN algorithm. The authors also combine image features with non-image features (patient records) to improve the accuracy of retrieval queries.

In chapter 15, Sung Wook Lee et al. describe how to aggregate agile methodologies into CMMI frameworks and suggest process-for-process appraisal. They also identify and define the processes for CMMI and extreme programming through the many existing comparison data-sets.

In chapter 16, Toshiyuki Maeda et al. present an e-mail-based lecture support system, consisting of an attendance management system, an attendance history management subsystem, a short examination management subsystem, a questionnaire subsystem and an assignment delivery subsystem. Because this system uses only email, the authors explain, it can be accessed through any machine with email capabilities. The authors' full work outlines this system and its functions and effects.

In chapter 17, Maruf Monwar and Marina Gavrilova present a system that uses multiple biometrics to overcome the drawbacks associated with monomodal biometric systems.

In chapter 18, Amel Borgi et al. focus on approximate reasoning in a symbolic framework, and more precisely on multi-valued logic. The authors propose a generalization of the approximate reasoning axiom introduced by Fukami, and show the weakness of Generalized Modus Ponens (GMP) approaches in the multi-valued context toward this axiom. Moreover, they propose two rules of symbolic GMP that check the axiom.

In chapter 19, Hyun-joong Kang et al. present a simulation environment that was made using a shadowing model for the position of a sensor node in order to test the communication impact by crop in observing the crop's growth.

In chapter 20, Emad Ghosheh et al. introduce new design metrics for measuring the maintainability of web applications from class diagrams. The metrics used are based on Web Application Extension (WAE) for UML and measure the design attributes for size, complexity and coupling. The authors describe an experiment carried out using a CVS repository from a US telecoms web application.

In chapter 21, Hye-Kyeong Jo and In-Young Ko propose an ontologybased model by which developers can semantically describe and find artifacts based on the common aspects and properties of component-based software development processes.

In chapter 22, Bindu Goel and Yogesh Singh present their investigation of the relationship between OO metrics and the detection of the faults in object–oriented software. Fault prediction models are made and validated using regression methods for detecting faulty classes. The authors compare and present the results of the models, as well as an investigation on the metrics.

In chapter 23, the final chapter of the book, Nitin et al. propose a new method which merges two communications systems, local communication between the Intellectual Property (IP) cores in Networks-on-Chips (NoC), and global communication between NoCs, in Networks-in-Package (NiP). In addition, the authors propose two $O(n^2)$ time Fault–tolerant Parallel Algorithms for the setting of Inter NoC Communication in NiP, which allows different NoCs in NiP to communicate in parallel using either fault-tolerant Hexa Multi–stage Interconnection Network (HXN) or fault tolerant Penta Multi–stage Interconnection Network (PNN).

It is our sincere hope that this volume provides stimulation and inspiration that we may all take with us as we continue our work in this exciting field of cutting edge technology.

Roger Lee

Contents

List of Contributors

Saioa Arrizabalaga
University of Navarra, Spain
sarrizabalaga@ceit.es

Takayuki Asada
Osaka University, Japan
asada@econ.osaka-u.ac.jp

Sussy Bayona Luz
Polytechnic University of Madrid, Spain
sbayona@zipi.fi.upm.es

Saoussen Bel Hadj Kacem
National School of Computer Science, Tunisia
saoussen.bhk@fst.rnu.tn

Sue Black
University of Westminster, UK
s.e.black@westminster.ac.uk

José Roberto Blaschek
State University of Rio de Janeiro, Brazil
blaschek@attglobal.net

Amel Borgi
National Institute of Applied Science and Technology, Tunisia
amel.borgi@insat.rnu.tn

Pablo Cabezas
University of Navarra, Spain
pcabezas@ceit.es

Jose Antonio Calvo-Manzano
Polytechnic University of Madrid, Spain
jacalvo@fi.upm.es

Ho-Jin Choi
Information and Communications University, Korea
hjchoi@icu.ac.kr

Gonzalo Cuevas
Polytechnic University of Madrid, Spain
gcuevas@fi.upm.es

Pooya Khosravyan Dehkordy
Islamic Azad University (Arak branch), Iran
pooya_khd@iaun.ac.ir

Esfandiar Eslami
University of Shahid Bahonar Kerman, Iran
eeslami@mail.uk.ac.ir

Marcus Vinícius Mazega Figueredo
Pontifícia Universidade Católica do Paraná, Brazil
marcus@ppgia.pucpr.br

Takayuki Fujimoto
Sonoda Women's University, Japan
me@fujimotokyo.net

Yae Fukushige
Osaka University, Japan
ppr-mint@r7.dion.ne.jp

Marina Gavrilova
University of Calgary, Canada

Khaled Ghedira
National School of Computer Science, Tunisia
khaled.ghedira@isg.rnu.tn

Emad Ghosheh
AT&T, USA
eg2534@att.com

Bindu Goel
Guru Gobind Singh Indraprastha University, India
bindu_delus@yahoo.com

Gongzhu Hu
Central Michigan University, USA
hu1g@cmich.edu

Xiaohui Huang
Central Michigan University

Harald Hungenberg
Friedrich-Alexander University, Germany
harald.hungenberg@wiso.unierlangen.de

Chan Kyou Hwang
KT Network Technology Lab, Korea
ckhwang@kt.com

Hye-Kyeong Jo
Information and Communications University, Korea
hgcho@icu.ac.kr

Hyun-joong Kang
School of Information and Communications Engineering, Korea
hjkang@mail.sunchon.ac.kr

Sungwon Kang
Information and Communications University, Korea
kangsw@icu.ac.kr

Hamid Khosravi
University of Shahid Bahonar Kerman, Iran
hkhosravi@mail.uk.ac.ir

Haeng-Kon Kim
Catholic University of Daegu, Korea
hangkon@cu.ac.kr

In-Young Ko
Information and Communications University, Korea
iko@icu.ac.kr

Sehgal Vivek Kumar
Jaypee University of Information Technology, India
sehgal.vivek@juit.ac.in

Farshad Kyoomarsi
University of Shahid Bahonar Kerman, Iran
Kumarci_farshad@graduate.uk.ac.ir

Jon Legarda
University of Navarra, Spain
jlegarda@ceit.es

Dan Hyun Lee
Information and Communications University, Korea
danlee@icu.ac.kr

Hee Won Lee
KT Network Technology Lab, Korea
hotwing@kt.com

Meong-hun Lee
School of Information and Communications Engineering, Korea
leemh777@sunchon.ac.kr

Roger Y. Lee
Central Michigan University, USA
lee1ry@cmich.edu

Sung Wook Lee
Catholic University of Deagu
sojiro@cu.ac.kr

Toshiyuki Maeda
Hannan University, Japan
maechan@hannan-u.ac.jp

Antonio Ruiz Martínez
University of Murcia, Spain
arm@um.es

Fabio Perez Marzullo
Federal University of Rio de Janeiro, Brazil
fpm@cos.ufrj.br

Tokuro Matsuo
Yamagata University, Japan
tokuro@tokuro.net

Md. Maruf Monwar
University of Calgary, Canada

Júlio César Nievola
Pontifícia Universidade Católica do Paran, Brazil

Alfredo Beckert Neto
Pontifícia Universidade Católica do Paraná, Brazil

Nitin
Jaypee University of Information Technology, India
delnitin@juit.ac.in

Tadayuki Okamoto
Ehime University, Japan
tadayuki@ll.ehimeu.ac.jp

Eun-Ju Park
Catholic University of Daegu, Korea
ejpark@cu.ac.kr

Huy Nguyen Anh Pham
Louisiana State University, USA
hpham15@lsu.edu

Rodrigo Novo Porto
Federal University of Rio de Janeiro, Brazil
rodrigo@cos.ufrj.br

Jihad Qaddour
Illinois State University, USA
jqaddou@ilstu.edu

Óscar Cánovas Reverte
University of Murcia, Spain
ocanovas@um.es

Sérgio Renato Rogal Jr.
Pontifícia Universidade Católica do Paraná, Brazil

Antonio Salterain
University of Navarra, Spain
asalterain@ceit.es

Tomás San Feliu
Polytechnic University of Madrid, Spain
tsanfe@fi.upm.es

Angel Sánchez
Polytechnic University of Madrid, Spain
Angel.Sanchez@everis.com

Thomas Seemann
Friedrich-Alexander University, Germany
thomas.seemann@wiso.unierlangen.de

Seung Woo Shin
Catholic University of Daegu, Korea
selab@cu.ac.kr

Geraldo Zimbrão da Silva
Federal University of Rio de Janeiro, Brazil
zimbaro@cos.ufrj.br

Chauhan Durg Singh
Jaypee University of Information Technology, India
ds.chauhan@juit.ac.in

Yogesh Singh
Guru Gobind Singh Indraprastha University, India
ys66@rediffmail.com

Antonio F. Gómez Skarmeta
University of Murcia, Spain
skarmeta@um.es

Jano Moreira de Souza
Federal University of Rio de Janeiro, Brazil
jano@cos.ufrj.br

Evangelos Triantaphyllou
Louisiana State University, USA
trianta@lsu.edu

Akiyoshi Wakatani
Konan University, Japan
wakatani@konan-u.ac.jp

Hyun Yoe
School of Information and Communications Engineering, Korea
yhyun@sunchon.ac.kr

Jae-Hyoung Yoo
KT Network Technology Lab, Korea
styoo@kt.com

An MDA Approach for Database Profiling and Performance Assessment

Fabio Perez Marzullo[1], Rodrigo Novo Porto[1], Geraldo Zimbrão da Silva[1], Jano Moreira de Souza[1], and José Roberto Blaschek[2]

[1] Federal University of Rio de Janeiro - UFRJ/COPPE
{fpm,rodrigo,zimbrao,jano}@cos.ufrj.br
[2] State University of Rio de Janeiro - UERJ
blaschek@attglobal.net

Summary. This paper describes a Model Driven Architecture (MDA) approach for assessing database performance. The increase in the area of component-based development is reaching a point where performance issues are critical to successful system deployment. It is widely accepted that the model development approach is playing an important role on IT projects; therefore the profiling technique discussed here presents a way to assess performance and identify flaws while performing software construction activities. The approach is straightforward: using new defined MDA stereotypes and tagged values in conjunction with profiling libraries, the proposed MDA extension enables code generation to conduct a thorough set of performance analysis elements. This implementation uses the well-known MDA framework AndroMDA [2], the profiling libraries JAMon [11] and InfraRED [15], and creates a Profiling Cartridge, in order to generate the assessment code.

1 Introduction

Research conducted in recent years has shown that model driven approaches are becoming essential tools in software development [1]. This new paradigm implies on new ways of analyzing overall software performance. Attempts to integrate performance analysis with MDA have been made and are still in course [3]. However, such attempts present mostly ways to generate test code for general software engineering aspects.

The relevant scenario here is: how does one assess performance on database specific aspects, such as joins, storage, memory load and so forth? And mostly, how does one measure such database aspects to guide overall software architecture design, improving model and generated code?

The lack of an efficient representation for Database profiling has motivated us to extend the MDA models in order to cope with undesirable performance situations along software execution. Therefore, the idea relies on an MDA extension with the ability to mark business domain models with stereotypes and tagged values, seeking to understand database performance aspects.

Since long the OMG has been revising requests and proposals on Software Profiling. As an example, the UML 2.0 Testing Profile Specification [4] defines a test modelling language that can be used with all major UML objects and component technologies, and applied to testing systems in various application domains. However, its objective is to create a testing environment that focuses on software engineering aspects.

R. Lee and H.-K. Kim (Eds.): Computer and Information Science, SCI 131, pp. 1–10, 2008.
springerlink.com

Despite what is being done, there are still gaps involving proper database analysis. Still, different approaches focusing on benchmarking MDA techniques [5], or on the testing and verification of model transformation [6] are currently in activity, and our work comes to join all efforts to upgrade the MDA world.

Although MDA promises to implement a self-sufficient model-driven development theory, supported now by a few important companies and tools, it still needs to address several specific development aspects, including an efficient and complete set of Meta Object Facility [7] models and objects to address database specifics.

2 Problem Definition

Since the beginning of computational growth, there have been efforts to address performance tuning issues. Regardless of the hardware or software involved the pursuit of a systematic approach on writing code which efficiently implements software functionality lingers on in today's projects. Few techniques are available to help developers to write better code; as a result of that, one that has been gaining visibility is the profiling technique. Many companies are now engaged in creating tools to aid software profiling, enabling developers to identify system flaws and rewrite problematic code [8, 9, and 10].

Consequently, the challenges faced by IT managers when developing software in this scenario have only increased. Due to painstaking, detailed domain-specific functionalities, and an ever growing knowledge consumption, it became urgent to invest in new ways to cope with the time to market information systems without lacking proper product quality [11].

The promise of a model development approach has gained many adepts. The AdroMDA project [2] is on the top of model development tools, based on the OMG MDA specification [1]. From single CRUD application to complex enterprise applications, it uses a set of ready-made cartridges (which implements a highly widespread Java development API) to automatically generate up to 70% of the software source code. It is expected that all efforts should point to an interesting future where the modelling and coding stages will be merged without significant losses.

The same horizon is faced by database experts. The old promise of uniting Object-Oriented development and Relational data representation is now very well taken care of by the Hibernate API [13, 14]. Also, different persistence technologies have been developed to relieve the ordinary developer from the overheads in dealing with database laborious queries and the joining of structures.

Nevertheless, in spite of all the promises of database abstraction, it is not always possible to guarantee that what you see is what you get. For example, Hibernate is a powerful, high-performance, object/relational persistence and query service [13]; however, using it in its default configuration might prove that working with huge amounts of data is cumbersome. It lacks the ability to identify and target inefficient table joins which may imply in performance degradation. It is known that after a few adjustments in configuration values it may become more reliable; however the work load might be significant.

Given the foregoing, it is necessary to explain the problematic scenario that we have been faced with in our project. The AndroMDA environment comprises a set of specialized cartridges that are used to generate the software code. One of its cartridges is the Hibernate Cartridge, which is responsible for generating all the database manipulation code. As for database code generation it all seemed perfect, and the development database was used to test the system while as it was being constructed. Only three to five users were simultaneously accessing the database and development seemed to go smoothly. When the system was deployed and put into production, a small set of functionalities presented longer response times than we had anticipated. The production environment needed to be accessed by 10 to 20 users simultaneously and had to deal with thousands, even millions of records to be queried or joined. It suffices to say that performance degradation perception was immediate, and as such, something should be done to locate and solve the root problem.

To solve it we began by stating three topics that needed to be immediately addressed:

1 - How to isolate the functionalities and their problematic attributes?
2 - How to solve these problems in an elegant and transparent way for the end user?
3 - How to automate the solution, extending the MDA objects to address Database Testing and Analysis?

Using the current infra-structure to address the database performance analysis was not the best way, since we wanted to create a systematic and self-sustained analysis approach. It was necessary to extend the MDA infra-structure, more precisely the Hibernate Cartridge, and implement a new Profiling Cartridge as explained in the next section.

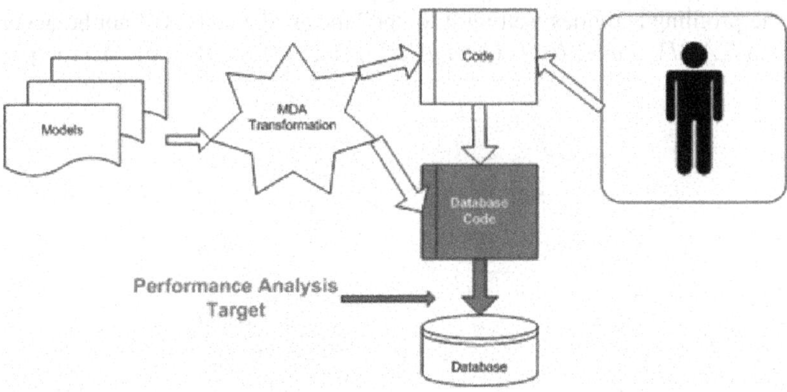

Fig. 1. Schematics of the performance analysis target

3 The Profiling Extension

The profiling extension process was straightforward. We needed to create new design elements to indicate when it would be necessary to embed monitoring code inside the software. For that purpose we followed a five-step design process, in order to extend the MDA framework:

1 - Defining the profiling library: This stage was responsible for identifying the profiling libraries that would be used. Our objective was to create an infra-structure that could cope with any profiling library available, so we decided to use design patterns to enable the ability to plug the library as necessary. This technique is explained in the following sections. However, after analyzing several libraries, two appeared to be the best choices: the JAMon Library [11] and the InfraRED Library [15].

2 - Defining Stereotypes: This stage was responsible for creating all the stereotypes necessary for model configuration. For each feature available a stereotype was created:

(a) - API timing: Average time taken by each API. APIs might be analyzed through threshold assessment, and first, last, min, and max execution times:

$\langle\langle APIView \rangle\rangle$: which enables api monitoring;

(b) - JDBC and SQL statistics: Being the goal of our research, we created the following stereotype for accessing JDBC and SQL statistics:

$\langle\langle SQLQueryView \rangle\rangle$: which enabled the displaying of the created hibernate queries;

(c) - Tracing and Call Information: Responsible for showing statistics for method calls. The following stereotype was created to mark/enable this option:

$\langle\langle TraceView \rangle\rangle$: which enabled a detailed call tracing for method calls;

3 - Defining Tagged Values: This stage was responsible for creating all tagged values necessary to support and configure stereotypes:

(a) - @profiling.active: Defines whether profiling activities are going to be executed. Setting its value to "yes" implies generating the profiling code and consequently enabling the profiling activities; setting it to "no" means the code will not be generated. Applies to $\langle\langle APIView \rangle\rangle$,$\langle\langle SQLQueryView \rangle\rangle$ and $\langle\langle TraceView \rangle\rangle$ stereotypes;

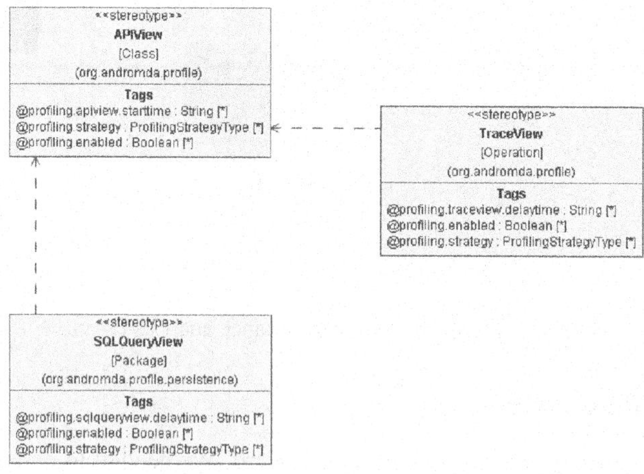

Fig. 2. Stereotypes and tagged values hierarchy

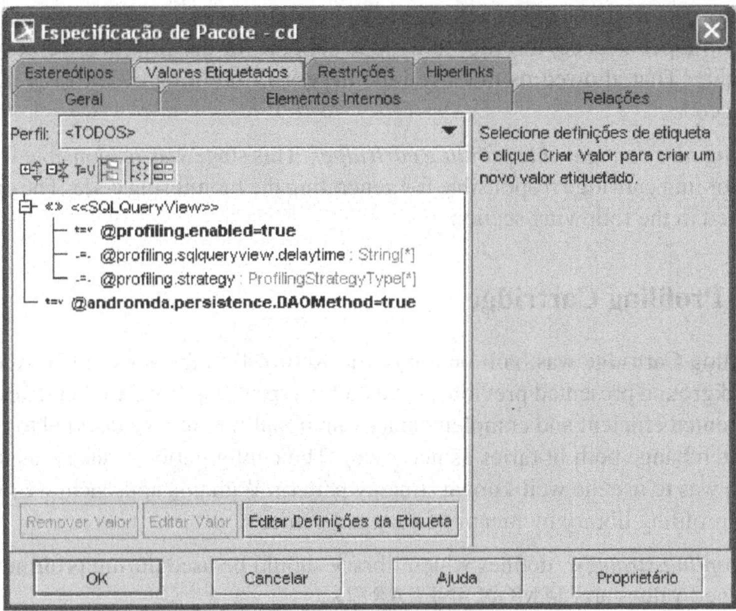

Fig. 3. Associating the stereotype to the persistence package

(b) - @profiling.apiview.starttime: Defines the starting time in which the profiling code should start monitoring. For example: it is not interesting to monitor every aspect of the system, and we therefore indicate the minimum limit in which the profiling code should initiate the logging process. Applies to the $\langle\langle APIView \rangle\rangle$ stereotype;

(c) - @profiling.sqlqueryview.delaytime: Defines the starting time in which the profiling code should initiate SQL monitoring. For example: it is not interesting to monitor all query execution in the system. It is only necessary to assess queries that are over a certain delay threshold. Applies to the $\langle\langle SQLQueryView \rangle\rangle$ stereotype;

(d) - @profiling.traceview.delaytime: Defines the starting time in which the profiling code should start Trace monitoring. For example: it is not interesting to monitor all method calls in the system. It is only necessary to assess the calls that are over a certain delay threshold. Applies to the $\langle\langle TraceView \rangle\rangle$ stereotype;

4 - Adjusting the Hibernate Cartridge: This stage was responsible for extending the hibernate cartridge so it would be able to cope with the new stereotypes. The Hibernate API works through the concept of Sessions, where a set of instructions, referring to database interactions, creates the notion of an unique Session. Normally, when a database connection is opened, a Hibernate Session is opened to accommodate that database connection. It stays open as the hibernate continues to access the database. Given those rules, we realized that we would have not only to analyze the software according to a class-based scope but we needed a broader and more complete approach in which the profiling would be able to gather information globally. The profiling scope should reach the entire persistence layer. This constraint motivated us to design the stereotype as having a package scope and not only class scope.

The solution involved the extension of the AndroMDA persistence base package (andromda-profile-persistence) in order to include support for our profiling stereotype and tagged value. That allowed us to embed the necessary profiling code in the hibernate generated code.

5 - Creating new configurable profiling cartridge: This stage was responsible for creating the profiling cartridge responsible for generating the monitoring code. The cartridge is explained in the following section.

4 The Profiling Cartridge

The Profiling Cartridge was built on top of the AndroMDA framework [2]. According to the background presented previously, we had two profiling library candidates to use. Both presented efficient and complementary functionalities and we decided to design a way to interchange both libraries as necessary. The configuration strategy, as the term describes, was to use the well-known strategy pattern. With this approach, we managed to set the profiling library by means of a tagged value:

(a) - @profiling.strategy: defines which library should be used during profiling analysis. Assigned values are: JAMON, INFRARED.

Figure 4 illustrates the strategy of interchanging profiling libraries:

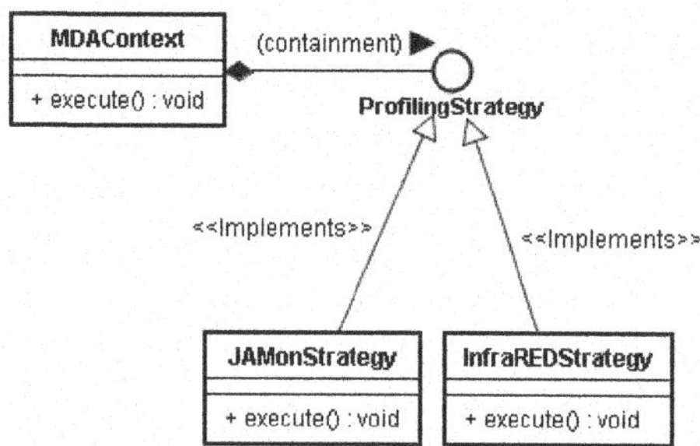

Fig. 4. Strategy pattern for profiling configuration

The specification of the profiling configuration details the functional relation between the MDA, as the source model, and the profiling technique, as the target implementation. This configuration method enabled us to interchange any profiling library of interest.

As for the cartridge implementation in AndroMDA we executed six steps, in which the basic process for cartridge development was fulfilled:

1 - Identify, design and generate PSM metaclasses: This first step was useful to plan how the profiling libraries would be used and mapped as stereotypes and tagged values;

2 - Identify transformation rules: After planning profiling attributes we created all transformation rules needed to generate the code for the corresponding stereotype and tagged value;

3 - Model, generate and write metafacades: Modelling stereotypes and tagged values were the first real difficulty found along the extension process. Deficiencies were uncovered and solved to match every PSM metaclass and transformation rule planned earlier;

4 - Write templates: This step was responsible for using Velocity as a template language for reference objects defined in Java code [18]. Templates for both profiling libraries were created; and Java code generation could be set according to the strategy assigned to the @profiling.strategy tagged value;

5 - Write deployment descriptors: The AndroMDA core uses cartridge descriptors to discover available capabilities such as: the supported metafacades, stereotypes, outlets, templates, property references, etc. The cartridge descriptor must reside in the META-INF/andromda subdirectory and must be named cartridge.xml [2].

6 - Deploy cartridge: This step is responsible for deploying the new cartridge into the AndroMDA environment.

5 Profiling Analysis and Results

For the definition of the analysis process, we must understand the execution scenario. The development was focused on creating a Web-based application, using Struts [19], running on the Jboss Application Server (JAS) [20], and accessing a Data Warehouse. System modelling was done using an UML tool, the AndroMDA, and the Maven tool [16], and for Java development we used the Eclipse IDE [17]. The data load was irrelevant at development time but it became crucial by the time the system was put into production. The generated hibernate code and configuration did not comprise with the system's response time non-functional requirements.

As soon as system performance started to interfere with overall system usability, we thought it urgent to locate the bottleneck source and re-factor all models involved. On the other hand, we needed to ensure that the new model would not create new bottleneck sources.

Each system functionality was assigned an estimated execution time, based on the system non-functional requirement, as shown in table 1. The system was deployed in a production environment and measured without the optimized code. Those functionalities that went over the estimated value by 20% were marked as problematic (that in our case easily overcame this requirement). We then regenerated the code and redeployed the system. Obtaining the actual execution times we observed that functionalities F2 and F3 were highly outside the required range.

The profiling technique enabled us to embed monitoring code that gathered all necessary information and pointed the features that needed to be remodelled.

Statistics were acquired along two different scenarios: (1) less than 10,000 registers in a non-optimized and optimized environments; (2) more than 2,000,000 registers with a non-optimized and optimized environments. After performing measurements in both scenarios we came up with the averages shown bellow. Tables 1 and 2 present three functionalities that were assessed to test the MDA-Profiling strategy:

Table 1. Identifying problematic functionalities. Functionality names were omitted for confidentiality purposes (Non-Optimized Environment). Time values averages were rounded up.

Functionalities	Expected Execution Time	Actual Execution Time	Problematic Functionality
Single Table (F1)	10 ms	30 ms	No
Join with 2 Tables (F2)	400 ms	50,000 ms	Yes
Join with 3 Tables (F3)	600 ms	10,000,000 ms	Yes

Table 2. Solving problematic functionalities. Functionality names were omitted for confidentiality purposes (Optimized Environment). Time values averaged were rounded up.

Functionalities	Expected Execution Time	Actual Execution Time
Single Table (F1)	10 ms	23 ms
Join with 2 Tables (F2)	400 ms	1000 ms
Join with 3 Tables (F3)	600 ms	2000 ms

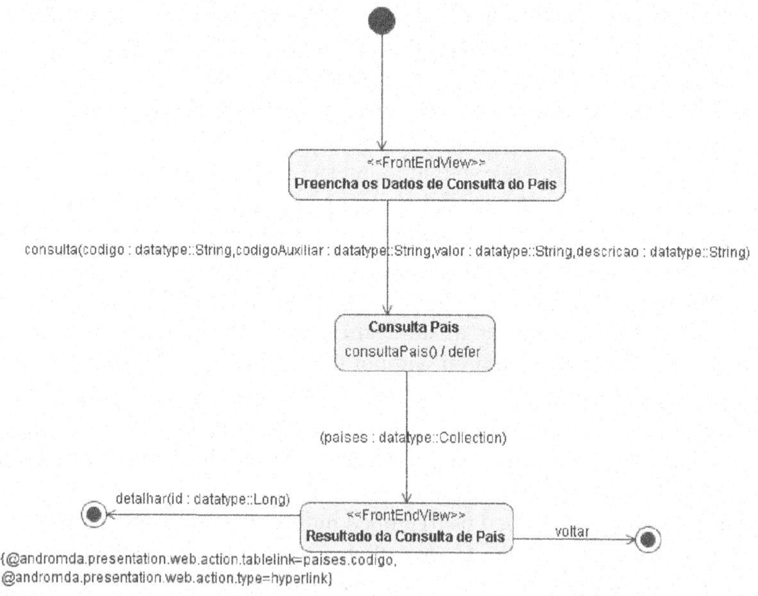

Fig. 5. Use Case used as Test Case

Analyzing the JAMon profiling results we discovered a number of items that could be re-factored in order to produce better optimized code. For instance, we detected that the Hibernate session was not reusing SQL statements, and was creating new SQL plans each time a query was submitted to the database.

Values presented in table 2 show that the new optimized code significantly reduced execution time. It showed a large reduction in database resource consuming, taking into consideration that in a Web-based environment such optimization might prove to be extremely relevant.

6 Conclusions

This paper presented a new approach for embedding profiling techniques in MDA development. Our contribution aimed at creating a MDA extension to help identifying and solving performance problems regarding information system and database (data warehouse) communications. In addition, we defined an extensible, easy-to-use profiling infra-structure that can be configured to execute different profiling libraries and techniques, obtaining a more complete set of results.

The analysis results have validated the profiling approach and proved that the MDA extension might be used to analyze the code as soon as it is deployed. The initial effort to create the infra-structure proved laborious although following developments shall not suffer from the same problems as it has already been implemented and added to the AndroMDA features.

Finally, the intention was to obtain development information in order to allow developers and analysts to make proper decisions regarding software design. According to the analysis results, the extension was able to expose flaws and delays during system execution, and, consequently promote the necessary corrections to ensure that the generated code was reliable in both scenarios.

Future works point to the use of aspect based development to add considerable new techniques on profiling the system and using autonomic computing, in conjunction with the profiling libraries, in order to detect the problems, understand the root problems and use embedded intelligence to allow the system to self correct and recover from eventual failures.

References

1. Model Driven Architecture, `http://www.omg.org/mda`
2. AndroMDA, v3.0M, `http://www.andromda.org/`
3. Zhu, L., Liu, Y., Gorton, I., Bui, N.B.: MDAbench, A Tool for Customized Benchmark Generation Using MDA. In: OOPSLA 2005, San Diego, California, October 16-20 (2005)
4. UML 2.0 Testing Profile Specification, `http://www.omg.org/cgi-bin/doc?formal/05-07-07`
5. Rodrigues, G.N.: A Model Driven Approach for Software System Reliability. In: Proceedings of the 26th International Conference on Software Engineering (ICSE 2004). IEEE, Los Alamitos (2004)
6. Lamari, M.: Towards an Automated Test Generation for the Verification of Model Transformations. In: SAC 2007. ACM, New York (2007)

7. Meta Object Facility, http://www.omg.org/mof
8. Eclipse Test & Performance Tools Platform Project, http://www.eclipse.org/tptp/
9. ej-technologies JProfiler, http://www.ej-technologies.com/products/jprofiler/
10. JAMon (Java Application Monitor), http://jamonapi.sourceforge.net/
11. NetBeans Profiler, http://profiler.netbeans.org/
12. Frankel, D.S.: Model Driven Archtecture - Applying MDA to Enterprise Computing. OMG Press, Wiley Publications (2003)
13. Hibernate, http://www.hibernate.org
14. Bouer, C., King, G.: Hibernate in Action. Manning Publications Co. (2004)
15. InfraRED - Perfromance and Monitoring Tool for Java, http://sourceforge.net/projects/infrared/
16. Maven project management and comprehension tool, http://maven.apache.org
17. Eclipse Project., http://www.eclipse.org
18. Velocity Project, http://velocity.apache.org/
19. Struts Project, http://struts.apache.org/
20. JBoss Application Server, http://www.jboss.org/

Prediction of Diabetes by Employing a New Data Mining Approach Which Balances Fitting and Generalization

Huy Nguyen Anh Pham and Evangelos Triantaphyllou

Department of Computer Science, 298 Coates Hall, Louisiana State University,
Baton Rouge, LA 70803
hpham15@lsu.edu, trianta@lsu.edu

Summary. The Pima Indian diabetes (PID) dataset [1], originally donated by Vincent Sigillito from the Applied Physics Laboratory at the Johns Hopkins University, is one of the most well-known datasets for testing classification algorithms. This dataset consists of records describing 786 female patients of Pima Indian heritage which are at least 21 years old living near Phoenix, Arizona, USA. The problem is to predict whether a new patient would test positive for diabetes. However, the correct classification percentage of current algorithms on this dataset is oftentimes coincidental. The root to the above critical problem is the overfitting and overgeneralization behaviors of a given classification algorithm when it is processing a dataset. Although the above situation is of fundamental importance in data mining, it has not been studied from a comprehensive point of view. Thus, this paper describes a new approach, called the Homogeneity- Based Algorithm (or HBA) as developed by Pham and Triantaphyllou in [2-3], to optimally control the overfitting and overgeneralization behaviors of classification on this dataset. The HBA is used in conjunction with traditional classification approaches (such as Support Vector Machines (SVMs), Artificial Neural Networks (ANNs), or Decision Trees (DTs)) to enhance their classification accuracy. Some computational results seem to indicate that the proposed approach significantly outperforms current approaches.

1 Introduction

Insulin is one of the most important hormones in the body. It aids the body in converting sugar, starches and other food items into the energy needed for daily life. However, if the body does not produce or properly use insulin, the redundant amount of sugar will be driven out by urination. This phenomenon (or disease) is called diabetes. The cause of diabetes is still a mystery, although obesity and lack of exercise appear to possibly play significant roles.

According to the American Diabetes Association [4] in November 2007, 20.8 million children and adults in the United States (i.e., approximately 7% of the population) were diagnosed with diabetes. Thus, the ability to diagnose diabetes early plays an important role for the patient's treatment process. The World Health Organization [5] proposed the eight attributes, depicted in Table 1, of physiological measurements and medical test results for the diabetes diagnosis.

Furthermore, one of the many applications of data mining involves the analysis of data for which we know the class value of each data point. We wish to infer some

R. Lee and H.-K. Kim (Eds.): Computer and Information Science, SCI 131, pp. 11–26, 2008.
springerlink.com

Table 1. The eight attributes for the diabetes diagnosis

No.	Attribute
1	Number of times pregnant
2	Plasma glucose concentration in an oral glucose tolerance test
3	Diastolic blood pressure (mm/Hg)
4	Triceps skin fold thickness (mm)
5	2-hour serum insulin (μU/ml)
6	Body mass index (kg/m^2)
7	Diabetes Pedigree function
8	Age (years)

patterns from these data which in turn could be used to infer the class value of new points for which we do not know their class values. For instance, a doctor could be interested in knowing whether a patient would test positive for diabetes based on the above eight attributes. This kind of data mining analysis is called classification or class prediction of new data points.

The PID dataset [1], originally donated by Vincent Sigillito from the Applied Physics Laboratory at the Johns Hopkins University, is one of the most wellknown datasets for testing classification algorithms. This dataset consists of records describing 768 female patients of Pima Indian heritage which are at least 21 years old living near Phoenix, Arizona, USA. From the 768 patients in the PID dataset, classification algorithms used a training set with 576 patients and a testing dataset with 192 patients. However, the correct classification percentage of current algorithms on this dataset is oftentimes coincidental.

For instance, Smith et al. in [6] used an early neural network to diagnose the onset of diabetes mellitus. Their approach yielded 76.0% accuracy. Similarly, Jankowski and Kadirkamanathan in [7] developed a radial basis function network suite called IncNet which used 100 neurons and trained for 5,000 iterations. This approach yielded 77.6% accuracy. Au and Chan in [8] attempted to improve the correct classification percentage on the PID dataset by using a fuzzy approach. Au and Chan first represented the revealed regularities and exceptions using linguistic terms, and then mined interesting rules for the classification based on membership degrees. Their approach yielded 77.6accuracy. Rutkowski and Cpalka in [9] introduced a new neural-fuzzy structure called a flexible neuralfuzzy inference system (FLEXNFIS). Based on the input and output data, they proposed the parameters of the membership functions and the type of the neuron systems (Mamdani or logical). However, their correct classification percentage on the PID dataset was 78.6%. Davis in [10] developed a fuzzy neural network by using the BK-Square products. This fuzzy neural network was then tested on the PID dataset. The result of his approach yielded 81.8the results obtained from the StatLog project [11] when evaluating for many different classification algorithms on the PID dataset showed that their correct classification percentage was less than 78%.

The root to the low accuracies is the overfitting and overgeneralization behaviors of a given classification algorithm when it is processing this dataset. Although the above situation is of fundamental importance in data mining, it has not been studied from a comprehensive point of view. Thus, the main goal of this paper is to apply a new approach, called the Homogeneity-Based Algorithm (or HBA), as described in [2-3], to optimally control the overfitting and overgeneralization behaviors on the PID dataset. That is, the HBA would minimize the total misclassification cost in terms of the false-positive, false-negative, and unclassifiable rates. By doing so, it is hoped that the classification/prediction accuracy of the inferred models will be very high or at least as high as it can be achieved with the available training data.

The next section is a brief description of the HBA and it is adopted from [2-3]. That section shows how a balance between fitting and generalization has the potential to improve many existing classification algorithms. The third section discusses some promising results. These results give an indication of how this methodology may improve the classification/prediction accuracy. Finally, this chapter ends with some conclusions.

2 Description of the HBA

2.1 Problem Description

As described in [2-3], many real-life applications have the following three different penalty costs:

- A cost when a true-positive point is classified as negative.
- A cost when a true-negative point is classified as positive.
- A cost when a data point cannot be classified by any of the classification patterns.

The first case is known as *false-negative*, while the second case is known as *false-positive*. The last case is known as *unclassifiable*. Furthermore, [2-3] showed that attempts to minimize any of the previous rates might affect to the other rates. Thus, we cannot separate the control of fitting and generalization into two independent studies. That is, we need to find a way to simultaneously balance fitting and generalization by adjusting the inferred systems (i.e., the positive and the negative systems) obtained from a classification algorithm. The balance of the two systems will attempt to minimize the total misclassification cost of the final system.

In particular, let us denote C_{FP}, C_{FN}, and C_{UC} as the unit penalty costs for the false-positive, the falsenegative, and the unclassifiable cases, respectively. Let $RATE\_FP, RATE\_FN, and\ RATE\_UC$ be the falsepositive, the false-negative, and the unclassifiable rates, respectively. Then, the problem is to achieve a balance between fitting and generalization that would minimize, or at least significantly reduce, the total misclassification cost denoted as TC. Thus, the problem is defined as in the following expression:

$$TC = min(C_{FP} \times RATE\_FP + C_{FN} \times RATE\_FN + C_{UC} \times RATE\_UC) \quad (1)$$

This methodology may assist the data mining analyst to create classification systems that would be optimal in the sense that their total misclassification cost would be minimized. As mentioned in [2-3], there are two key issues regarding the HBA:

- The accuracy of the inferred classification systems can be increased if the derived patterns are, somehow, more compact and *homogenous*. A pattern C of size n_C is a homogenous set if the pattern can be partitioned into smaller bins of the same unit size h and the density of these bins is almost equal to each other.
- The accuracy of the inferred classification systems may also be affected by a *density* measure. Such a density could be defined as the number of data points in each inferred pattern per unit of area or volume. Therefore, this density will be called the *homogeneity degree*. Suppose that a homogenous set C is given. Then, $HD(C)$ will denote its homogeneity degree.

2.2 Non-parametric Density Estimation

As seen in Section 2.1, the density estimation of a typical bin plays an important role in determining whether a set is a homogenous set. One of the most appropriate approaches for the non-parametric density estimation is Parzen Windows [12]. That is, the Parzen Windows approach temporarily assumes that the bin R is a D-dimensional hypercube of unit size h. To find the number of points that fall within this bin, the Parzen Windows approach defines a kernel function $\varphi(u)$ as follows:

$$\varphi(u) = \begin{cases} 1, & \mid u \mid \leq 1/2, \\ 0, & otherwise \end{cases} \tag{2}$$

It follows that the quantity $\varphi(\frac{x-x_i}{h})$ is equal to unity if the point x_i is inside the hypercube of unit size h and centered at x, and zero otherwise. In the D-dimensional space, the kernel function can be presented as follows:

$$\varphi(\frac{x - x_i}{h}) = \prod_{m=1}^{D} \varphi(\frac{x^m - x_i^m}{h}) \tag{3}$$

Let n_C be the number of points in C and $d(x)$ denote the x's density, then:

$$d(x) \approx \frac{1}{n_C \times h^D} \sum_{i=1}^{n_C} \prod_{m=1}^{D} \varphi(\frac{x^m - x_i^m}{h}) \tag{4}$$

Choosing a value for h plays the role of a smoothing parameter in the Parzen Windows approach. We propose a way for finding an appropriate value for h as follows:

Heuristic Rule 1: *If h is set equal to the minimum value in set S and this value is used to compute d(x) by using Equation (4), then d(x) approaches to a true density.*

For instance, suppose that we determine all distances between all possible pairs formed by taking any two points from pattern C of size 5 (i.e., there are five points in C). Thus, there are 10 distances totally. For easy illustration, assume that these distances are as follows: 6, 1, 2, 2, 1, 5, 2, 3, 5, 5. Then, we define S as the set of the distances which have the highest frequency. For the previous illustrative example, we have set S equal to 2, 5 as both distances 2 and 5 occur with frequency equal to 3 (which is the highest frequency). By using the concept of the previous set S, Heuristic Rule 1 proposes an appropriate value for h which is equal to 2. The following section briefly provides the key details of the HBA [2-3].

2.3 The HBA

There are five parameters which are used in the HBA and are computed by using a Genetic Algorithm (GA) approach:

- Two expansion threshold values α^+ and α^- to be used for expanding the positive and the negative homogenous sets, respectively.
- Two breaking threshold values β^+ and β^+ to be used for breaking the positive and the negative patterns, respectively.
- A density threshold value γ to be used for determining whether either a positive or a negative hypersphere is approximately a homogenous set.

The HBA depicted in Figure 1 is summarized in terms of the following six phases:

- **Phase #1:** Randomly initialize the threshold values. Assume a training dataset T is given. We divide T into the two random sub-datasets: T_1 whose size is equal to, say 90%, of $T's$ size and T_2 whose size is equal to 10% of T's size (these percentages are determined empirically).
- **Phase #2:** Apply a classification approach (such as SVMs, ANNs, or DTs) on the training dataset T_1 to infer the two classification systems (i.e., the positive and the negative classification systems). Suppose that each classification system consists of a set of patterns. Next, break the inferred patterns into hyperspheres.
- **Phase #3:** Determine whether the hyperspheres derived in Phase #2 are homogenous sets or not. If so, then compute their homogeneity degree and go to Phase #4. Otherwise, break a non-homogenous set into smaller hyperspheres. Repeat Phase #3 until all of the hyperspheres are homogenous sets.
- **Phase #4:** For each homogenous set, if its homogeneity degree is greater than a certain breaking threshold value, then expand it. Otherwise, break it into smaller homogenous sets. Phase #4 stops when all of the homogenous sets have been processed.
- **Phase #5:** Evaluate the classification models (i.e., the homogenous sets processed in Phase #4) by using the dataset T_2 as a calibration dataset. The evaluation returns the value of Equation (1). Next, apply a genetic algorithm (GA) with the expression in Equation (1) as the fitness function to find the new threshold values (α^+, α^*, β^+, β^+) and then go to Phase #4. After a number of iterations, Phase #5 returns the optimal threshold values (α_*^+, α_*^-, β_*^+, β_*^-) and the classification model S_1 (i.e., the positive and the negative classification models) with the best value for Equation (1).
- **Phase #6:** Suppose that the calibration dataset T_2 can be divided into the two sub-datasets: $T_{2,1}$, which consists of the points classified by S_1, and $T_{2,2}$, which includes the unclassifiable points by S_1. We apply Phases #2 to #4 on the sub-dataset $T_{2,2}$ with the optimal threshold values (α_*^+, α_*^-, β_*^+, β_*^-). This phase infers the additional classification model S_2. The final classification model is the union of S_1 and S_2.

The six phases described earlier lead to the formulation of six sub-problems as follows:

- **Sub-Problem #1:** Apply a data mining approach to infer the two classification systems.

Input:
- The training dataset T with the positive and the negative points.
- A given classification algorithm.
- The density threshold value γ.

1. Divide T into T_1 and T_2 as described in Phase #1.
2. Randomly initialize the values of the control parameters α^+, α^-, β^+, and β^-.
3. Call **Sub-Problem #1** with the training dataset T_1 to infer the two classification models.
4. Call **Sub-Problem #2** to form the hyperspheres from the inferred patterns.
5. For each hypersphere C, do:
   Call **Sub-Problem #3** with inputs C and γ to determine whether C is a homogenosu set.
   If C is a non-homogenous set, then call **Sub-Problem #4** to break it and go to Step 5.
6. Sort the homogeneity degrees in decreasing order.
7. For each homogenous set C, do:
   If $HD(C) \geq \beta^+$ (for positive sets) or $HD(C) \geq \beta^-$ (for negative sets), then
   Call **Sub-Problem #5** with inputs $HD(C)$ and α^+ or α^- to expand C.
   Else, Call **Sub-Problem #6** to break C.

Notes:
- Apply a GA approach on Steps 5 to 7 by using Equation (1) as the fitness function and T_2 as a calibration dataset to find the classification model S_1 and the optimal threshold values $(\alpha_*^+, \alpha_*^-, \beta_*^+, \beta_*^-)$.
- For the unclassifiable points by S_1 in T_2, we use Steps 3 to 7 with the optimal threshold values $(\alpha_*^+, \alpha_*^-, \beta_*^+, \beta_*^-)$ to infer the additional classification model S_2.

10. Let $S = S_1 \cup S_2$.

Output: A new classification system S.

Fig. 1. The HBA

- **Sub-Problem #2:** Break the inferred patterns into hyperspheres.
- **Sub-Problem #3:** Determine whether a hypersphere is a homogenous set or not. If so, then its homogeneity degree is estimated.
- **Sub-Problem #4:** If a hypersphere is not a homogenous set, then break it into smaller hyperspheres.
- **Sub-Problem #5:** Expand a homogenous set C by using HD(C) and the corresponding expansion threshold value plus some stopping conditions.
- **Sub-Problem #6:** Break a homogenous set C into smaller homogenous sets.

To solve Sub-Problem #1, one simply applies a traditional classification algorithm and then derives the classification patterns. Furthermore, a solution to Sub- Problem #2 is similar to solutions for Sub-Problem #4. Therefore, the following sections present some summary procedures for solving Sub-Problems #2, #3, #5, and #6 and the GA approach. The explanation and the illustrative examples for Sub-Problems #1 to #6 are described in more detail in [2-3].

2.4 Solving Sub-Problem #2

The goal of Sub-Problem #2 is to find the minimum number of hyperspheres that can cover a pattern C of size n_C. A heuristic algorithm for Sub-Problem #2 is proposed as depicted in Figure 2.

Input: Pattern *C* of size n_C.
1. Estimate the densities of the n_C points by using Equation (4).
2. For *K*=1 to n_C do
   Pick *K* points in *C* with the highest densities.
   Use the *K*-means clustering approach to find *K* hyperspheres.
   If the *K* hyperspheres cover *C*, then STOP.
   Else, *K* = *K* + 1.

Output: *K* hyperspheres.

Fig. 2. The algorithm for Sub-Problem #2

The algorithm starts by first estimating the densities of the n_C points by using Equation (4). Assume that the value for K is going from 1 to n_C. The algorithm will pick K points in C with the highest densities. Next, it uses these K points as centroids in the K-means clustering approach. If the K hyperspheres which are obtained from the clustering approach cover C, then the algorithm will stop. Otherwise, we repeat the algorithm with the value for K increased by one. Obviously, the algorithm will stop after some iterations because a hypersphere of size one is a homogenous set.

2.5 Solving Sub-Problem #3

Let us consider hypersphere C of size n_C. Sub- Problem #3 determines whether or not hypersphere C is a homogenous set as follows. Hypersphere C is first divided into a number of small bins of unit size h and then approximates the density at the center x of each bin. If the densities at the centers are approximately equal to each other, then C is a homogenous set.

A softer condition can be applied instead of requiring exactly the same density at the centers of the bins. That is, if the standard deviation of the densities at the centers of the bins is approximately less or equal to γ, say for $\gamma = 0.01$, then hypersphere C may be considered to be a homogenous set. The algorithm for Sub-Problem #3 is given in Figure 3.

Input: Hypersphere *C* and density threshold value *γ*.
1. Compute the distances between all pairs of points in *C*.
2. Let *h* be the distance mentioned in Heuristic Rule 1.
3. Superimpose *C* into hypergrid *V* of unit size *h*.
4. Approximate the density at the center *x* of each bin.
5. Compute the standard deviation of the densities at the centers of the bins.
6. If the standard deviation is less than or equal to *γ*, then
   C is a homogenous set and its homogeneity degree *HD(C)* is computed by using Equation (5).
   Else, *C* is not a homogenous set.

Output: Decide whether *C* is a homogenous set.

Fig. 3. The algorithm for Sub-Problem #3

As seen above, the homogeneity degree $HD(C)$ is a factor that may affect the total misclassification cost of the inferred classification systems. If an unclassified point is covered by a homogenous set C which has a higher homogeneity degree, then it may more accurately be assumed to be of the same class as the points covered by the homogenous set C. Thus, a definition for $HD(C)$ is an important step in improving the accuracy of the classification systems. Pham and Triantaphyllou in [2-3] have proposed a way for computing $HD(C)$ as follows:

$$HD(C) = \frac{1n(n_C)}{h}. \tag{5}$$

Intuitively, $HD(C)$ depends on the value h defined in Heuristic Rule 1 and the number of points n_C. If n_C increases, then $HD(C)$ would slightly increase since the volume of C does not change and C has more points. Furthermore, if h increases, then the average distance between pairs of points in homogenous set C increases. Obviously, this leads to $HD(C)$ decreases. Hence, $HD(C)$ is inversely proportional to h while $HD(C)$ is directly proportional to n_C. We use the function $1n(n_C)$ to show the slight effect of n_C on $HD(C)$.

2.6 Sub-Problem #5

Suppose that we are given a positive homogenous set F with its homogeneity degree $HD(F)$, the breaking threshold value β^+, and the expansion threshold value α^+. A similar definition exists for a negative homogenous set. According to the main algorithm depicted in Figure 1, if $HD(F)$ is greater than or equal to β^+, then the homogenous set F will be expanded by using the expansion threshold value α^+. Otherwise, we will break the homogenous set F into smaller hyperspheres.

There are two types of expansion for F: a radial expansion in which a homogenous set F is expanded in all directions and a linear expansion in which a homogenous set F is expanded in a certain direction. The following section explains in detail these two expansion types [2-3].

2.6.1 Radial Expansion

Let M be a region expanded from F. Let R_F and R_M denote the radiuses of F and M, respectively. The radial expansion algorithm is depicted in Figure 4.

```
Input: Homogenous set F with HD(F), R_F, and α+
1. Set M = F (i.e., R_F = R_M).
2. Set hypersphere G covering M with radius R_G = 2 ×
       R_M.
3. Repeat
       Set E = M (i.e., R_E = R_M).
       Expand M by using Equation (10).
   Until (R_M satisfies stopping conditions discussed in
       Section 2.6.3 or R_M = R_G).
4. If R_M satisfies stopping conditions, then STOP.
   Else, go to Step 2.
Output: An expanded region E.
```

Fig. 4. The algorithm for the radial expansion

The idea of this algorithm is to expand a homogeneous set F as much as possible by using a dichotomous search methodology. That is, R_F is increased by a certain amount denoted as T, called a *step-size increase,* in each iteration. Thus, one gets:

$$R_M = R_F + T. \tag{6}$$

A value for T is determined as follows. We first assume that there exists a hypersphere G which covers the homogenous set F. Without loss of generality, let us assume that the radius R_G may be computed by:

$$R_G = 2 \times R_F \tag{7}$$

By using R_G and R_F, we can derive the step-size increase T. That is, T must depend on the difference between R_G and R_F. One of the ways that T may be determined is as follows:

$$T = \frac{R_G - R_F}{2} \tag{8}$$

At the same time, T should depend on $HD(F)$ because of the dichotomous search methodology. That is, if $HD(F)$ gets higher, then T should get smaller. This means that $HD(F)$ is inversely proportional to T. We may use a threshold value L to ensure that $HD(F)$ is always greater than one. Thus, the value for T may be defined as follows:

$$T = \frac{R_G - R_F}{2} \times \frac{1}{L \times HD(F)} \tag{9}$$

If we substitute back into Equation (6), R_M becomes:

$$R_M = R_F + \frac{R_G - R_F}{2} \times \frac{1}{L \times HD(F)} \tag{10}$$

2.6.2 Linear Expansion

The linear approach expands a homogenous set F in a certain direction. There is a difference between the method presented in the previous section and the one presented in this section (i.e., linear vs. radial). That is, now the homogenous set F is first expanded to hypersphere M by using the radial expansion. Then, hypersphere M is expanded in a given direction by using the radial approach until it satisfies the stopping conditions mentioned next in Section 2.6.3. The final region is the union of all the expanded regions.

2.6.3 Description of Stopping Conditions

As described in [2-3] the stopping conditions of the radial expansion approach for homogenous set F of size n_F must satisfy the following requirements:

- Depend on the homogeneity degree $HD(F)$. This has been mentioned in Section 2.1.
- Stop when an expanded region reaches other patterns. However, this condition can be relaxed by accepting several noisy data points in the expanded region. If the homogeneity degree $HD(F)$ is high, then the expanded region can accept more noisy data.

To address the first stopping condition, an upper bound for R_M should be directly proportional to the homogeneity degree $HD(F)$, the expansion threshold value α^+, and the original radius R_F. The second stopping condition can be determined while expanding. Furthermore, an upper bound on the number of noisy points should be directly proportional to $HD(F)$ and n_F. The stopping conditions are summarized as follows:

$$R_M \leq HD(F) \times R_F \times \alpha^+ \text{ and}$$
$$\textit{the number of noisy points} \leq \frac{HD(F) \times \alpha^+}{n_F} \tag{11}$$

Similar conditions exist for the expansion threshold value α^-.

2.7 Sub-Problem #6

Suppose a given positive homogenous set F is available. Recall that if its homogeneity degree $HD(F)$ is less than β^+, then the homogenous set F is broken into sub-patterns. The sub-patterns are also homogenous sets. Thus, they can be expanded or broken down even more.

2.8 A Genetic Algorithm (GA) Approach for Finding the Threshold Values

Recall that the main algorithm depicted in Figure 1 uses the four threshold values α^+, α^-, β^+, and β^- to derive a new classification system. If the breaking threshold values (i.e., β^+, and β^-) are too high, then this would result in the overfitting problem. On the other hand, too low breaking threshold values may not be sufficient to overcome the

α^+	α^-	β^+	β^-

Fig. 5. An illustrative example of a chromosome consisting of the four genes

Chromosome A	α_1^+	α_1^-	β_1^+	β_1^-
Chromosome B	α_2^+	α_2^-	β_2^+	β_2^-
Chromosome C	α_1^+	α_2^-	β_2^+	β_1^-

Fig. 6. An illustrative example of the crossover function

overgeneralization problem. The opposite situation is true with the expansion threshold values (i.e., α^+ and α^-).

Since the ranges for the threshold values depend on each individual application, the search space may be large. In this investigation an exhaustive search would be impractical. Thus, we propose to use a GA approach to find approximate optimal threshold values as follows. The HBA uses Equation (1) as the fitness function and the dataset T_2 as a calibration dataset. The GA approach has been applied here because Equation (1) is not unimodal. Furthermore, each chromosome consists of four genes corresponding to the four threshold values (α^+, α^-, β^+, β^-) as depicted in Figure 5. The initial population size is 20 (this size was determined empirically).

The algorithm creates the crossover children by combining pairs of parents in the current population. At each coordinate of the child's chromosome, the crossover function randomly selects the gene at the same coordinate from one of the two parents and assigns it to the child.

In order to help motivate the crossover function, we consider the two chromosomes A and B depicted in Figure 6. Assume that the chromosomes A and B consist of the four genes (α_1^+,α_1^-,β_1^+,β_1^-) and (α_2^+,α_2^-,β_2^+,β_2^-), respectively. The algorithm randomly selects the gene at the same coordinate from one of the chromosomes A and B and then assigns it to child C. Thus, the chromosome C may be (α_1^+,α_2^-,β_2^+,β_1^-).

The algorithm creates the mutation child (g_1, g_2, g_3, g_4) by randomly changing the genes of the parent chromosome (α^+, α^-, β^+, β^-). Suppose that the first two genes α^+ and α^- are in the range $[a, b]$, while the last two genes β^+ and β^- are in the range $[c, d]$. The algorithm first randomizes a chromosome (t_1, t_2, t_3, t_4) by using the Gaussian distribution. Next, one would prefer that the genes in the mutation child are also in the corresponding ranges. Thus, for each gene at the same coordinate from the parent, the algorithm uses either one of the following Equations (12) or (13) to create the corresponding gene for the mutation child:

$$g_1 = ((\alpha^+ \text{ or } t_1) \text{ or } a) \text{ and } b,\ g_2 = ((\alpha^- \text{ or } t_2) \text{ or } a) \text{ and } b \tag{12}$$

$$g_3 = ((\beta^+ \text{ or } t_3) \text{ or } c) \text{ and } d,\ g_4 = ((\beta^- \text{ or } t_4) \text{ or } c) \text{ and } d \tag{13}$$

In order to help motivate the mutation function, let us consider a parent chromosome, say, (2, 1, 5, 7). Assume that (α^+, α^-) are in the range [0, 3], while (β^+, β^-) are in

g_1	g_2	g_3	g_4
((2 or 10) or 0) and 3 = 2	((1 or 6) or 0) and 3 = 3	((5 or 3) or 0) and 10 = 2	((7 or 7) or 0) and 10 = 2

Fig. 7. An illustrative example of the mutation function

the range [0, 10]. Also suppose that the chromosome, which is created by using the Gaussian distribution, is (10, 6, 3, 7). The mutation child is presented in Figure 7. The GA stops if there is no improvement in the fitness function during successive iterations.

3 A Computational Study

3.1 Experimental Methodology

From the 768 patients, the HBA divided the PID dataset into a training dataset T with 576 patients and a testing dataset with 192 patients.

Suppose that we are given a certain 3-tuple of the unit penalty costs (C_{FP}, C_{FN}, C_{UC}). The experiments were done as follows:

Step 1: The original algorithm was first trained on the training dataset T and then derived the value for TC by using the testing dataset.
Step 2: The HBA was trained on the training dataset T_1 as described in Section 2.2 and then derived the value for TC by also using the testing dataset. It was assumed that β^+ and β^- were in [0, 2] while α^+ and α^- were in [0, 20].
Step 3: Compare the two values for TC returned in steps 1 and 2, respectively.

On the other hand, if we are given different values for the 3-tuple (C_{FP}, C_{FN}, C_{UC}), then we expect that the value for TC after controlling the fitting and generalization problems would be less than or at most equal to what was achieved by the original algorithms.

3.2 Experimental Results

The experiments were run on a PC with 2.8GHZ speed and 3GB RAM under the Windows XP operating system. The original classification algorithms used in these experiments are based on SVMs, ANNs, and DTs. There were more than 54 experiments done on the PID dataset with different values for the 3-tuple (C_{FP}, C_{FN}, C_{UC}). Furthermore, we used the libraries in Neural Network Toolbox 6.0, Genetic Algorithm and Direct Search Toolbox 2.1, and Statistics Toolbox 6.0 [13] for implementing the classification algorithms, the GA approach, and the density estimation approach. The experimental details are as follows:

Case 1: At first we studied the case of a 3-tuple (C_{FP}, C_{FN}, C_{UC}) in which the application would not penalize for the unclassifiable cases while the application would penalize at the same cost, say one unit, for the other two types of error. Under this scenario, the problem is equivalent to the evaluation of the current classification algorithms which

Table 2. Results for minimizing $TC = 1 \times RATE\_FP + 1 \times RATE\_FN$ on the PID dataset

Algorithm	*RATE_FP*	*RATE_FN*	*RATE_UC*	*TC*	% of improvement
SVM	0	74	0	74	
DT	27	36	0	63	
ANN	22	39	0	61	
SVM-HBA	0	10	0	10	86.49%
DT-HBA	0	16	0	16	74.60%
ANN-HBA	0	10	0	10	83.61%

Table 3. Results for the PID dataset

Algorithm	% Accuracy	% of improvement
[6]	76.0%	
[7]	77.6%	
[8]	77.6%	
[9]	78.6%	
[10]	81.8%	
[11]	77.7%	
SVM-HBA	94.79%	16.57%
ANN-HBA	94.79%	16.57%
DT-HBA	91.67%	13.45%

require either positive or negative outputs (see Table 4). Thus, the objective function in this case was assumed to be:

$$TC = 1 \times RATE\_FP + 1 \times RATE\_FN.$$

The results are presented in Table 2. In this case, Table 2 shows the three rates and the value of TC obtained from the algorithms. The notation "SVMHB" means that the HBA used the classification models first obtained by using the SVM algorithm before controlling the fitting and generalization problems. Two similar notations are used for DT-HBA (the Decision Tree algorithm and the HBA) and ANN-HBA (the Artificial Neural Network algorithm and the HBA). Table 2 presents that after 100 generations, SVM-HBA, DT-HBA, and ANN-HBA found the optimal TC to be equal to 10, 16, and 10 units, respectively. These values of TC were less than the average value of TC achieved by the original algorithms (i.e., the SVM, DT, and ANN) by about 81.57%. The values for α^+, α^-, β^+, β^- when ANN-HBA found the optimal TC by using the GA approach are 0.39, 18, 0.23, and 0.35, respectively.

Table 3 presents a comparison between the achieved classification percentages of the different classification algorithms. Clearly, the results by the HBA when it was combined with the traditional approaches were more accurate than those by the stand alone algorithms.

Case 2: Now we consider a case in which the application would penalize the same way, say three units, for the false-positive, the false-negative, and the unclassifiable cases. Thus, the objective function in this case was assumed to be:

$$TC = 3 \times RATE\_FP + 3 \times RATE\_FN + 3 \times RATE\_UC.$$

The results are presented in Table 4. In this case, Table 4 shows that after 100 generations, SVM-HBA, DT-HBA, and ANN-HBA found the optimal value for TC which was less than the value of TC achieved by the original algorithms by about 50.48%.

Table 4. Results for minimizing $TC = 3 \times RATE\_FP + 3 \times RATE\_FN + 3 \times RATE\_UC$ on the PID dataset

Algorithm	*RATE_FP*	*RATE_FN*	*RATE_UC*	*TC*	% of improvement
SVM	0	74	109	549	
DT	27	36	118	543	
ANN	22	39	118	537	
SVM-HBA	2	40	54	288	47.54%
DT-HBA	1	61	24	258	52.49%
ANN-HBA	1	57	29	261	51.40%

Case 3: Now we consider a case in which the application would penalize much more for the falsenegative cases than for the other types of error. Thus, the objective function in this case was assumed to be:

$$TC = 1 \times RATE\_FP + 20 \times RATE\_FN + 3 \times RATE\_UC.$$

The results are presented in Table 5. In this case, Table 5 shows that after 100 generations, SVM-HBA, DT-HBA, and ANN-HBA found the optimal value for TC which was less than the value of TC achieved by the original algorithms by about 51.59%.

Table 5. Results for minimizing $TC = 1 \times RATE\_FP + 20 \times RATE\_FN + 3 \times RATE\_UC$ on the PID dataset

Algorithm	*RATE_FP*	*RATE_FN*	*RATE_UC*	*TC*	% of improvement
SVM	0	74	109	1,807	
DT	27	36	118	1,101	
ANN	22	39	118	1,156	
SVM-HBA	0	16	105	635	64.86%
DT-HBA	5	10	136	613	44.32%
ANN-HBA	0	10	143	629	45.59%

We also experimented with the following different objective functions on this dataset:

$$TC = 20 \times RATE\_FP + 2 \times RATE\_FN + 1 \times RATE\_UC,$$
$$TC = 20 \times RATE\_FP + 20 \times RATE\_FN + 1 \times RATE\_UC,$$
$$TC = 20 \times RATE\_FP + 1 \times RATE\_FN + 20 \times RATE\_UC,$$
$$TC = 1 \times RATE\_FP + 20 \times RATE\_FN + 20 \times RATE\_UC,$$
and $TC = 3 \times RATE\_FP + 6 \times RATE\_FN$.

In all these tests we concluded that the HBA always found the optimal combinations of α^+, α^-, β^+, and β^- by using the GA approach in order to minimize the value of TC. Furthermore, the value for TC in all these cases was significantly less than or at most equal to what was achieved by the original algorithms.

4 Conclusions

Millions of people in the United States and the world have diabetes. Many of these people do not even know they have it. The ability to predict diabetes early plays an important role for the patient's treatment process. However, the correct prediction percentage of current algorithms is oftentimes low. Thus, this chapter applied a new approach, called the Homogeneity-Based Algorithm (HBA), for enhancing the diabetes prediction. That is, the HBA is first used in conjunction with traditional classification approaches (such as SVMs, DTs, ANNs). A GA approach was then used to find optimal (or near optimal) values for the four parameters of the HBA. The Pima Indian diabetes dataset was used for evaluating the performance of the HBA. The obtained results appear to be very important both for accurately predicting diabetes and also for the data mining community, in general.

References

[1] Asuncion, A., Newman, D.J.: UCI-Machine Learning Repository. School of Information and Computer Sciences. University of California, Irvine, California, USA (2007)

[2] Pham, H.N.A., Triantaphyllou, E.: The Impact of Overfitting and Overgeneralization on the Classification Accuracy in Data Mining. In: Maimon, O., Rokach, L. (eds.) Soft Computing for Knowledge Discovery and Data Mining, Part 4, ch. 5, pp. 391–431. Springer, Heidelberg (2007)

[3] Pham, H.N.A., Triantaphyllou, E.: An Optimization Approach for Improving Accuracy by Balancing Overfitting and Overgeneralization in Data Mining (January 2008) (submitted for publication)

[4] American Diabetes Association (2007), `http://www.diabetes.org/home.jsp`

[5] World Health Organization, Diabetes Mellitus: Report of a WHO Study Group. Geneva: WHO, Technical Report Series 727 (1985)

[6] Smith, J.W., Everhart, J.E., Dickson, W.C., Knowler, W.C., Johannes, R.S.: Using the ADAP learning algorithm to forecast the onset of diabetes mellitus. In: Proceedings of 12th Symposium on Computer Applications and Medical Care, Los Angeles, California, USA, pp. 261–265 (1988)

[7] Jankowski, N., Kadirkamanathan, V.: Statistical control of RBF-like networks for classification. In: Gerstner, W., Hasler, M., Germond, A., Nicoud, J.-D. (eds.) ICANN 1997. LNCS, vol. 1327, pp. 385–390. Springer, Heidelberg (1997)

[8] Au, W.H., Chan, K.C.C.: Classification with degree of membership: A fuzzy approach. In: Proceedings of the 1st IEEE Int'l Conference on Data Mining, San Jose, California, USA, pp. 35–42 (2001)

[9] Rutkowski, L., Cpalka, K.: Flexible neuro-fuzzy systems. IEEE Transactions on Neural Networks 14, 554–574 (2003)

[10] Davis IV, W.L.: Enhancing Pattern Classification with Relational Fuzzy Neural Networks and Square BKProducts. PhD Dissertation in Computer Science, pp. 71 - 74 (2006)

[11] Michie, D., Spiegelhalter, D.J., Taylor, C.C.: Machine Learning, Neural and Statistical Classification, ch. 9. Series Artificial Intelligence, pp. 157–160. Prentice Hall, Englewood Cliffs (1994)

[12] Duda, R.O., Hart, P.E.: Pattern Classification and Scene Analysis, pp. 56–64. Wiley Publisher, Chichester (1973)

[13] Artificial Neural Network Toolbox 6.0 and Statistics Toolbox 6.0, Matlab Version 7.0, http://www.mathworks.com/products/

Compression of Electrocardiogram Using Neural Networks and Wavelets

Marcus Vinícius Mazega Figueredo, Júlio César Nievola, Sérgio Renato Rogal Jr., and Alfredo Beckert Neto

Programa de Pós-Graduação em Informática
Pontifícia Universidade Católica do Paraná
marcus@ppgia.pucpr.br

Summary. The real-time transmission of the electrocardiogram (ECG) in urgent situations can improve the chances of the patient. However, one of the greatest problems involving this kind of telemedicine application is the leakage of network bandwidth. ECG exams may generate too much data, which makes difficult to apply telecardiology systems in real life. This problem motivated many authors to look for efficient techniques of ECG compression, such as: transform approaches, 2-D approaches, similarity approaches and generic approaches. The present work proposes a new hypothesis: neural networks may be applied together with Wavelet transforms to compress the ECG. In this approach, the Wavelet transform acts as a pre-processor element for a multilayer perceptron neural network, trained with the backpropagation algorithm. The original signal was divided in two parts: the "plain" blocks and the "complex" ones. The "plain" blocks were compressed with a 40:1 ratio while the "complex" blocks were compressed with a 5:1 ratio. The use of both compressors guaranteed a compression rate of 28:1, approximately. The process obtained good grades in the quality aspect: percent root mean squared difference (1.846%), maximum error (0.1789) and standard derivation of errors (0.1044).

Keywords: ECG compression, data compression, wavelets, neural networks.

1 Introduction

The recent advances in telecommunications stimulated the development of telemedicine applications, such as telecardiology. The real-time transmission of electrocardiogram (ECG) in emergency situations can considerably improve the survival chances of the patient. For example, several studies showed that a 12-lead ECG exam, when realized inside an ambulance, can anticipate the necessity of the thrombolytic therapy avoiding a cardiac attack [1].

Nevertheless, one of the greatest problems of telecardiology involves the high utilization of telecommunications bandwidth. ECG exams can generate a significant amount of data, which makes unviable many daily applications. This problem motivated many authors to look for efficient ECG compression techniques [2, 3, 4, 5, 6, 7, 8, 9, 10, 11]. Generally, these many techniques can be grouped in four large collections: transform approaches, 2-D approaches, similarity approaches and generic approaches. In the last years, the methods utilizing Wavelet transforms have been largely investigated, because of their promising results.

R. Lee and H.-K. Kim (Eds.): Computer and Information Science, SCI 131, pp. 27–40, 2008.
springerlink.com

In general, this kind of method involves the following steps during the compression: the signal segmentation; the application of a Wavelet transform; the thresholding of the Wavelet coefficients; and, the application of an efficient compression method such as Huffman coding or Golomb coding. On the other hand, the expansion of the signal involves the decoding of the encoded Wavelet coefficients and the application of an inverse Wavelet transform.

The present work investigates the utilization of artificial neural networks and Wavelet transforms in ECG compression. The main concept is to replace the traditional coding algorithms for a non-linear compression method, based in a multilayer perceptron trained with the backpropagation algorithm.

2 Revision

2.1 Electrocardiogram

According to [12], electrocardiography is the study of the electrical activity of the heart, the record of this activity is the electrocardiogram (ECG) and the equipment that records it is the electrocardiograph. The ECG is one of the most important resources in the cardiology because it is cheap, very informative and non-invasive. Indeed, the Nobel Prize winner of medicine Bernardo A. Houssay affirmed in 1946: "a physician does not consider an exam complete without analyzing the electrical activity of the heart".

Figure 1 shows the schematic representation of a normal electrocardiogram highlighting its main components and patterns. It is relevant to consider that ECG is a non-stationary signal, where the high frequencies take place locally and the low frequencies happen during the entire signal.

2.2 ECG Compression

The telecardiology is one of the more mature applications of telemedicine. It can involve the real-time broadcast of the waveform or the transmission of single exams. Historically, the last one was more investigated by science. The reason for this is due to the bandwidth limitations, in the past. The recent advances in telecommunications, however, instigated the development of real-time applications [1].

In paper [2], it is presented an ECG compression method based in Wavelet transforms. The signal is segmented, separating each QRS complex. Each segment is subtracted from a template signal selected from a pattern dictionary. The residual signal and the template are coded with a Coiflet Wavelet expansion. The compression is obtained with the thresholding of the Wavelet coefficients. Normally, the ECG compression techniques utilize thresholding to remove the coefficients near to zero. Then, the significant coefficients are compressed using Huffman coding, for example. The employment of Huffman coding, however, needs the previous knowledge of the symbols that will be compressed. This characteristic is negative in situations where the ECG waveform is much different from the normal.

To avoid the problems of Huffman coding, some authors [3] proposed a compression schema with a uniform scalar dead zone quantiser for encoding the significant coefficients. The method also utilizes the Golomb coding for the zero sequences and the

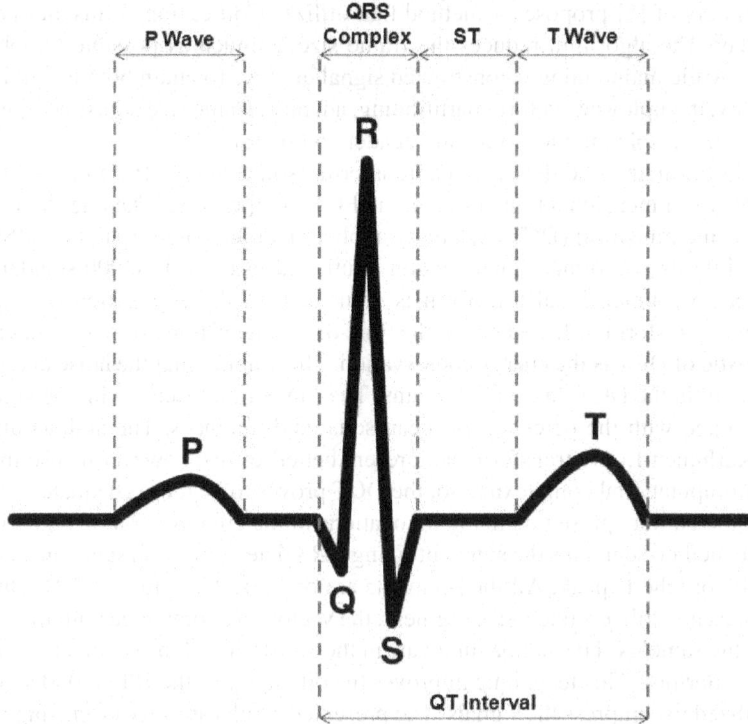

Fig. 1. A schematic representation of a regular electrocardiogram signal. The main components of the signal are highlighted: P wave, QRS complex, ST segment and T wave. All these components are clinically important since they are utilized during diagnosis. Therefore, the compression method cannot compromise their morphology.

Golomb-Rice coding for the non-zero sequences. As these algorithms do not need a previous knowledge of the symbols, the method becomes more robust. Additionally, the great computational efficiency of Golomb coding algorithm turns it into a good approach to real-time applications.

An ECG compression method without thresholding was proposed by the authors of [4]. They utilized a reversible round-off no recursive one-dimensional (1-D) discrete periodized Wavelet transform, executing a staged decomposition that is more resistant to errors. Normally, the increase of the compression rate implicates in the increase of the error. To avoid this, the authors tried to stay in the linear operation zone, reaching the desired compression rate with many stages.

In real-time applications, the transmission delay is a very relevant characteristic to be analyzed. Generally, high compression rates present high delays, because they utilize large portions of the waveform. The use of large time windows, however, makes this kind of technique unviable to applications where time is crucial; such is the case of telecardiology in ambulances.

The authors of [5] proposed a method that utilizes Coiflet transforms and presents a low delay. The algorithm reduces the frame size as much as possible to achieve a low delay, while maintaining reconstructed signal quality. To attain both low delay and high quality, it employs waveform partitioning, adaptive frame size adjustment, wavelet compression, flexible bit allocation, and header compression [5].

Instead of utilizing one-dimensional transforms, some authors [6] proposed the utilization of two-dimensional transforms in the ECG compression. The one-dimensional discrete cosine transform (DCT) is largely applied in audio compression like ISO/IEC MPEG, while its two-dimensional version is utilized in the JPEG2000 standard. Basically, the two-dimensional transform is equivalent to the application of one one-dimensional transform followed by another one-dimensional transform. One interesting characteristic of DCT is the energy conservation. This implies that the noise energy will be equivalent in the DCT and time domains. The same thing occurs with the signal-to-noise ratio and with the percent root-mean-squared differences. The authors affirmed that the Karhunen-Loeve transform can present better results, however its use involves a higher computational complexity. So, the DCT proves to be more adequate.

In [7], the authors presented many innovations in the compression of very irregular ECG. The method segments the signal utilizing QRS detection. Each segment starts 130 samples before the R peak. All the segments are organized in a matrix 2-D. Once the method does not utilize a fixed-size segment, the vectors are completed with the average values of the samples. The authors innovate in the reordering of the segments according to their similarities. This technique improves the efficiency of the JPEG2000 algorithm that is utilized to compress the signals. The presented results are very promising and the method is very efficient in the compression of irregular waveforms. The methodology is very efficient in the transmission of complete records, but not for real-time applications, because it utilizes large time windows (approximately, 10 minutes).

The great regularity of ECG suggests that similarity-based methods can achieve good results. In this kind of approach, the signal is usually segmented according the RR interval, utilizing one of the many available QRS detection algorithms. Nevertheless, this kind of approach is not very adequate in situations where the QRS detection does not works efficiently, such as the cases of ventricular tachycardia and ventricular fibrillation. In order to overcome this inconvenience, the authors of [8] proposed the utilization of Bezier cubic curves to identify the important points of the signal. These points simplify the signal, allowing the utilization of k-means algorithm to group similar waves. The patterns are stored in a template dictionary, avoiding the repetition of similar waves. This approach presents results similar to the methods based in Wavelets.

Recently, a multidimensional multiscalar parser (MMP) was successfully utilized in the ECG compression [9]. The periodic nature of ECG makes this kind of technique a good alternative; however, some ECG anomalies can affect the performance of MMP. So, they must be preprocessed to avoid the loss of relevant information.

An interpolation method was proposed in [10]. The signals are segmented so each cardiac beat is related to a template stored in a patterns dictionary. This method presented results that are similar to the Wavelets [11] and JPEG2000 [12] approaches.

A predictive method was proposed in [13]. An auto-regression technique is utilized to predict a k-th sample based in the k-1 samples before it. The method presented a low error rate against many types of abnormal signals, like arrhythmias, fibrillations, and others. Nevertheless, the compression rate was inferior to results described in the state of art.

Artificial neural networks were trained with backpropagation algorithm to compress ECG [14]. The signal was segmented with a fixed size and the training aimed to predict some portions of the waveform. The technique achieved high compression rates: 47:1 and 155:1. Nonetheless, this approach also revealed to be very sensible to the presence of noise. Indeed, the measured signal-to-noise ratio was unsatisfactory. An important contribution of this work was to show the promising compression potential of artificial neural networks. A previous work [15] with neural networks also showed great compression rates. However, the technique was very sensible to the variability of ECG waves. The Hebbian training was applied in the development of another ECG compression method based in neural networks [16]. The work achieved a compression rate superior to 30:1 and percent root-mean-squared differences inferior to 5%.

3 Methodology

3.1 Hypothesis

This work raises the following hypothesis: a neural network can be trained to compress Wavelet coefficients of an ECG signal, achieving results similar to the state of art.

3.2 Database

The MIT-BIH Arrhythmia Database [17] was utilized in this work. This database was created from an arrhythmias research conducted by the Boston's Beth Israel Hospital [18] and the MIT [19], between the years of 1975 and 1979. Totally, forty-seven subjects were monitored and analyzed by BIH Arrhythmia Laboratory, generating 48 hours and 30 minutes of ECG records. Twenty-three records were selected from a collection of 4.000 24 hours exams of intern (60%) and extern (40%) patients. Other 25 records were selected to represent important but rare arrhythmias. The records were digitalized with 360 samples per second, using 11 bits to map 10 mV. Two or more specialists analyzed independently each cardiac beat, generating a total of 110.000 annotations.

3.3 Segmentation

An algorithm sliced the ECG in segments, which start 250 ms before the QRS complex. This time interval was selected to guarantee that each segment will contain just one cardiac beat. A segment could have the maximum length of 1024 samples, when the QRS complex is not present. This technique was firstly proposed by [2].

3.4 Classification

Each segment was sliced in the middle, generating two blocks with the same size, as proposed by [5]. The standard deviation was calculated for both blocks, as in Eq. (1), where xi represents an ECG sample, represents the average value of the samples and N

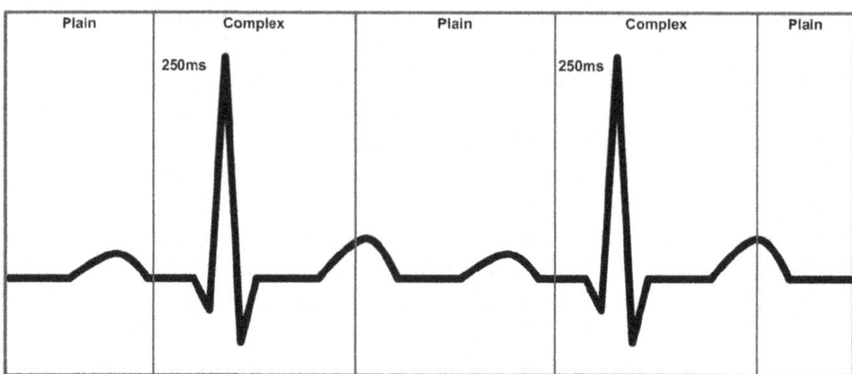

Fig. 2. Segmented and labeled ECG signal, as defined in methodology. The ECG is sliced 250 ms prior to the cardiac peak. Each segment is sliced in 2 blocks, which were labeled as "plain" or "complex". The "complex" blocks contain more relevant information, so they should not lose much information during compression. On the other hand, the "plain" blocks can lose more information without compromising the clinical validity.

represents the total number of samples. If the deviation is bigger than a defined threshold, the block is classified as "complex". If it is not, it is classified as "plain". The threshold limit is calculated empirically. Figure 2 shows the processed ECG according to these steps.

$$\textbf{Deviation} = \sqrt{\frac{{}^{\mathrm{N}}_{\mathrm{i=1}}\sum\left(\mathrm{x}-\overline{X}\right)^2}{N}} \tag{1}$$

3.5 Wavelet Transform

A discrete Wavelet transform was utilized to process each block with ECG samples, as shown in figure 3. The Coiflet Wavelet was selected due to promising results of its application in other studies [2]. The resulting Wavelet coefficients were preprocessed in the following way:

1. The signal of each coefficient is removed by the application of an "absolute" function;
2. The coefficients are reordered in a descendent way;
3. The first 40 coefficients are selected;
4. The meta-information (signal +/- and original position) of each selected coefficient is stored in an auxiliary data file.

3.6 Artificial Neural Networks

Two different neural networks were trained to compress and expand the ECG signals. The first network was trained only with "complex" blocks, while the second one worked just with "plain" blocks. The main concept behind this strategy was proposed by [5].

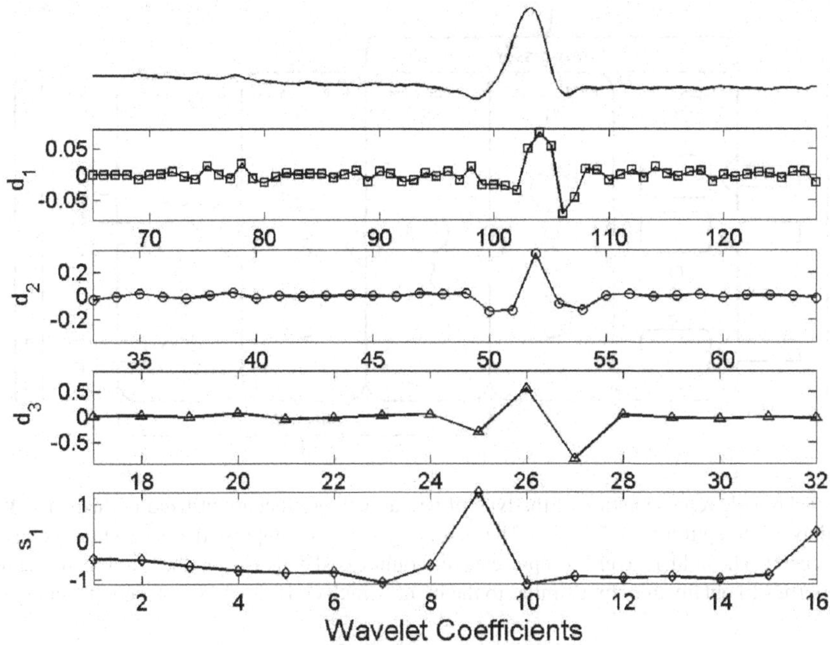

Fig. 3. ECG blocks are decomposed in Wavelet coefficients by a Coiflet transform

Generally, the "plain" blocks have less information than the "complex" blocks, so they may be compressed with a more aggressive compression rate without compromising the quality of the process. On the other hand, the "complex" blocks should not receive a high compression rate. The final compression rate of the method is obtained by the weighted average of both compressors.

Both neural networks are multilayer perceptrons with 5 layers and were trained with the backpropagation learning algorithm, as shown in figure 4. The number of neurons and the compression rates of the networks are shown in table 1.

Table 1. Number of Neurons in Each Layer

Network	Compression Rate	X	Y1	Y2	Y3	Z
"Plain" blocks	40:1	40	39	1	39	40
"Complex" blocks	5:1	40	26	8	26	40

3.7 Metrics

Four metrics are normally utilized to analyze the ECG compression techniques: compression rate (CR), percent root mean square difference (PRMSD), maximum error (ME) and standard deviation of errors (SDE), where y is an original sample, ŷ is a processed sample and N is the total number of samples.

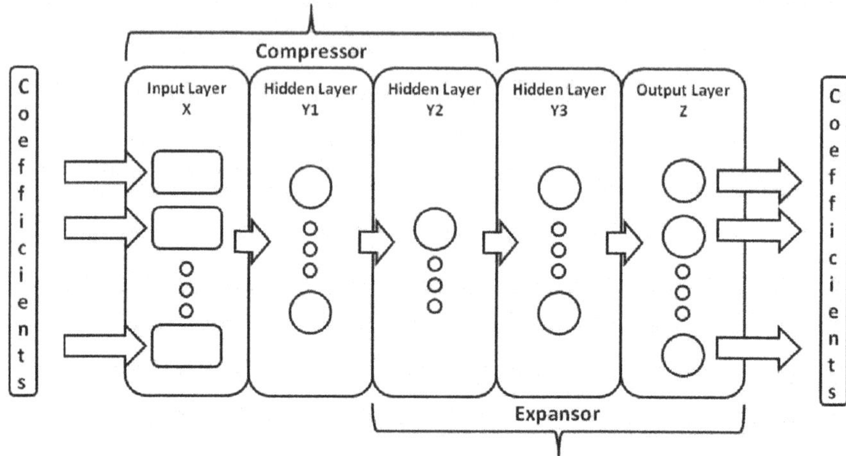

Fig. 4. Schematic representation of the type of neural network that are utilized in this work. It is a multilayer perceptron with 5 layers. The first 3 layers does compress the data while the last 3 reconstruct it. The hidden layer Y2 represents the compressed data. During the training, the neural network tries to obtain an output similar to the input, which consists of Wavelet coefficients.

The compression rate (2) is calculated as the number of bits in the original ECG signal in relation to the total number of bits in the compressed data:

$$\mathbf{CR} = \mathbf{a} \div \mathbf{b} \tag{2}$$

The percent root mean square difference (3) measures normalized root mean square difference between the compressed and original ECG sequences. There are many forms to calculate this metric. The formula that is utilized in this work was selected because it is the most common in the literature, which facilitates the comparison with other works:

$$\mathbf{PRMSD} = \sqrt{\frac{{}_{n=1}^{N}\sum\left[\mathbf{y(n)}-\hat{y}(\mathbf{n})\right]^2}{{}_{n=1}^{N}\sum\left[\mathbf{y(n)}-\mathbf{1024}\right]^2}} \times \mathbf{100\%} \tag{3}$$

The maximum error (4) is the largest absolute value of all errors, where an error (5) is the difference between the sample in the original ECG signal and the compressed one:

$$\mathbf{error(n)} = \mathbf{y(n)}-\hat{y}(\mathbf{n}) \tag{4}$$

$$\mathbf{ME} = \mathbf{max}_{n=1}^{N} \left|\mathbf{error(n)}\right| \tag{5}$$

The standard deviation of errors (6) is calculated by:

$$\mathbf{SDE} = \sqrt{\frac{{}_{n=1}^{N}\sum\left[\mathbf{error(n)}-\overline{error}\right]^2}{\mathbf{N-1}}} \tag{6}$$

Table 2. Results

Metric	Object	Measure
CR	"Plain" blocks	40:1
	"Complex" blocks	5:1
	Both blocks	28:1
PRMSD	"Plain" blocks	2.312%
	"Complex" blocks	0.387%
	Both blocks	1.846%
ME	"Plain" blocks	0.1789
	"Complex" blocks	0.0232
	Both blocks	0.1789
SDE	"Plain" blocks	0.1038
	"Complex" blocks	0.0442
	Both blocks	0.1044

4 Results

The proposed method was tested against the MIT-BIH Arrhythmia Database, obtaining the results that are shown in the table 2. The compression algorithm is shown in figure 5. The results from the proposed methodology were compared to others available in the scientific literature [3, 4, 5, 7, 8, 10, 11, 20]. This analysis can be observed in table 3, where the compression rate and the percent root mean square difference of each method

Table 3. Comparison with Other Methods

Author	CR	PRMSD	Author	CR	PRMSD
Proposed Method	28:1	1.85	Proposed Method	28:1	1.85
[3]	20:1	6.13	[8]	33.4:1	7.6
[3]	16:1	4.67	[8]	30.7:1	7.1
[3]	12:1	3.46	[8]	25.7:1	7.2
[3]	10:1	2.93	[8]	22.6:1	7.8
[3]	8:1	2.39	[10]	20:1	3.41
[4]	20:1	6.10	[10]	16:1	2.95
[5]	24.77:1	7.32	[10]	10:1	2.14
[5]	5.63:1	7.33	[10]	8:1	1.83
[7]	24:1	4.06	[10]	4:1	1.02
[8]	13.7:1	6.6	[11]	20:1	7.52
[8]	14.4:1	4.6	[11]	16:1	5.46
[8]	9.3:1	5.7	[11]	12:1	3.82
[8]	8.6:1	3.9	[11]	10:1	2.11
[8]	8.5:1	4.0	[11]	8:1	2.5
[8]	40.2:1	7.9	[20]	24:1	8.10

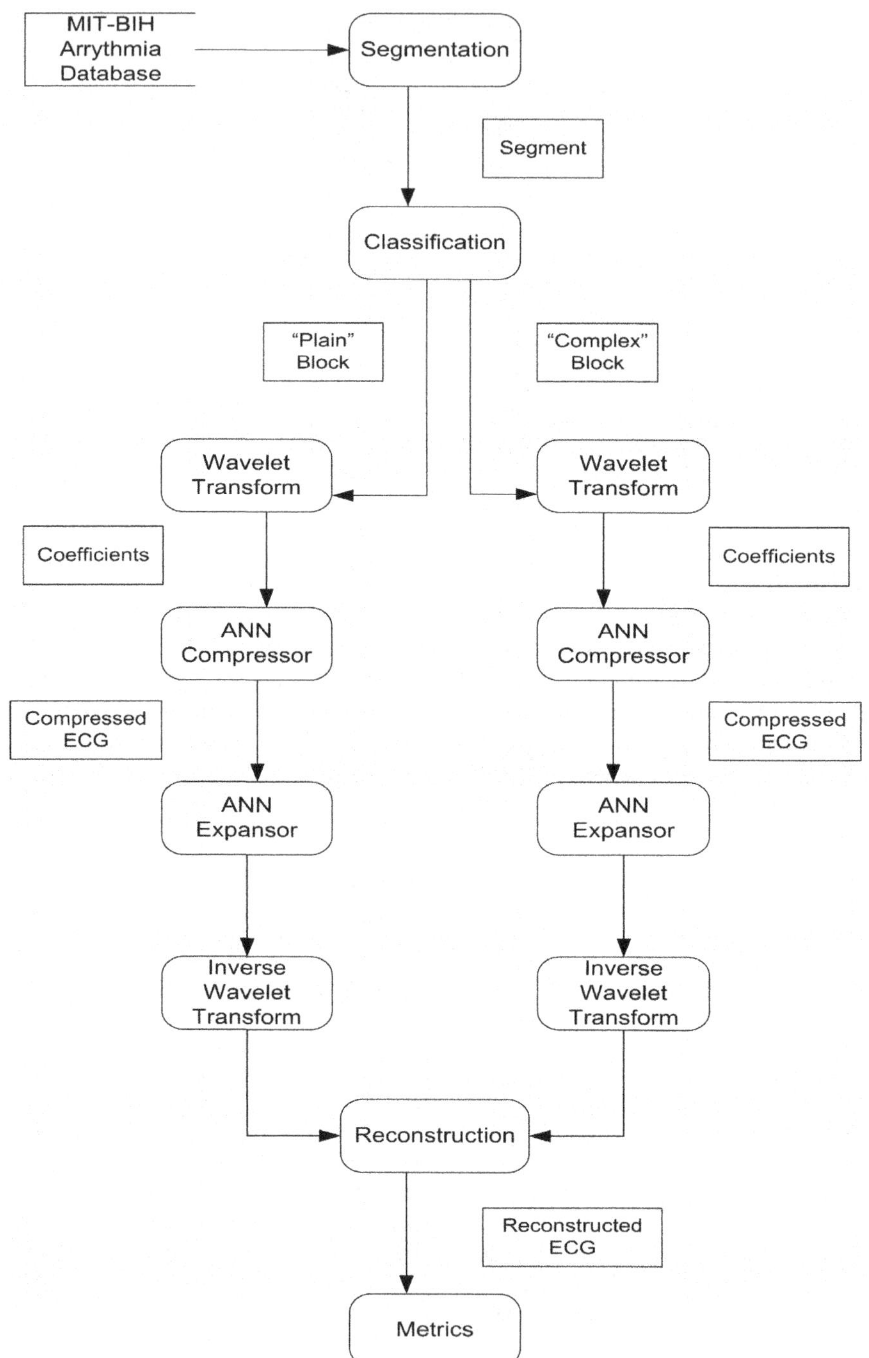

Fig. 5. Overview of the compression algorithm, during tests

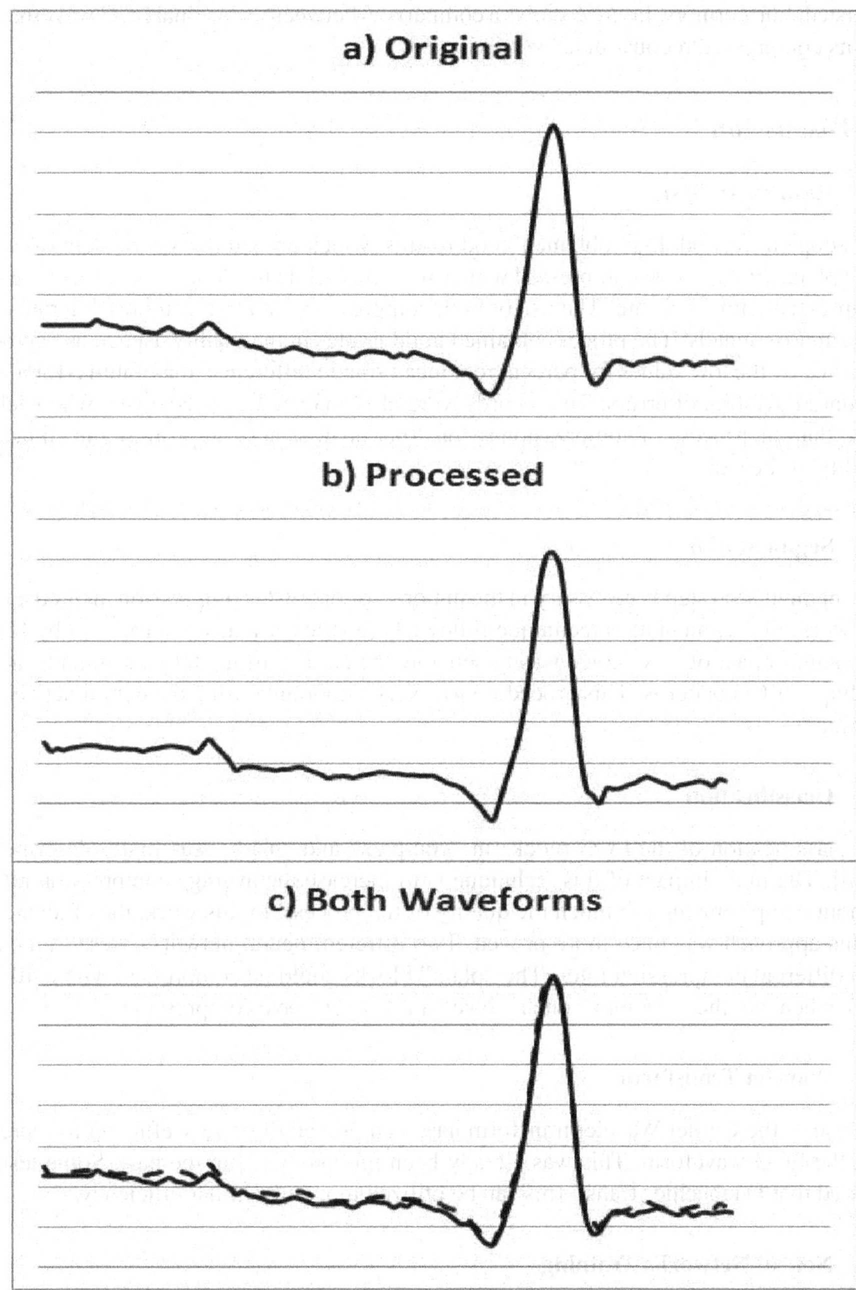

Fig. 6. Processed ECG waveform. In (a), it can be seen the original waveform, without processing. In (b), there is the same waveform, after compression and reconstruction. The great similarities between the two waveforms can be observed in (c), where both signals are plotted together.

are listed. Furthermore, figure 6 shows a comparison between an original ECG waveform and its compressed/reconstructed version.

5 Discussion

5.1 Results Analysis

The adopted methodology obtained good results, which are similar to the state of art. The "plain" blocks were compressed with a 40:1 rate, while the "complex" blocks were compressed with a 5:1 rate. The use of both compressors lead to a compression rate of 28:1, approximately. The process obtained good grades in the quality aspect, as can be observed by the low values for percent root mean square difference, maximum error and standard deviation of errors. The records were also analyzed by specialists, who evaluated that the process of compression/reconstruction does not compromise the clinical validity of the exam.

5.2 Segmentation

The segmentation step is necessary in the majority of the ECG compression methods. In this work, the segmentation technique followed the strategy that was proposed by [2]. The main impact of this choice was to separate the cardiac beats in two segments, according to ECG patterns. This procedure was very important during the neural network training.

5.3 Classification

The classification of the ECG blocks in "complex" and "plain" was firstly described by [5]. The main impact of this technique is to increase the average compression rate without compromising too much the quality of the process. In this work, the efficiency of this approach was once more proved. Two different neural networks were trained, with different compression rates. The "plain" blocks could be compressed with a 40:1 ratio, whenever the "complex" one received a less aggressive compression.

5.4 Wavelet Transform

The use of the Coiflet Wavelet transform has been proved to be very efficient to represent the ECG waveform. This was already been proven by [2] in the past. Some tests showed that Daubechies transforms can be utilized too, with similar efficiency.

5.5 Neural Networks Training

Many approaches were tried to make a multilayer perceptron able to compress Wavelet coefficients. Firstly, a 3 layers MLP was trained, however the results were not good. This happened due to the great disparity between the number of input neurons and the number of hidden neurons. Indeed, the initial networks had not sufficient neurons and connections to establish a reasonable compression/rebuilding process. Nevertheless, the

5 layers architecture proved to be capable to compress and reconstruct the coefficients efficiently. The 2 additional inner layers made the network more robust during the learning process.

Also, it is important to salient the importance of the two steps in the preprocessing: the ordering and the elimination of the negative signal from the coefficients. In fact, in the beginning of the investigation, the negatives coefficients were not normalized, which affected badly the training process. The application of an "absolute" function, however, proved to solve this issue. In the same way, the descendent ordering of the coefficients highlighted the ECG pattern, which improved the effectiveness of the learning process.

6 Conclusion

The main contribution of this work is to demonstrate that neural networks can be applied with Wavelets in the ECG compression successfully. The use of Wavelets transforms influences positively the learning process, because the time-frequency domain is more robust to signal noises and makes the ECG patterns more evident.

The obtained results are similar to the state of art, what indicates that better results may be found in future studies involving neural networks and Wavelets in the ECG compression. Other Wavelet families may be tested, besides the Coiflets. Also, other learning algorithms can be utilized instead of backpropagation. Indeed, one of the main objectives of this work was to demonstrate the feasibility of this approach and to stimulate more advanced studies in this way.

Finally, it is relevant to verify whether this same approach may be effective in the compression of other waveforms, besides the ECG. For example, they can be cited signals like the electroencephalogram, the electromyogram, audio, and others. Indeed, one of the next steps of this work is to test and evaluate the effectiveness of the proposed method in these areas.

Acknowledgment

We are pleased to acknowledge the funding support from the Coordernação de Aperfeiçoamento de Pessoal de Nível Superior (CAPES) - Brazil, which provided scholarships for the authors.

References

1. Sedgewick, M.L., Dalziel, K., Watson, J., Carrington, D.J., Cobbe, S.M.: Perfomance of an established system of first responder out-of-hospital defibrillation: the results of the second year of the heartstar Scotland project in the Ulstein style, vol. 6, pp. 75–78 (1993)
2. Alesanco, A., Olmos, S., Istepanian, R.S.H., García, J.: Enhanced Real-Time ECG Coder for Packetized Telecardiology Applications. IEEE Transactions on Information Technology in Biomedicine 10(2), 229–236 (2006)
3. Chen, J., Zhang, Y., Shi, X.: ECG Compression Based on Wavelet Transform and Golomb Coding. Electronic Letters 42(2), 322–324 (2006)

4. Ku, C.T., Wang, H.S., Hung, K.C., Hung, Y.S.: A Novel ECG Data Compression Method Based on Nonrecursive Discrete Periodized Wavelet Transform. IEEE Transactions on Biomedical Engineering 53(12), 2577–2583 (2006)
5. Kim, B., Yoo, S.K., Lee, M.H.: Wavelet-Based low-Delay ECG Compression Algorithm for Continuous ECG Transmission. IEEE Transactions on Information Technology in Biomedicine 10(1), 77–83 (2006)
6. Alexandre, E., Pena, A., Sobreira, M.: On the Use of 2-D Coding Techniques for ECG Signals. IEEE Transactions on Information Technology in Biomedicine 10(4), 809–811 (2006)
7. Chou, H.H., Chen, Y.J., Shiau, Y.C., Kuo, T.S.: An Effective and Efficient Compression Algorithm for ECG Signals with Irregular Periods. IEEE Transactions on Biomedical Engineering 53(6), 1198–1205 (2006)
8. Henriques, J., Brito, M., Gil, P., Carvalho: Searching for Similarities in Nearly Periodic Signals with Applications to ECG Data Compression. In: 18th International Conference on Pattern Recognition, Hong Kong, pp. 942–945 (2006)
9. Carvalho, M.B., Silva, A.B., Finamore, W.: Multidimensional Signal Compression using Multiscale Recurrent Patterns. Signal Processing Image and Video Coding beyond Standards 82, 3201–3204 (2002)
10. Filho, E.B.L., Eduardo, A.B., Junior, W.S.S., Carvalho, M.B.: ECG Compression using Multiscale Recurrent Patterns with Period Normalization. In: Proceedings of the IEEE International Symposium, Greece, p. 4 (2006)
11. Lu, Z., Kim, D.Y., Pearlman, W.A.: Wavelet Compression of ECG Signals by the Set Partioning in Hierarchial Trees (SPIHIT) Algorithm. IEEE Transactions on Biomedical Engineering 47(7) (2000)
12. Bilgin, A., Marcellin, M.W., Altbach, M.I.: Compression of Electrocardiogram Signals using JPEG2000 49(4) (November 2003)
13. Brito, M., Henriques, J., Antunes, M.: A Predictive Adaptive Approach to Generic ECG Data Compression. In: IEEE International Symposium on Intelligent Signal Processing, Portugal, pp. 32–37 (2005)
14. Kannan, R., Eswaran, C., Sriraam, N.: Neural Networks Methods for ECG Data Compression. In: Proceedings of the 9th International Conference on Neural Information Processing ICONIP 2002, vol. 5 (November 2002)
15. Zhao, Y., Wang, B., Zhao, W., Dong, L.: Applying incompletely connected feedforward neural network to ambulatory ECG data compression. Electronic Letters 33, 220–221 (1997)
16. Al-Hukazi, E., Al-Nashash, H.: ECG data compression using Hebbian neural networks. Journal of Medical Engineering & Technology (November 1996)
17. Physyonet. MIT-BIH ECG Arrythmia Database (2007) (Accessed 20 April 2007), `http://www.physionet.org/physiobank/database/mitdb`
18. Beth Israel Hospital Inc (2007) (Accessed 21 April 2007), `http://www.bih.harvard.edu`
19. Massachussets Institute of Technology (2007) (Accessed 21 April 2007), `http://www.mit.edu`
20. Lee, H., Buckley, K.: ECG Data Compression using Cut and Align Beats Approach and 2-D Transforms. IEEE Transactions on Biomedical Engineering 46(5), 556–564 (1999)

Multi-step Parallel PNN Algorithm for Distributed-Memory Systems

Akiyoshi Wakatani

Faculty of Science and Engineering
Konan University
8-9-1, Okamoto, Higashinada, Kobe, 658-8501, Japan
wakatani@konan-u.ac.jp

Summary. Parallel system with distributed memory is a promising platform to achieve a high performance computing with less construction cost. Applications with less communications, such as a kind of parameter sweep applications (PSA), can be efficiently carried out on such a parallel system, but some applications are not suitable for the parallel system due to a large communication cost. We focus on *PNN (Pairwise Nearest Neighbor)* codebook generation algorithm for VQ (Vector Quantization) compression algorithm and propose a parallel version of the PNN algorithm suitable for the parallel system with distributed memory, called "multi-step parallel PNN".

The multi-step parallel PNN is a modified version of the PNN algorithm that creates a different codebook than the original PNN does, thus the quality of a codebook created by using the multi-step parallel PNN may be worse than that of a codebook by the original PNN. However, our experimental results show that the quality of the codebook is almost same as that of the original one. We also confirm the effectiveness of the multi-step parallel PNN by the evaluation of the computational complexity of the algorithm and the preliminary experiment executed on a PC cluster system.

1 Introduction

VQ (vector quantization) is a lossy compression algorithm that consumes vast computing resources in encoding process[1]. Especially, the codebook generation is a key issue in order to achieve a high compression rate with keeping an excellent image quality. PNN (pairwise nearest neighbor) is one of the codebook generation methods, which produces an excellent codebook, but consumes vast computing time[2]. Therefore, PNN should be implemented with a parallel processing approach in order to reduce the computing time.

Recently, most parallel processing platforms equipped with distributed memory configurations. For example, Cell BE (Broadband Engine) is considered as a prominent next generation processor having a distributed memory system. Although it was originally designed for a processor suitable for streaming applications such as media processing, it also can be applied to other high performance computing applications. The Cell BE is a heterogeneous multi-core architecture that consists of a PPE processor (Power 6 architecture) and 8 SPE processors. The SPE cannot access the main memory directly, but it can access a Local Storage (LS) of 256 Kbyte. The SPE sends a DMA request to Memory Flow Controller (MFC) to fetch data from the main memory to the

R. Lee and H.-K. Kim (Eds.): Computer and Information Science, SCI 131, pp. 41–50, 2008.
springerlink.com

LS and then accesses the LS directly after completing the DMA request[3]. So, the LS of each SPE is considered as a kind of distributed memory.

On the other hand, many researchers have studied the PC Grid system and applications since the marvelous success of SETI@home[4]. The PC Grid can achieve a high performance computing by organizing a large number of idle computers in an university or a tremendous number of idle computers on Internet and consists of a master computer, which executes a main routine and controls others, and many worker computers, which execute divided slave tasks independently. In general, the worker computers do not exchange data with each other and the only master computer sends and receives data to/from the worker computers. However, since the data transfer is very expensive, only a kind of parameter sweep applications (PSA) is suitable for the PC Grid system, which iterates the same task with different data set[5].

Even on other shared memory platforms except for the Cell BE and the PC Grid, most applications should be implemented in a distributed memory manner in order to avoid cache miss penalties and achieve a high performance computing. Thus, we propose a modified PNN method suitable for distributed memory platforms.

In general, parallel algorithms require a variety of data exchange between computers, including local data exchange between neighboring computers, broadcast communication from a computer to other computers and reduction communication of data generated on all computers. Thus, parallel algorithms with such data exchanges cannot be efficiently implemented on the distributed memory system.

In order to overcome such a difficulty, we should consider an algorithm with a slight amount of data exchange that can be carried out on the distributed memory system. Although the quality of the algorithm, such as the quality of compressed image for the compression algorithm, may be degraded due to the modification of the algorithm, the execution time of the modified algorithm can be drastically improved by using plural computers on the parallel system and thus the increase of the application size can avoid the degradation of the quality of the algorithm. Therefore we aim to produce a new parallel algorithm suitable for the distributed memory system to reduce the execution time with keeping the quality of the algorithm.

Namely, the key to the integrated performance improvement of the algorithm on the parallel system is the following: 1) the efficient modification of the algorithm for the distributed memory system, 2) the scalability of the modified algorithm, 3) the reduction of the communication overhead between the master computer and the worker computers, 4) the minimization of the degradation of the quality of the modified algorithm.

2 PNN Algorithm

The VQ compression is one of the most important methods for compressing multimedia data, including images and audio, at high compression rate. The key to achieving high compression rate is to build an efficient codebook that represents the source data with the least quantity of bit stream.

Methods of generating a codebook for the VQ compression include the PNN[6] and the *LBG*[1] algorithms. Both methods require the vast computing resource to determine an efficient codebook. Although LBG algorithm has been studied more frequently than the PNN algorithm, we focus on the PNN algorithm because LBG algorithm sometimes finds non-optimal codebook with only a local minimum and then the algorithm must be iterated by beginning with a different initial codebook in order to avoid capturing the local minimum.

The squared Euclidean distance value of training vectors $vector_a$ and $vector_b$ weighted by the number of vectors, the number of vectors of a merged vector and the merged vector of $vector_a$ and $vector_b$ are defined as

$$d(S_a, S_b) = \frac{n_a \cdot n_b}{n_a + n_b} \cdot ||S_a - S_b||^2, \tag{1}$$

$$n_{a+b} = n_a + n_b, \tag{2}$$

$$S_{a+b} = \frac{n_a S_a + n_b S_b}{n_a + n_b} \tag{3}$$

where S_i is a training vector of $vector_i$ and n_i is the number of vectors of S_i if $vector_i$ is a merged vector[2]. Note that n_i is 1 if $vector_i$ is not a merged vector.

The PNN algorithm mainly consists of four steps: 1) select initial training vectors from source data, 2) calculate all the distance (or distortion) value between two vectors, 3) select the minimum distance from the calculated distance values and merge a vector pair with the minimum into one vector, and 4) iterate the above steps 2 and 3 until the number of vectors equals the size of codebook (=K). This algorithm can determinately find the optimal codebook after several iterations of the above steps, however the computational complexity of step 2 is very computationally expensive with the cost of $O(T^2)$ where T is the number of initial training vectors.

In order to alleviate the computational complexity of the algorithm, Franti proposed the variation of the PNN algorithm called *tau-PNN*[2]. A nearest neighbor vector (`nn`) of each vector (or training vector) is calculated in advance. Let `(imin, jmin)` be the pair selected at the third step of the PNN algorithm. The distance should be recalculated only for a vector whose `nn` is `imin` or `jmin` at the second step, so the computational complexity can be reduced dramatically with the cost of $O(T)$.

3 Related Works

The codebook generation algorithms for the VQ compression are mainly classified into non-hierarchical clustering algorithms and hierarchical clustering algorithms. One of the most popular method using the former approach is *k-means algorithm*. Dhillon and Modha described a parallel version of the k-means algorithm and implemented it on a distributed memory multicomputers[7]. Garg et al. presented a parallel version of BIRCH that utilized parallel k-means algorithm and showed the speedup of the parallel algorithm was up to 4.3 on an 8 CPU SMP system[8].

The PNN codebook generation is one of the typical hierarchical clustering algorithm using the modified weighted centroid metric. Olsen studied parallel implementations of hierarchical clustering algorithms with several distance metrics and evaluated their computational complexities on a PRAM[9].

To the best of our knowledge, a small number of prior studies have dealt with parallel versions of the PNN algorithm on SMP systems and PC cluster systems[10, 11]. This paper mentions a parallel approach suitable for a PC Grid system that requires the large communication cost for message exchanges.

4 Parallel PNN Algorithm

4.1 Base Algorithm

In this paper, we focus on the original PNN algorithm and propose a modified algorithm suitable for the distributed memory system. Let P be the number of worker computers in the parallel system. A naive version of parallel PNN is as follows:

1. Set n to T
2. On P worker computers in the parallel system, calculate the distances of $\frac{n}{P}$ training vectors to all n training vectors, find the minimum and send it to a master computer
3. On the master computer, find the global minimum among received data and broadcast it to the worker computers
4. On the worker computers, merge the pair with the minimum distance and set n to $n-1$
5. if $n = K$, stop the procedure, otherwise go back to step 2

This algorithm ("naive parallel PNN") seems to provide enough parallelism, but this algorithm is not appropriate for the distributed memory system because the step 3 requires the communication between the master and the worker computers and the communication cost in the distributed memory system is generally large. Note that the step 3 is within the iterative loop. In order to cope with this difficulty, a modified algorithm without expensive communication part is proposed as follows:

First of all, the intermediate value L is decided appropriately ($\frac{K}{P} \leq L \leq \frac{T}{P}$).

1. Set n to T/P
2. On P worker computers in the parallel system, calculate the distances between n training vectors, find the minimum and merge the pair with the minimum distance
3. Set n to $n-1$ and go to step 2 if $n \neq L$
4. From the worker computers, send each merged L vectors to a master computer
5. On the master computer, receive $P \times L$ vectors from the worker computers in total and set n to $P \times L$
6. On the master computer, calculate the distances between n training vectors, find the minimum and merge the pair with the minimum distance
7. Set n to $n-1$ and go to step 6 if $n \neq K$

This algorithm is called "multi-step parallel PNN". Although the naive parallel PNN requires a large amount of communication, the multi-step parallel PNN needs communication only in the step 4 and then the communication cost is slight because the step 4 is out of the iterative loop. Note that the worker computers mainly iterate the steps 2 and 3 and then the master computer iterates the steps 6 and 7. Although the computational complexity of the master computer in the steps 6 and 7 affects the total performance drastically, the parallelism of the steps 2 and 3 is large and thus the multi-step parallel PNN is suitable for the parallel system.

On the other hand, since the effective parallelism of the multi-step parallel PNN may be degraded due to the latter sequential part of the master computer, we also propose alternative version of the algorithm to distribute the latter part to the worker computers. First of all, we appropriately choose intermediate values $L_1, L_2, , , L_{\log P-1}$ ($\frac{K}{P} \leq L_0, L_1, , , L_{\log P-1} \leq \frac{T}{P}$ and $L_0 = L, \ L_{i+1} \leq 2L_i$). Then, $P/2$ worker computers merge $2L$ vectors into L_1 vectors in a PNN way, second, $P/4$ worker computers merge $2L_1$ vectors into L_2 vectors, and so on. Then, the master computer merges $2L_{\log P-1}$ vectors into K vectors, finally. This procedure consist of the PNN merge steps of $\log P$ in total. This is called "reduced (multi-step) parallel PNN". Although the reduced parallel PNN can enhance the parallelism of the latter part using worker computers, the communication between the master and the worker computers also increases. Therefore, the effectiveness and feasibility of the reduced parallel PNN should be confirmed empirically.

Moreover, since the multi-step parallel PNN is inherently different from the naive parallel PNN, the algorithm creates a different codebook from the original PNN. The multi-step parallel PNN merges T/P vectors into L vectors on every worker computer in the former part and then the master computer gathers all data from the worker computers, but the created codebook by the multi-step PNN may be worse than that by the naive parallel PNN. Therefore, the quality of the codebook generated by the multi-step parallel PNN should be evaluated empirically.

4.2 L Value

T training vectors should be distributed onto P computers, which create L merged vectors. Then each computer sends the L merged vectors to the master computer, which creates K vectors from received $L \times P$ vectors in a PNN way.

From the performace point of view, L should be as small as possible because the cost of the second PNN, which is a sequential part done on the master computer, can be slight or nothing. However, when L is small, the quality of the final K vectors may be

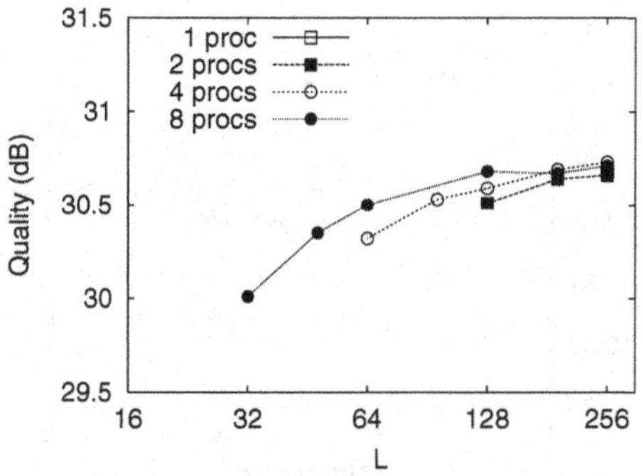

Fig. 1. The effect of the size of L ($K = 256$)

worse than when large L, bacause the quality of the codebook can be improved in the second PNN step in accordance with the size of L.

Figure 1 shows the relation of the size of L and the quality (PSNR) of the generated codebook. We compress the 8 bit gray-map image of the natural face ("lenna") with 512×512 pixels by using the multi-step parallel PNN with $T = 4096 (= 2^{12})$, $K = 256 (= 2^8)$ and varying $P = 1, 2, 4, 8$ and $L = K/P$ to K. The figure shows PSNR of the compressed image. Note that $L = 256$ and $P = 1$ is PSNR of the compressed image by using the original PNN algorithm.

As shown in the figure, when $L = K$, the quality of $P = 2, 4, 8$ is the same as that of $P = 1$, but as L decreases, the quality becomes worse. Although small L reduces the computational complexity as described later, we choose K as L in order to keep the quality of the compression algorithm. Hereafter, L is K as long as not specified.

5 Evaluation

5.1 Image Quality

As mentioned in the previous section, by using the multi-step parallel PNN, the parallelism can be enhanced but the quality of the codebook may be degraded. In order to evaluate the quality of the codebook, we compare the quality of face image (lenna.pgm)) by the multi-step parallel PNN with that by the original PNN. Both images are 8 bit graymap with the size of 512×512 and the size of a vector is 4×4. It is assumed that the multi-step parallel PNN with the number of computers of $P = 1, 2, 4, 8$ creates the codebook of the size of $K = 2^8, 2^9$ from the training vectors of the size of $T = 2^{12}$. Note that the multi-step parallel PNN with $P = 1$ means the original PNN.

The quality of the compressed images is shown Figure 2. The y axis stands for PSNR of compressed image compared with the original image. The computational complexity of the latter part of the multi-step parallel PNN on the master computer exceeds that of the original PNN when $K = 2^9$ and $P = 8$. So the cases of $K = 2^9$ and $P = 8$ are omitted. As shown in the figure, the difference between $P = 1$ and others is very slight for any cases. For example, the PSNR of the image compressed by the codebook of the

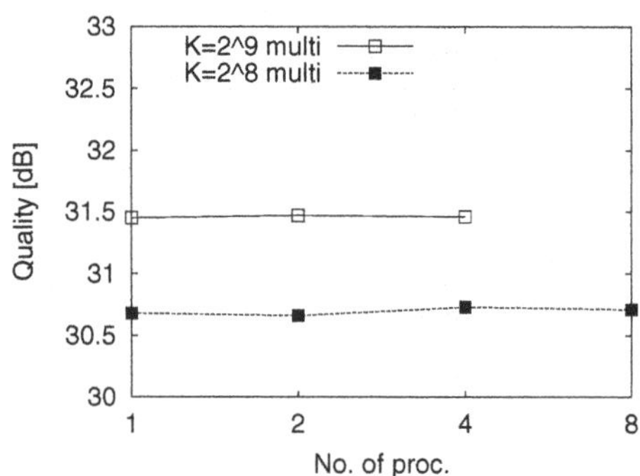

Fig. 2. Quality of images compressed by Grid PNN algorithm (Lenna)

size of $K = 2^8$ is around 30.7 dB with any P. Therefore, since our results show that the quality of the codebook generated by the multi-step parallel PNN is not worse than that by the original PNN, it is rational that the multi-step parallel PNN is adopted as the parallel PNN algorithm on the parallel system.

5.2 Computational Complexity

Let α be the computational cost of calculating the distance between vectors. Also let β be the communication cost of one codeword in the intermediate result between the master and the worker computer. Since the computational cost of merging i vectors into $i-1$ vectors in a PNN way is described as $C_i = \frac{\alpha}{2}(i^2 - i)$, the cost of the original PNN, the multi-step parallel PNN and the reduced parallel PNN (L_i is constant(=K)) is as follows:

$$C_{PNN} = \sum_{i=K+1}^{T} C_i = \frac{\alpha}{6}(T^3 - T - K^3 + K) \tag{4}$$

$$\begin{aligned} C_{multi} &= \sum_{i=L+1}^{\frac{T}{P}} C_i + \sum_{i=K+1}^{P\cdot L} C_i = \frac{\alpha}{6}((\frac{T}{P})^3 - \frac{T}{P} - L^3 + L \\ &+ (PL)^3 - PL - K^3 + K) + \beta PL \end{aligned} \tag{5}$$

$$\begin{aligned} C_{reduced} &= \sum_{i=K+1}^{\frac{T}{P}} C_i + \log P \times \sum_{i=K+1}^{2K} C_i \\ &= \frac{\alpha}{6}((\frac{T}{P})^3 - \frac{T}{P} - K^3 + K + \log P(7K^3 - K)) \\ &+ \beta \cdot P \cdot K \cdot 2 \end{aligned} \tag{6}$$

As shown in Equation 5, C_{multi} is minimized when

$$L = \sqrt{\frac{\alpha(P-1) - 6\beta P}{3\alpha(P^2-1)}}.$$

Thus, as L decreases, the computational complexity decreases. However, we choose K instead of $\frac{K}{P}$ as L value due to the quality of the codebook as mentioned earlier.

When $\frac{T}{P}$ is extremely larger than K, $C_{multi,reduced}$ are nearly equal to $\frac{\alpha}{6}(\frac{T}{P})^3$. Namely, the cost of the multi-step parallel PNN is $\frac{1}{P^3}$ of that of the original PNN, so the cost can be alleviated drastically. For example, the cost can be reduced to $\frac{1}{1000}$ of the original when $P = 10$. On the other hand, when $\frac{T}{P}$ and K are the same order, the dominant of C_{multi} is the cost of the latter part, that is, $\frac{\alpha}{6}((PK)^3 - PK - K^3 + K)$ and thus C_{multi} is in proportion to P^3. Namely, as the number of worker computers increases, the computational cost increases. Therefore, in order to implement the multi-step parallel PNN on the parallel system efficiently, it is important to choose the optimal P, which depends on T and K.

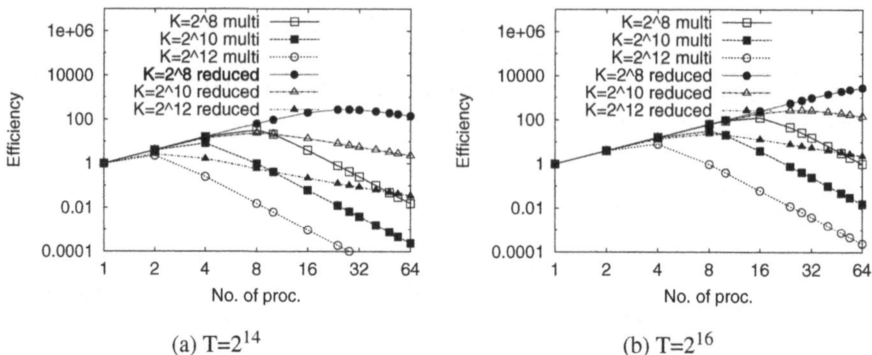

(a) T=2^{14} (b) T=2^{16}

Fig. 3. Efficiency of multi-step parallel PNN algorthm

By using Eqs. 4, 5 and 6, the efficiency of those ($L = K$ and $L_i = K$) is shown in Figure 3. The efficiency is defined as follows:

$$Efficiency = \frac{Sequential\ time}{Parallel\ time \times P.}$$

Note that a parallel algorithm is good if the efficiency is close to or larger than 1. As shown in the figure, the reduced parallel PNN is superior to the multi-step parallel PNN. For example, when $T = 2^{16}$ and $K = 2^8$, the efficiency of the multi-step parallel PNN is largest with $P = 4$ but that of the reduced parallel PNN is a monotone increase for P. This trend depends on K, but we expect that the reduced parallel PNN is always superior to the one-step.

6 Preliminary Experiment and Discussion

We implement the multi-step parallel PNN to create a codebook of the size of $K = 2^8, 2^9, 2^{10}$, from $T = 2^{12}$ training vectors using $P = 1, 2, 4, 8$ computers. The PC cluster system consists of 8 PCs with Celeron processor (1GHz) and 512 Mbyte memory under Linux 2.4 and MPICH 1.2.4. Note that the multi-step parallel PNN does not utilize data exchange function except for that between the master computer and the worker computers.

As mentioned earlier, the number of computers available depends on T and K, so we measure the efficiency on $P = 1, 2$ for $K = 2^{10}$, $P = 1, 2, 4$ for $K = 2^9$ and $P = 1, 2, 4, 8$ for $K = 2^8$. Experimental results of the multi-step parallel PNN with $T = 2^{12}$ are shown in Figure 4. When $K = 2^{10}$, the efficiency on $P = 2$ is slightly larger than 1 because the cost of the latter part of the master computer is very large and thus the effective parallelism can be degraded. The efficiency is enough large for P when $K = 2^8$ and the maximum efficiency is over 8. However, the efficiency decreases after the specific P because the cost of the master computer is in proportion to P^3, as the evaluation of the computational complexity. In order to alleviate this difficulty, the reduced parallel PNN should be considered for an efficient implementation on the parallel system.

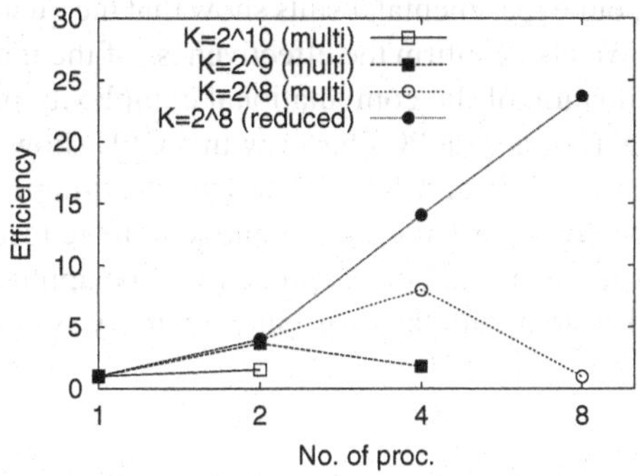

Fig. 4. Experimental results (Lenna)

When the number of worker computers is 8, the latter part of the algorithm is as follows: 1) the worker computers with the task number of 1, 3, 5 and 7 send their intermediate results to the master computer, 2) the master send the result of 1 to the 0th worker computer, the result of 3 to the 2nd worker computer, the result of 5 to the 4th worker computer and the result of 7 to the 6th worker computer, 3) the worker computers with the task number of 2 and 6 send their intermediate results to the master computer, 4) the master send the result of 2 to the 0th worker computer and the result of 6 to the 4th worker computer, 5) the worker computers with the task number of 0 and 4 send their intermediate results to the master computer, 6) the master computer does the final the PNN procedure with the results of 0 and 4. The results of the experiment are also shown in Figure 4 . It is shown that the efficiency of the reducde parallel PNN algorithm goes to 14.07 from 8.01 and to 23.71 from 0.962 compared with the multi-step parallel PNN when P is 4 and 8, respectively, so the effectiveness of the multi-step algorithm is confirmed. Note that α and β is 0.7×10^{-6} sec and 15.6^{-6} sec, respectively, according to our results.

On the other hand, we will extend our approach and results to general clustering problems because the PNN algorithm is a kind of hierarchical clustering algorithms as mentioned before. The prior art of parallel clustering algorithms has been done only on SMP systems and cluster systems and thus requires lots of communication exchanges within an iterative loop, so a different approach should be invented for the parallel system with a high communication latency. We will carry out these further studies in the near future.

7 Conclusion

We focus on the PNN codebook generation algorithm for the VQ compression algorithm, and propose a parallel version of the PNN algorithm suitable for the distributed memory parallel system, called "multi-step parallel PNN".

The multi-step parallel PNN is a modified version of the PNN algorithm that creates a different codebook than the original PNN does, thus the quality of a codebook created by using the multi-step parallel PNN may be worse than that of a codebook by the original PNN. However, our experimental results show that the quality is almost same as that of the original one. We also confirm the effectiveness of the multi-step and reduced parallel PNN by the evaluation of the computational complexity of the algorithm. Our preliminary experimental results on a PC cluster with 8 CPU show that the efficiency is enough large for P when $T = 2^{12}$ and $K = 2^8$ and the maximum efficiency is over 8.

In the near future, we will confirm the effectiveness of the reduced parallel PNN empirically using larger experiments and our approach will be applied to other clustering problems on the parallel system with the distributed memory system.

Acknowledgments

This was supported by MEXT ORC (2004-2008), Japan.

References

1. Gersho, A., Gray, R.: Vector Quantization and Signal Compression. Kluwer Academic Publishers, Boston (1992)
2. Franti, P., Kaukoranta, T., Shen, D., Chang, K.: Fast and Memory Efficient Implementation of the Exact PNN. IEEE Tran. on Image Processing 9(5), 773–777 (2000)
3. Sony Computer Entertainment Inc., Cell Broadband Engine, `http://cell.scei.co.jp/index_e.html`
4. Anderson, D., Cobb, J., Korpela, E., Lebofsky, M., Werthimer, D.: SETI@home: an experiment in public-resource computing. Communications of the ACM 45(11), 56–61 (2002)
5. Huedo, E., Montero, R., Llorente, I.: Experiences on Adaptive Grid Scheduling of Parameter Sweep Applications. In: Proc. of the 12th Euromicro Conference on Parallel, Distributed and Network-Based Processing (PDP 2004), pp. 28–33 (2004)
6. Equitz, W.: A new vector quantization clustering algorithm. IEEE trans. on Acoustics, Speech and Signal Processing 37(10), 1568–1575 (1980)
7. Dhillon, I.S., Modha, D.S.: A data-clustering algorithm on distributed memory multiprocessors. In: Proc. of Large-scale Parallel KDD Systems Workshop, ACM SIGKDD (1999)
8. Garg, A., Mangla, A., Gupta, N., Bhatnagar, V.: PBIRCH: Acalable parallel clustering algorithm for incremental Data. In: Proc. of the 10th Int'l Data Engineering and Applications Symp. (2006)
9. Olsen, C.F.: Parallel algorithms for hierarchical clustering. Parallel Computing 21, 1313–1325 (1995)
10. Wakatani, A.: Evaluation of parallel VQ compression algorithms on an SMP system. In: Proc. of IEEE CCECE 2006, pp. 1899–1904 (2006)
11. Wakatani, A.: A VQ compression algorithm for a multiprocessor system with a global sort collective function. In: Proc. of IEEE/ACIS ICIS 2006, pp. 11–16 (2006)

Software Repository for Software Process Improvement

Eun-Ju Park[1], Haeng-Kon Kim[2], and Roger Y. Lee[3]

[1] Department of Computer Information & Communication Engineering,
Catholic University of Daegu, Korea
ejpark@cu.ac.kr

[2] Department of Computer Information & Communication Engineering,
Catholic University of Daegu, Korea
hangkon@cu.ac.kr

[3] Software Engineering & Information Technology Institute,
Central Michigan University, USA
leelry@cmich.edu

Summary. A large number of organizations are trying to acquire CMMI certification for software process improvement for their organization. But, the existing solutions are used only in a limited fashion in specific process area. Because of this, created assets in each solution hard to utilize and reuse in software process improvement. One of reason is that they do not have specialized repository to support CMMI.

In this paper, we suggest SIR(SPIC Integration Repository) prototyping system for effective management and support of heterogeneous assets and tools of SPIC(Software Process Improvement Center in Korea). We provide the main design and modeling issues for organizing and managing of SIR for software process improvement. The most important advantage of SIR is flexibility and portability. SIR will provide API that is suitable to the manipulation of the SPICs meta-assets, related tools and models for helping the small and medium enterprise to achieve the CMMI.

Keywords: Software Process Improvement, Repository, CMMI.

1 Introduction

Organizations in today's global world are competing in highly complex and dynamic environments. Organizations are trying to acquire quality certification of software processes. Organizations are objectively verifying software development and system development capabilities with quality certification acquisition. Consequently, it enhances the competitive power of the organization, and thus it can improve its profitability. CMMI model [1] and SPICE (ISO 15504) model [2] in software process improvement model are recognized internationally [3]. Generally, CMMI model is used in America and the SPICE model is used in Europe. CMMI has became the standard in software process improvement, and organizations are trying to acquire CMMI certification. But, Most organizations do not understand the application and data which will be stored because there are so many defined processes and practices in CMMI. Moreover, theres no reference about the scale of the project in CMMI. Therefore, we will study these models for application of CMMI in the world [4, 5, 6, 7].

R. Lee and H.-K. Kim (Eds.): Computer and Information Science, SCI 131, pp. 51–64, 2008.
springerlink.com

Organizations use tools and assets which support CMMI for acquisition of CMMI certification. To reuse the valuable assets and tools, related assets and tools in the process must be stored and managed in a special repository for adapting the heterogeneous data. Therefore, we need integrated repository that can store and manage heterogeneous assets and tools to achieve the software process improvement certifications.

In this paper, we suggest a repository of software process improvement for effective management and support of heterogeneous assets and tools that were produced by a different organizer in SPIC in Korea. We finally developed the SIR Prototyping. We also described the main design and modeling issues for organizing and managing of SIR. It enables effective assets presentation with configuration management under the constraints of different SPIC documents as a reference for developers. In the long run, SIR could serve as a basis for standardization of software processes and interaction techniques, creation of assets and corresponding guidelines for SPIC.

The following section briefly describes related works with CMMI and measurement repository. In section 3, we define and suggest the metadata of SIR-CM, Organizing and managing of assets and SIR-CM modeling. And section 4 presents an execution example of SIR-CM and conclude in section 5.

2 Related Work

2.1 CMMI

The Capability Maturity Model Integration (CMMI) for software developed by the Software Engineering institute (SEI) has had a major influence on software process and quality improvement around the world. CMMI is a process improvement approach that provides organizations with the essential elements of effective processes. It can be used to guide process improvement across a project, a division, or an entire organization. CMMI helps integrate traditionally separate organizational functions, set process improvement goals and priorities, provide guidance for quality processes, and provide a point of reference for appraising current processes [8, 9, 10, 11].

CMMI is a model that consists of the best practices for system and software development and maintenance. The CMMI model provides guidance to use when developing system and software processes. The model may also be used as a framework for appraising the process maturity of the organization [8, 12, 13].

The CMMI is best known for its five levels of organizational maturity (see figure 1). Each level represents a coherent set of the best practices organizations are expected to implement as they become better at what they do. To each maturity level are associated a number of related process areas. The process areas can be viewed as very detailed checklist of what goals need to be achieved, what activities performed, and what artifact created and maintained to satisfy the requirements for a specific part of the overall development process [14, 15, 16, 17].

CMMI has the features as follows [1, 18]:

- Integration of software engineering and system engineering
- Treating an each process very minutely
- focusing on continuous improvement
- Two Representation-Continuous Representation, Staged Representation

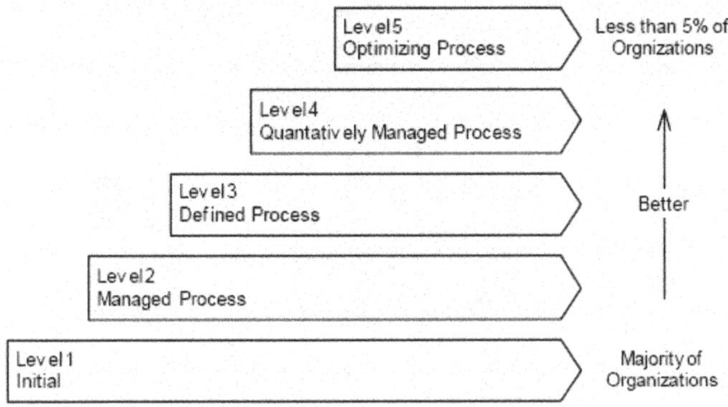

Fig. 1. CMMI maturity levels

2.2 Measurement Repository

When metrics are collected pertaining to a software product or process, the measurements are usually stored for later retrieval. Such a collection of metrics data is known as a metrics repository. The importance of repositories has been recognized in the past. A recommendation from the Workshop on Executive Software Issues held in 1988 stated that "Software organizations should promptly implement programs to: Define, collect, store in databases, analyze, and use process data" [19]. All the data collected on the project should be stored in a computerized data base. Data analysis routines can be written to collect derived data from the raw data in the data base. The data collection process is clearly iterative. The more we learn, the more we know about what other data we need and how better to collect it. Even then he clearly envisioned a data collection process that would change over time [20].

A neglected aspect of software measurement programs is what will be done with the metrics once they are collected. As a consequence, databases of metrics information tend to be developed as an afterthought, with little, if any concessions to future data needs, or long-term, sustaining metrics collection efforts. A metric repository should facilitate anon-going metrics collection effort, as well as serving as the "corporate memory" of past projects, their histories and experiences [21].

3 SIR Modeling and Design

3.1 An Overview

There are various SPIC supporting tools for software process improvement. Produced assets during the lifecycle of each tool must utilized each other. They can not utilize others heterogeneous asset and tools easily. Thus, we need integrated repository that can store and manage assets for adapting the heterogeneous data.

SIR is an integration repository which can store and manage heterogeneous software process assets and tools. The assets are produced from much different software

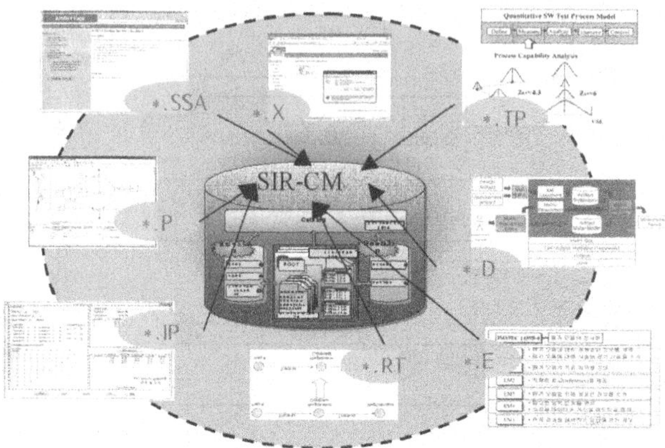

Fig. 2. Relationship between SIR and assets

development tools that support software process improvement. It is controlled and managed at the different data and forms on each development phase.

The design requirements are considered a sharable and a reusable service unit or component for each application. It also allows interoperability different type data and offer quick responses for dynamical changes. The relationship between SIR-CM and assets is shown in figure 2. It consists of many different categories of software process improvement assets. That is heterogeneous assets from several organizations that can afford to create centralized organizational SIR. SIR is managed and controlled by different tools on external of SIR.

SIR includes the configuration management with workspace and process management. It also allows the user to dynamically manage behavior for historical process information.

3.2 SIR Architecture

Figure 3 illustrates the architecture of SIR. It has three different layers as a platform, a business and a presentation by determining interactions between the roles. SIR stores all assets of tools which supports software process improvement as input data and is serviced or retrieved by each different tool. It consists of the guidelines of CMMI, measurement data about each process, methodologies and tools. Stake holders for SIR are the examiners of CMMI, project managers, developers and system operators. The SIR data is stored and provided for a browser through configuration management. To develop the SIR, we used s server platform from the x86 series, windows server from Microsoft and .Net framework as the based development language for SIR system.

There are three subsystems and several components as in figure 4. SIR simply stores the guideline or asset which is requested by developer and users in each process area. It also reports the requested data mining and search related processes. Related assets in each process area should be standardized, transformed and provided guidelines that are

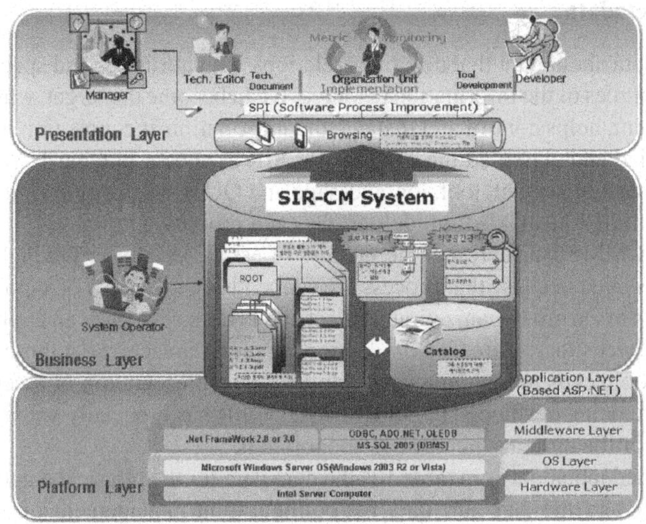

Fig. 3. SIR architecture

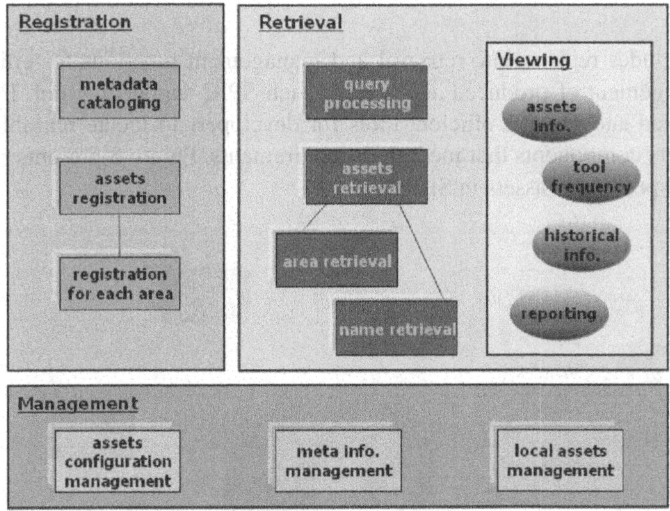

Fig. 4. Subsystem in SIR architecture

used for organizations. It also includes a functions to transform the assets type using each tool automatically.

SIR collects data for legacy tools and reprocess as a formed document type. Then, it stores produced metrics and confirms goals in the process. It improves the process through the process. Produced data ensures data integrity of repository through version management and workspace management.

3.3 SIR Metadata

Metadata is data about data that describes relations with attributes and special qualities and other resources of the information resource. It can become the target, which from the user's viewpoint, helps control and manage the information at connection institution.

We identified and classified the metadata of SIR as shown in table 1. We divide the main categories into artifact, tool, log, message and OU (Organization Unit). Basically, the metadata of SIR has CMMI guidelines for each process area.

Basically, CMMI can represent the process of assets serially or gradually, but SIR-CM is not classified because that process area depends on the assets of supporting tools. The categories in Ou are Organization, Project and Process. An OU can apply to CMMI. When an OU needs to acquire CMMI, supporting tools can achieve it through SIR. Assets of each tool needs to communicate with SIR. The presentation of assets and additional tools are being developed in guidelines OU. SIR has especially valuable assets in OU; CMMI certification has not always supported it.

The classified metadata as in table 1 is the core data for the physical database scheme design of SIR and the standard data for configuration management of SIR-CM assets. Also, it provides a standard format for retrieval and acquisition to understand the information.

3.4 SIR

The SIR includes registration, retrieval and management functions for efficient integrated management of produced assets from each SPIC supporting tool. It should be well-organized and provide efficient tools for developers to locate reusable software process assets components that meet their requirements. Figure 5 presents registration and retrieval process of assets in SIR-CM.

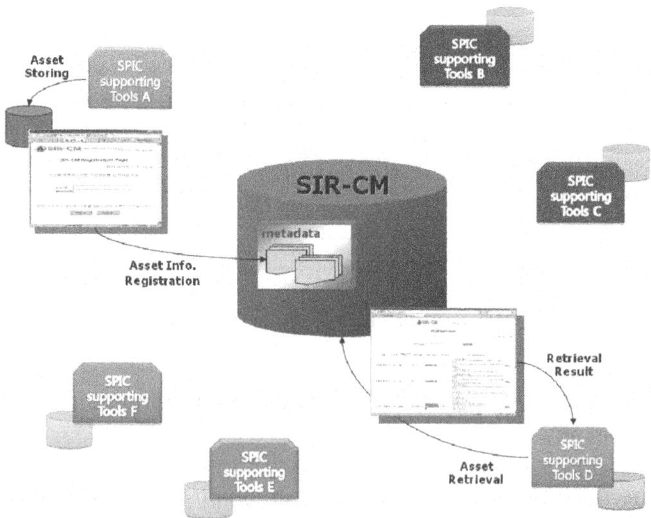

Fig. 5. Registration and retrieval of assets in SIR

Table 1. The metadata of SIR

Category	Metadata		Description
Artifact	Artifact_Id		Id of artifact
	Artifact_Name		Name of artifact
	Process_Area		It presents related process area of artifact
	Artifact_Type		Artifact type such as guideline, source, software, supporting tool and data, metric value
Tool	Tool_Id		Tool id
	Tool_Name		Tool name
	Tool_Url		It presents physical location information of artifact in tool.
Log	Artifact_Frequency		The use frequency of artifact
	Artifact_Relationship		It presents relationship between artifact and other artifact what frequently referred artifact.
	Last_Update_Date		Lastly updated date of artifact
	History_Info		It records all information such as registration, retrieval and modify of artifact
Message	Message_Create_Date		Initially created date of message that notified the related tools changed facts of tool
	Message_Date		Sent date of message that notified the related tools changed facts of tool.
	Received_Tool_Name		Name of related tool that sent message
OU	Organization	Organization_Id	Id of Organization
		Organization_Name	Name of organization
		CMMI_Level	CMMI Level
	Project	Project_Id	Id of project
		Project_Name	Name of project
		CMMI_Level	CMMI Level
		Related_Process_Area	Related process area
	Process	Process_Id	Id of process
		Process_Name	Name of process
		Related_Process_Area	Related process area

(1) Registration

Whenever assets are produced in each SPIC supporting tool, the assets must be registered SIR-CM. SIR-CM does not store all information about produced assets in each tool, but metadata of assets as table 1.

The detailed information of the asset produced in each tool is stored in local database at each SPIC tools. SIR user can browse the pre-stored asset items to select and register

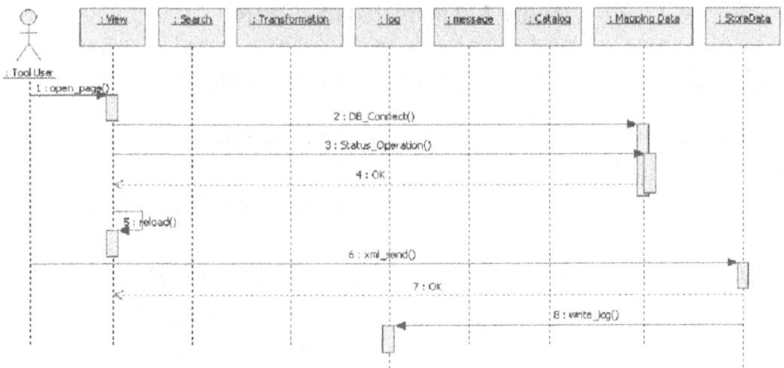

Fig. 6. Scenario diagram of asset registration

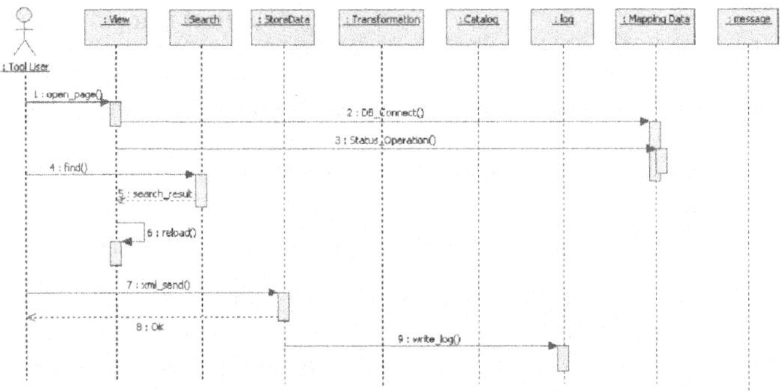

Fig. 7. Scenario diagram of asset retrieval

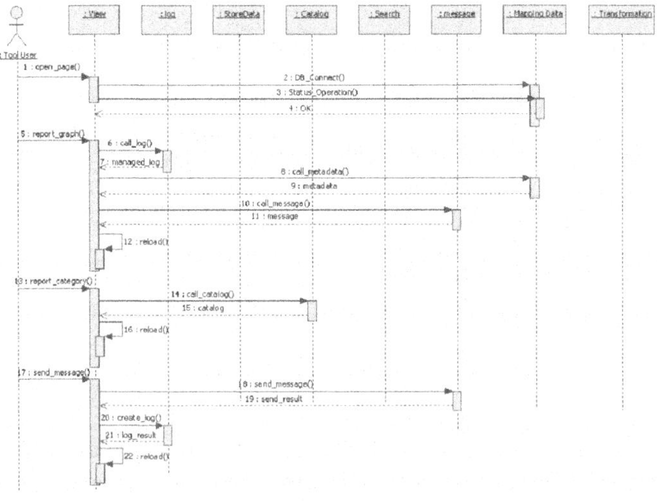

Fig. 8. Scenario diagram of asset management

in SIR. The user fills the dialog box in the browser and clicks the registration button to register it. Metadata is only stored in SIR. Scenario diagrams for assets registration is shown as figure 6.

(2) Retrieval

SPIC tool users can retrieve the assets from SIR. There are two ways to retrieve the required assets from tool; (a) by keyword matching of the name and, (b) by process area retrieval through matching each process area. We can retrieve the information such as the name of asset, the name of produced tool and related process areas as a result matching the candidates in the required assets. Figure 7 shows a scenario diagram of assets retrieval.

(3) Management

Management monitors the log records of assets, registers and retrieves in SIR, and manages the metadata. The log record presents use the frequency of the use assets and represents the relationship between assets and other assets that frequently are referred to as artifact. And this can derive relationships among tools through relationship among assets. Figure 8 shows a scenario diagram of asset management.

Table 2. Major function of SIR-CM

Function	Description
Messaging	When errors of management are discovered or new assets are updated or managers advise the user of each tool in current status, this function sends messages to the user of each tool.On the other hand, when errors or matters occur inside of the manager of each tool. SIR-CM manager function sends message to SIR-CM manager.
Reporting	Measures metric value, frequency of tools and progress of recent project visually shows through graphs or views.
Supporting	Supports standardized document forms such as XML and HTML.
Communicating	Communicates with legacy tools in standardized method.
Reusing	Reuses legacy tools and interoperability with it.
Measuring	Measures the value that is produced from each process area, include tools for measurement and tools for automatic or guideline.
Managing	Manages of metadata for each asset.
Confirming	Confirms guidelines and measures values through classification of each process area or level.
Reporting and searching	Reports and searches according to access area.
Ensuring reliability	Ensures reliability of the data when data of message format sends/receives from each tool.
Storing	Stores past results of measured metric value. If past results are need, it can confirm.

3.5 Subsystem Design

We analyze and classify three different CMMI; measurement tools for storing measurement values, automatic processing tools and tools for assets of document types. We suggest SIR for supporting of these tools and describing detail functions. There are 11 major functional requirements to develop the SIR-CM as in table 2.

All of these requirements are more focus on existing tools based on its value as UML modeling. Figure 9 shows the requirement specification for Usecase diagram.

SIR has many different component units. Each component unit introduces concepts of service oriented architecture, and remains a service except for configuration

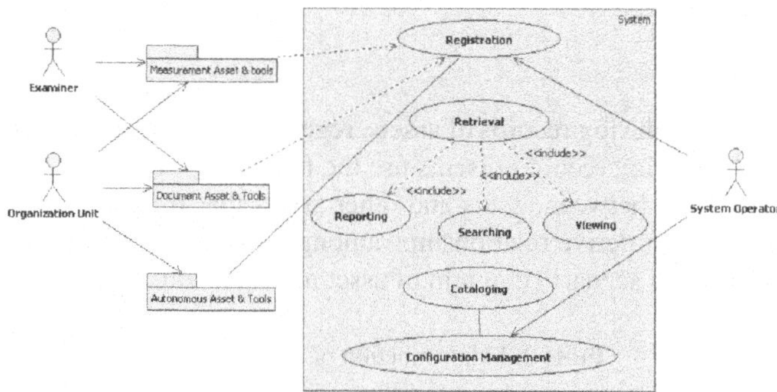

Fig. 9. Usecase diagram of SIR-CM

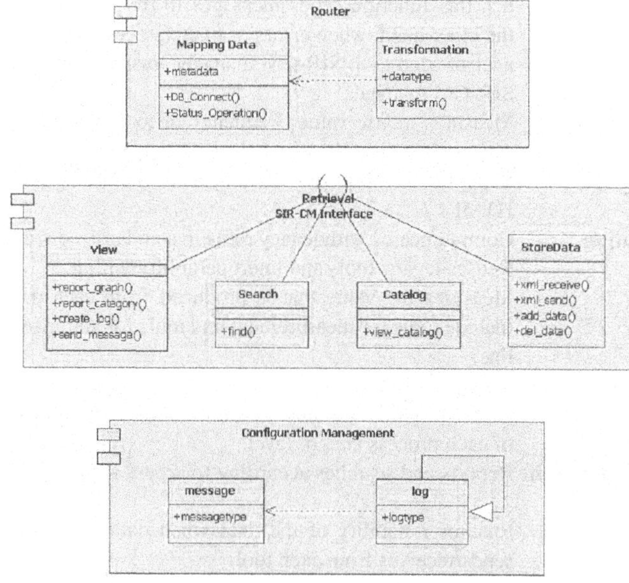

Fig. 10. Component diagram of SIR-CM

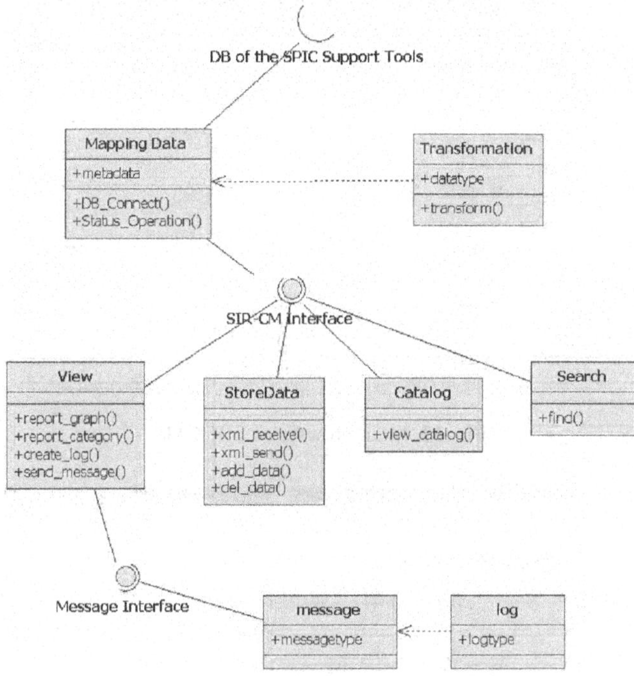

Fig. 11. Class diagram of SIR-CM

management and storing service excluding a direct connection. Figure 10 shows component diagrams using UML. Each component defines what the service is allowed to do; specifically, which information it is allowed to access. There are services that interact with other services by well-defined interfaces. Message units define the responsibility to transmitting the real data. Figure 11 shows the class diagram for detailed component diagrams.

4 Prototyping

We propose the modeling of the SIR prototype.

Figure 12 shows a main browsing screen. It contains three different areas such as menu area, tools area for SPIC supporting and area for Guidelines, assets and measurement data.

Figure 13 shows the registration browser of assets. The user can register asset in browser by filling it up. Metadata information about the assets is stored in SIR after confirming the information.

Figure 14 presents the retrieval browser of assets. It shows information such as the name of asset, the name of produced tool and the related process area as a result matched candidate assets.

Figure 15 shows dashboard browsing screen of SIR that presents recent work update, recent project progress chart, new message and tools frequency.

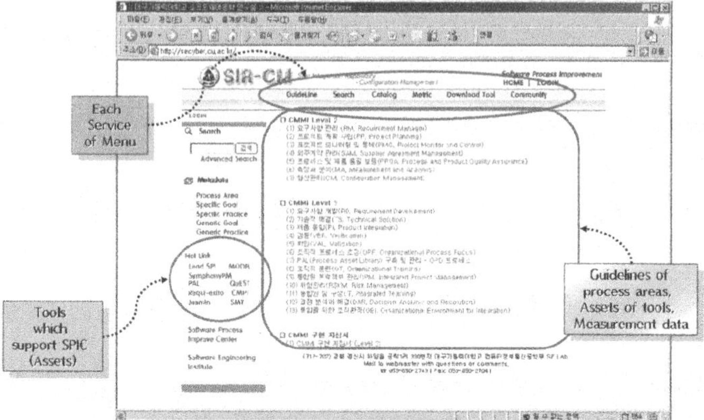

Fig. 12. Main browser of SIR-CM

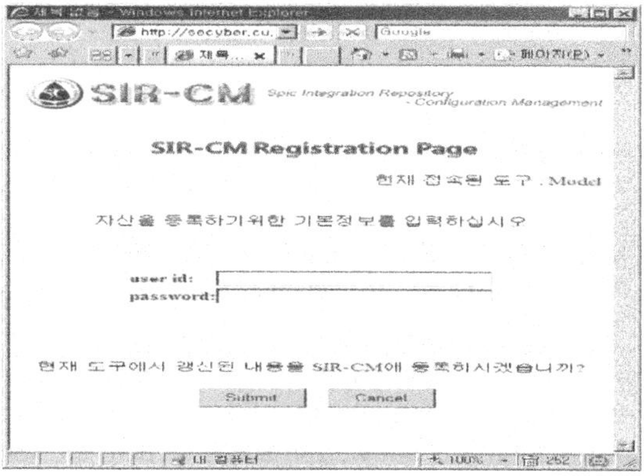

Fig. 13. Registration browser of assets in SIR-CM

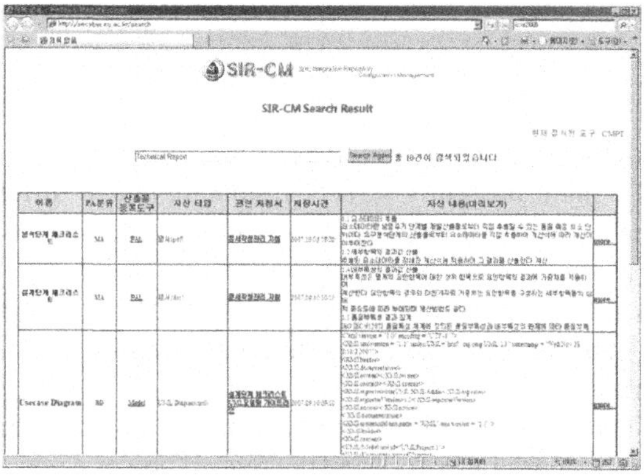

Fig. 14. Retrieval browser of assets in SIR-CM

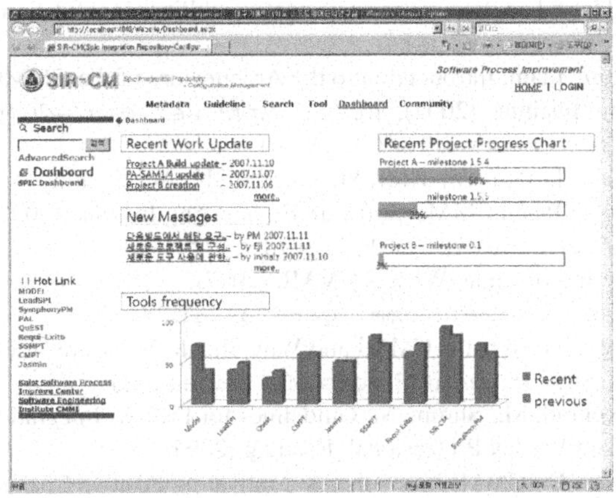

Fig. 15. Dashboard browser of SIR-CM

5 Conclusions and Future Work

Software process improvement forces in the organizing and managing the software assets. Existing software process improved tools have not been utilized and integrated for support small business organizations. SIR can be used for software process improvement modeling.

In this paper, we develop SIR prototyping system for effective management and support of heterogeneous assets and tools. And, we described organizing and managing of SIR for software process improvement.

We expect that SIR will provide an API that is suitable to the manipulation of the SPICs meta-assets and models to help the small and medium enterprise to achieve the CMMI, through this research. SIR approach in the study has advantages for acquiring certification of CMMI and supporting the quality and productivity of software process improvement.

In the future, we will implement the full set of SIR to cover our requirement in this paper. We also will contrive our work to use in the practical industry area. We will find some of the limitation from different SPIC tool site through a practical running environment.

References

1. Kulpa, M.K., Johnson, K.A.: Interpreting the CMMI: A Process Improvement Approach. Taylor & Francis (2006)
2. van Loon, H.: Process Assessment and ISO/IEC 15504: A Reference Book. Springer, Heidelberg (2004)
3. Lee, M.J.: CMMI as software international standard (2007), `http://www.dt.co.kr/dt_txt_see.htm?article_no=2007031402011060600001`
4. Werth, L.H.: Introduction to Software Process Improvement. Carnegie Mellon University (1993)

5. Lee, M.J.: CMMI based on Process Improvement for ability consolidation of IT Organization (2006), http://www.itjr.net/
6. CMMI Architecture Team, Introduction to the Architecture of the CMMI Framework. Software Engineering Institute (2007), http://www.sei.cmu.edu/publications/documents/07.reports/07tn009.html
7. Wall, D.S., McHale, J., Pomeroy-Huff, M.: Case Study: Accelerating Process Improvement by Integrating the TSP and CMMI. Software Engineering Institute (2007), http://www.sei.cmu.edu/publications/documents/07.reports/07tr013.html
8. Software Engineering Institute, What is CMMI. (2007), http://www.sei.cmu.edu/cmmi/general/index.html
9. Glazer, H.: Entinex, Inc, What Is CMMI and Why Should You Care? (2003), http://www.entinex.com/WhatIsCMMI_page1.cfm
10. Chrissis, M.B., Konrad, M., Shrum, S.: Guidelines for Process Integration and Product Improvement. Addison Wesley Professional, Reading (2003)
11. Bush, M., Dunaway, D.: CMMI Assessments: Motivating Positive Change. Addison-Wesley Professional, Reading (2005)
12. Software Engineering Institute, Capability Maturity Model Integration (CMMISM), Version 1.1. Carnegie Mellon University (2002)
13. Garcia, S., Cepeda, S., Miluk, G., et al.: CMMI in Small Settings Toolkit Repository from AMRDEC SED Pilot Sites. Software Engineering Institute (2004), http://www.sei.cmu.edu/cmmi/publications/toolkit/index.html
14. Process Academy's White Paper, What is the CMMI? Version: 1.00 (2003)
15. International Software Benchmarking Standards Group, CMMI and the ISBSG. International Software Benchmarking Standards Group (2007), http://www.isbsg.org/
16. Layman, B., Weber, C.: Measurement Maturity and the CMM: How Measurement Practices Evolve as Processes Mature. Software Quality Practitioner 4(3) (2002)
17. Bareisa, E., Karciauskas, E., Blazauskas, T.: Development of Case Tools for Software Process Improvement. Information Technology and Control 34(2A) (2005)
18. Chrissis, M.b., Konard, M., Shrum, S.: CMMI(R): Guidelines for Process Integration and Product Improvement. Addison-Wesley, Reading (2003)
19. Martin, R., Carey, S., Coticchia, M., et al.: Proceedings of the Workshop on Executive Software Issues. SEI Technical Report CMU/SEI-89-TR-006 (1988)
20. Basili, V.R.: Data collection, Validation, and Analysis. In: Tutorial on Models and Metrics for Software Management and Engineering. IEEE Computer Society Press, Los Alamitos (1981)
21. Harrison, W.: A Universal Metrics Repository. In: Proceedings of the Pacific Northwest Software Quality Conference, Portland, Oregon (2000)

Payment Annotations for Web Service Description Language (PA-WSDL)

Antonio Ruiz Martínez[1], Óscar Cánovas Reverte[2], and Antonio F. Gómez Skarmeta[1]

[1] Department of Information and Communication Engineering
[2] Department of Computer Engineering, University of Murcia, 30100 Espinardo (Murcia), Spain
{arm,ocanovas,skarmeta}@um.es

Summary. Nowadays the use of Web Services is based on two approaches. The first approach is the use of web services freely and the second one is a restricted use of services based on payment, subscription or any other method based on a commercial agreement. However, for this last approach, the information related to the payment, and therefore needed to use the service, is not provided along with the description of the service. This fact limits the discovery of web services in automated- way because this information is usually provided in web pages. In this paper, we describe how to include payment information in Web Services Description Language (WSDL), WS-Policy and Universal Description, Discovery and Integration (UDDI) for those services that require a payment.

1 Introduction

Web services facilitate the interoperability because they are independent of the underlying technology. This is the main reason many companies have started a migration process to provide their services as web services. This has also encouraged the development of new services as web services because the number of potential users can be significant, for example, Google offers some services as web services, such as Google Search, Adwords,...

These web services are described with Web Service Description Language (WSDL) [7]. These descriptions are usually published on the web or in repositories such as UDDI (Universal Description, Discovery and Integration). Furthermore, usually, the access to these services is free. However, more and more companies are developing web services based on payment because they want to get revenues from the services provided, e.g., the Google Adwords service is based on this approach and depending on the operation being carried out the fee is different.

The main problem that appears in these services is that there is no way to describe payment information with the elements defined in WSDL. For this reason, this payment information is provided on web pages and it is described in a natural language, that is, without using any machine-understandable language that facilitates its automatic processing. Therefore, the discovery of services and their requirements is more complex and cannot be automated by engines or intelligent agents as proposed in

R. Lee and H.-K. Kim (Eds.): Computer and Information Science, SCI 131, pp. 65–76, 2008.
springerlink.com

[11, 14]. Moreover, price comparison or negotiation between services with similar features is a difficult process.

As a solution to these problems, we present our proposal named *Payment Annotations for Web Service Description Language* (PA-WSDL). In this proposal we describe how to incorporate payment information with service description in WSDL for those web services that require a payment so that they can be accessed. However, with our proposal, we are not specifying the price of the different products that can be purchased by accessing to that service. Basically, we propose two ways to include payment information in WSDL: either extending it directly or by means of WS-Policy (see section 6). Moreover, we define how to include this information in UDDI specification (see section 7) to facilitate the discovery services, as well as, the payment information associated to these services that are annotated with payment information.

From our proposal, we can derive several potential benefits. First, the payment information is linked to the service description in a single document. Therefore, the process of publishing and discovering services and their payment information is simplified because, apart from the WSDL file, is not needed to surf on the net to find the payment information. Second, the automatic discovery of services based on intelligent agents [10, 11, 19] and UDDI [16] is facilitated because (almost) all the information about the service is provided in the description. Therefore, this information could be used with semantic information, in intelligent agents, to discover services that carry out operations expressed in a semantic way and that require a payment to be used. It is important to point out that our extension can be complementary with other extensions of WSDL such as P3P for privacy [8], SAWSDL [13] that facilitates the discovery of services based on those automated agents that we have just mentioned or WS-Agreement to negotiate Service Level Agreements (SLAs) between the service provider and the service consumer. Third, this solution provides a uniform way of publishing payment web services. This fact can encourage to service providers to publish their service descriptions in UDDI servers or any other kind of directory service. This facilitates to the clients to determine the best offer depending on their requirements in an easy way. Thus, it is promoted the competence and the development of services that offer a better quality.

2 WSDL and WS-Policy Overview

Web Service Description Language (WSDL) [7] is a language targeted to describe web services. The main element in the description of a web service is named description. It can contain six main elements: import, include, types, interface, binding and service that are used to describe the different operations that the service offers. WSDL is designed to be extensible. As a matter of fact, both description element and any of its elements can also be extended. Therefore, we can extend the description of the operations, the interfaces, the bindings and the services.

Web Services Policy (WS-Policy) framework [2] provides a general purpose model and machine-readable language to describe the capabilities, requirements, and general characteristics of entities in a WS-based system. WS-Policy attachment [3] specifies how policies are attached to a web service. WS-Policy is a simple language that has only four elements – *Policy, All, ExactlyOne* and *PolicyReference* – and only one attribute: *wsp:Optional.* Where Policy is the main element of a policy expression. Furthermore, *Policy, All, ExactlyOne* are operators that represent how the capabilities and requirements are combined. The *PolicyReference* element is used to reference a policy expression identified using either a name or a URI. Finally, the *wsp:Optional* attribute represents that an assertion is optional.

3 Payment-Based Web Service Scenario

Our goal is to describe the payment information associated to a payment-based web service in a machine-understandable way. Furthermore, we want to embed this information with the web service description to facilitate automated discovery and possible negotiations [11,14].

Figure 1 depicts a generic scenario. The first step is the publication of the Web Service with payment information. The different elements that we use to describe payment information in WSDL are commented in sections 5 and 6.

For the sake of simplicity, we are supposing the kind of service the user wants to use can only be accessed when a payment is made. Thus, when a client wants to make use of a specific type of service, he (or an intelligent agent on behalf of him) performs a search in the Web or in a UDDI server to discover Service Providers offering that kind of service (step 2). As a response, the client gets, from the UDDI server, the list of service providers (step 3). Next, the client requests the full WSDL description from the service provider itself (step 4). The information about the different bindings and operations can also be obtained from the UDDI server. The answer contains payment information expressed by the elements defined in the following sections (step 5).

From the WSDL files, the client can init a selection process to determine which service provider to use according to the personal preferences. For example, the client might discard those service providers requiring the use of unsupported payment protocols (step 6). Thus, using the whole set of service providers, the client might compare the prices and other payment information (step 7). Next, the client chooses the service provider that offers the best conditions (step 8). Then, the client begins the payment process with Payment Service Provider (PSP – an entity that accepts payments on behalf of other entity) (step 9) using some of the information provided in the WSDL extension, like prices, credentials, etc. Once the payment is made, the PSP provides a receipt so that the client can access the service. This receipt should be a signed statement based on XML. The receipt encodes statements such as "X paid 5$ to service provider SP through PSP Y to access to the operation O of service S".

After the payment, the client can invoke the service (step 11) using the receipt that PSP provided. Next, the service provider checks that the token comes from the PSP and that the amount and the operation selected are valid. Therefore, there is a trusted relationship between the PSP and the service provider. All payments made to the service provider through a PSP are received in the service providers account. This account was established when the service provider was registered with the PSP. In this scenario, the PSP role could be carried out by either the service provider or a bank or any other party.

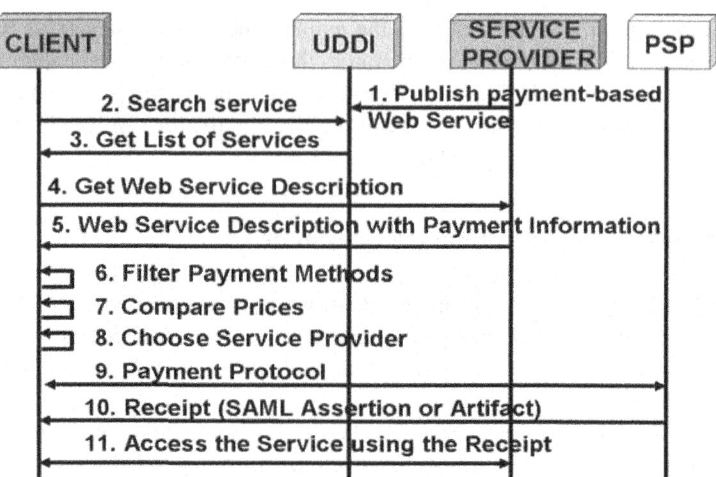

Fig. 1. Payment process

4 Requirements

This section defines the requirements we identified for describing payment information in web services:

- The payment information should be included with the WSDL description. Thus, we facilitate the discovery of payment information associated to a web service.
- We should take into account that some operations, interfaces or services could provide a free access.
- A web service might be paid with different payment protocols. Therefore, it should be possible to specify several payment protocols at the same time. Furthermore, for each protocol it should be possible to specify different versions of that protocol.
- Different brands can support the same payment protocol. Therefore, for each protocol we should be able to choose between different brands.
- The service provider should be able to reference the different PSPs who s/he usually operates with.
- The costs and commissions associated to each protocol and brand can vary depending on: the computational resources to use, payment models, on-line participation of a third party, etc. Therefore, we should be able to express different prices depending on the specific payment protocol, brand and payment service provider chosen. We should also be able to express these prices in different currencies to facilitate internationalization of the web services.
- The payment information should be expressed at several levels in the description of the WSDL.
- It should be supported the description of several payment models such as, a payment for the execution of an operation, a payment for a set of operations and a payment for the access during some period of time.
- It should be incorporated information related to the validity of the payment information to know when this should be updated.

- When a payment is required, sometimes it is needed that some additional conditions are satisfied, e.g. being older enough. For this reason, with the payment information, we should be able to express that the client can be required to present some credentials in order to authorize the payment.
- We should be able to express the loyalty information we accept and/or its benefits in a payment.

5 Payment Elements Definition

In this section we describe the elements that are needed to express payment information. These elements are used to include payment information in our proposal. For each element there is a XML type that represents it. Next, we describe each element. It is important to point out that each element defined has an attribute called ID which type is ID from XML Schemas. In the description of these elements this attribute will not be mentioned any longer.

5.1 Basic Elements

The *PaymentProtocol* element contains information about a specific payment protocol. Its *ProtocolID* attribute identifies the payment protocol with a URI. Its main elements are: *Name,* to name the protocol; *Version,* to specify the version of the protocol supported; and, *AdditionalInfo,* to include additional information of the protocol. Examples of payment protocols: Millicent [12], SET [20], or others more recent mentioned in [18].

The *Brand* element provides information about a specific brand. Its *BrandID* attribute identifies the brand with an URI. The information we can specify in a brand is: *Name,* the brand name; *Description,* a description of the brand; and, *Logo* which is the URL with the brand logo. Examples of brands are: Mastercard, Visa, etc.

The *PaymentServiceProvider* element provides information regarding the PSP that will receive the payment. The PSP is identified with an URI in the *PSPID* attribute. This role can be played by the service provider, or by other trusted third parties like a bank. This element can contain the following information: *Name,* the PSP's name; *Description,* to specify some information about the PSP; *Logo,* URL to a logo of the PSP; *PaymentURL,* URL where the payment messages has to be sent; *VendorID*, which represents the vendor's (service provider) identifier in that PSP. This information is used to associate the payment to that vendor; and, *AdditionalData* is used by vendor to provide to the PSP some additional data related to the payment. This information has to be sent by the client when he is making the payment.

The *Price* element is used to describe the amount to pay to access to the service. Its main elements are: *Amount,* i.e., the amount to pay; and, *Currency,* to indicate in which currency the amount is expressed.

The *PaymentModel* element is used to define a payment model. With this element we indicate if the payment gives the right to access once to an operation of the web service (option by default), several times or to several operations, or during a period of time.

The *LoyaltyInformation* element is used to express the loyalty schemes that are applicable to a payment transaction. To express this information we can use different languages since it is language-independent.

The *Credential* element is provided to specify the different credentials that might be required as additional information for the payment, as for instance the use of X.509 certificates, SAML Assertions [6].

5.2 Sets of Elements and References

For every previous element, we have defined a new XML element that represents a set of that individual element. For instance, a set of instances of the element *Brand* is named *Brands.* The rest of set of elements are *PaymentProtocols, PaymentServiceProviders, Credentials, PaymentModels, Prices,* and *LoyaltyInformations.* We have also defined elements that represent a set of references to the elements described previously.

5.3 PaymentDescription Element

The *PaymentDescription* element is used to join the previous sets of elements in one single element. This element represents a specific payment description establishing the conditions for a particular transaction. It is composed of (see Figure 2): *PaymentReference,* to identify univocally the payment with a reference; then, we have a set of elements to reference a set of instances of the *PaymentProtocol, Brand, PaymentServiceProvider* and *Price* elements. We also have other elements such as: *PaymentModel,* to specify the payment model followed; *CredentialsRequired* to express the set of credentials the client should present; *LoyaltyInformationAccepted,* to indicate the loyalty schemes and discounts to apply; and, *Expiration,* to indicate the time until the offer is valid.

This element represents the prices to pay using a set of protocols and brands, and satisfying a set of conditions such as presenting a set of credentials or providing some loyalty information. Then, in a payment description, any tuple that can be formed with a combination of one element of the set of protocols and one element of the set of

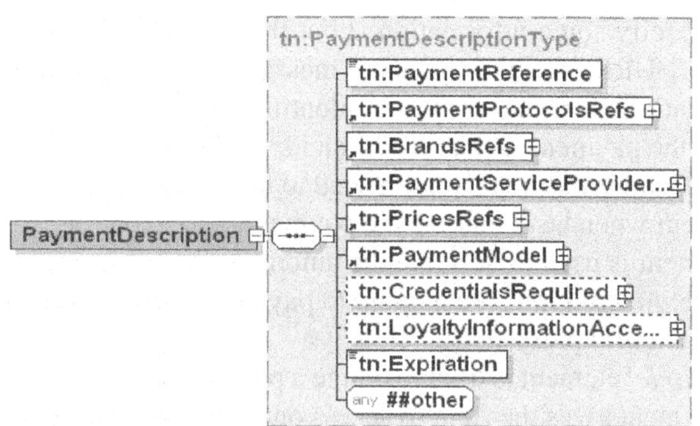

Fig. 2. *PaymentDescription* element

brands, represents the same price. If we wanted to express different conditions for the different combinations we should to create different payment descriptions simplifying the number of elements of each set. Therefore, this element is used to compare prices and conditions. Furthermore, this element defines an attribute called negotiable that indicates whether the price is negotiable or not. If the price is negotiable, the conditions could be negotiated with other protocols such as WS-Agreement.

5.4 PaymentInformation Element

The previous element is used to represent that the server provider is requesting a payment to access an operation or service. Thus, that structure contains the prices requested and the payment mechanisms supported. However, it does not contain the payment information itself. It only contains references where this information is located. Thus, we avoid the repetition of those elements.

The place to describe such payment information is the element named *PaymentInformation.* This element describes and groups all the information that describes the protocols, brands, PSPs, prices, credentials, loyalty information and the different payment descriptions. It contains the following elements (see Figure 3): *Payment- Protocols,* the set of the payment protocols supported; *Brands*; the set of brands supported; *PaymentServiceProviders,* the set of PSPs with the vendor works; Prices, the prices that are used to express the amount to pay; *Credentials,* the set of credentials that could be requested; *LoyaltyInformations,* the set about different loyalty schemes; *PaymentDescriptions,* that represents a set of instances of *PaymentDescription* element. This last element introduces the different prices that have to be paid depending on the protocols, brands and PSPs used. The *PaymentInformation* element also has an element named *Expiration* to indicate when this information should be updated, because it is not valid any more. Optionally, we provide an element named *Signature* to contain a XML digital signature for those cases when the vendor wants to guarantee that the information has not been modified.

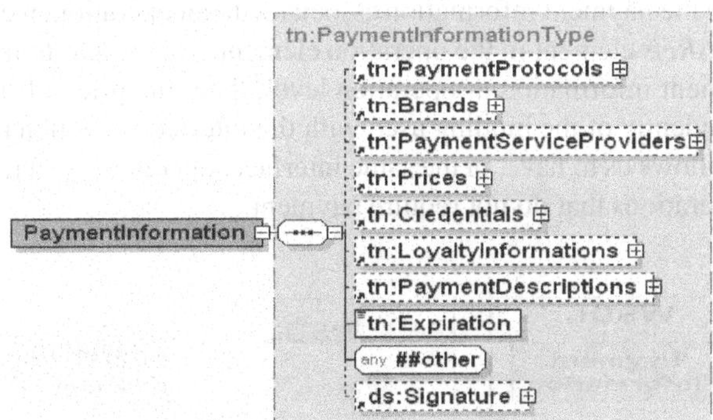

Fig. 3. *PaymentInformation* element

6 Payment Annotations for WSDL (PA-WSDL)

We can follow two approaches to include payment information in WSDL. First approach is to extend WSDL to include directly the information described in the previous sections. The second approach is to use WS-Policy extension to WSDL to include payment information as a requirement. The way in which we include payment information in both approaches constitutes *Payment Annotations* for WSDL (PA-WSDL). The first approach is used when service provider does not support WS-Policy or when he does not want to add an additional level of complexity to express payment information. The second approach is used for those service providers that already support WS-Policy. At the same they express other requirements such as security, reliability, etc. They describe payment information as another additional requirement. Therefore, the expression of all requirements is put in a uniform way. Next, we detail how PA-WSDL includes payment information in both approaches.

6.1 WSDL Extension

The *PaymentInformation* element contains information describing the payment protocols, brands and PSPs supported as well as the prices to pay. It also contains the different descriptions requesting a payment independently if payment is requested for an operation, a binding or a service. For this reason we decided to put this element at the same level as elements such as interface, binding and service (see Figure 4a) in WSDL. Additionally, this element could be placed in a document apart from the WSDL for reutilization purposes (Figure 4b) and thus, for this case, we defined an attribute named *PaymentInformationURI* to reference it. Hence, the elements defined throughout this document would be contained in it. Thus, this element is used to carry out the process of filtering information determining the protocols and brands supported.

To specify the payment information for operations, interfaces, bindings and services we use the *PaymentDescriptionsRefs* element. This element is defined as a set of references to PaymentDescription elements that are in the *PaymentInformation* element. Thus, to describe the payment information associated to an operation, we include a *PaymentDescriptionsRefs* element in the operation element. We want to point out that if we specify the payment information at operation level, then, the price of the operation is the same independently of the binding used with the interface in which the operation is described. This allows us to have, in the same interface, operations that require payment to access and operations that do not require payment.

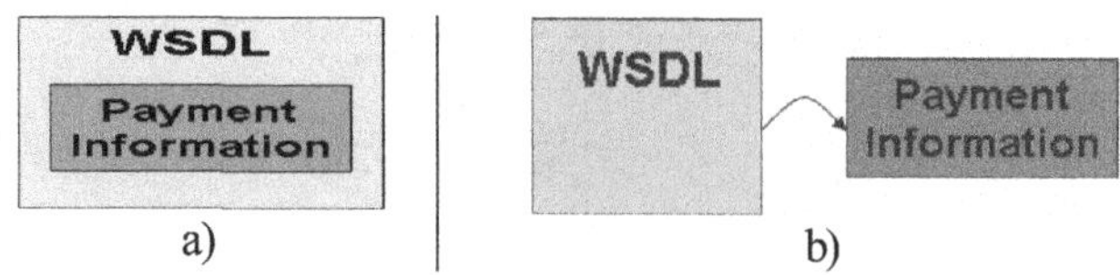

Fig. 4. a) Payment Information in WSDL. b) Payment Information referenced from WSDL.

To describe the payment information associated to an interface we put the *PaymentDescriptionRefs* element inside the interface element. Thus, we indicate that all operations described in that interface are based on payment and that all operations have the same cost independently of the binding used.

We can specify payment information at binding level also using the *Payment- DescriptionsRefs* element. If we specify it with the main elements of the binding element, then, the price of all the operations in this binding is the same. However, in this binding, we can also have operations with different prices, then, this information would be included in the operation element that is in the binding element.

Finally, at service level, we can specify payment information in a similar way to the previous elements with the same semantic.

If we specify the payment information at several levels, the total amount to pay is the sum of the quantities expressed at the different levels.

6.2 Payment Information in WS-Policy

We can include payment information as an additional requirement in WS-Policy. Thus, we can combine requirements of different domains (security, reliable messaging, etc) in a uniform way facilitating in that manner its subsequent processing.

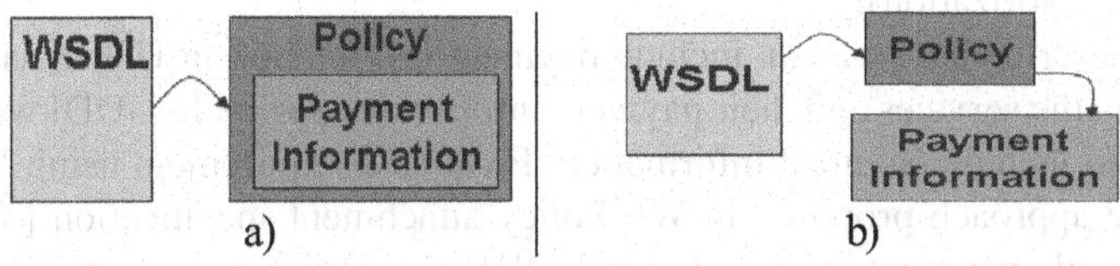

Fig. 5. a) WSDL references to a Policy Expression extended with Payment Information. b) Policy Expression references to Payment Information in other document.

WS-Policy also allows including policy expressions at several levels in WSDL. Therefore, we can include this policy, and thus, the payment information, at message, operation, binding and service levels. In this case, as policy assertion we include the element *PaymentDescriptionRef* instead the set of *PaymentDescriptionRef* since in WSDL policy is recommended the use of the normal form where each assertion is included in a different Policy element. This reference points a *PaymentDescription* element out. This element is included in the *PaymentInformation* element. In this case, this last element can be included either as an WSDL extension at the same level as the service element (see Figure 4a), or included inside the root element of *Policy* element, i.e., extending it (see Figure 5a). Needless to say, this element could be stored in other place while it could be referenced (see Figure 5b).

7 Payment Information in UDDI

UDDI (Universal Description, Discovery and Integration) is a specification to facilitate the publishing and discovery of web services. Its main components are: *businessEntity*

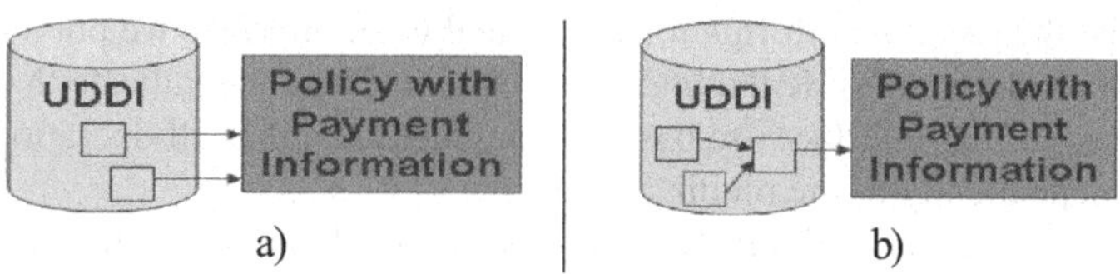

Fig. 6. a) UDDI references a policy. b) tModel references to a Policy.

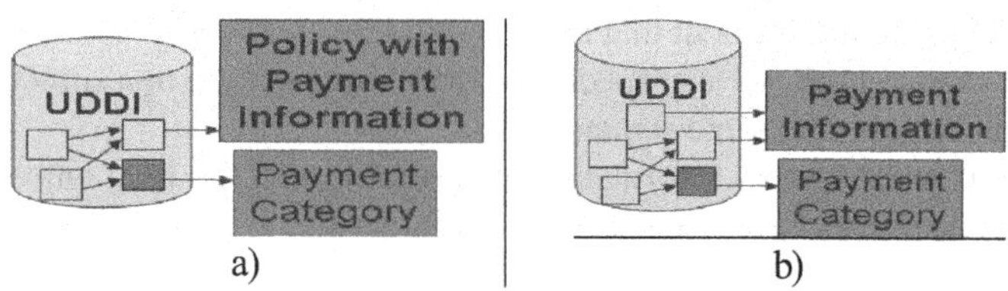

Fig. 7. Payment Information in UDDI with WS-Policy. b) WS-Policy from WSDL.

(to describe an enterprise or business), *businessService* (to describe the services offered by a business entity), *businessTemplate* (to describe the different endpoints of a service) and Technical Models or *tModels* (to support classification using various taxonomy systems for categorization).

Next we describe how we can include payment information in UDDI with the aim of discovering the services and their payment information easily. In UDDI we have two possibilities to include payment information. First, as a requirement using WS-Policy following the approach proposed in WS-Policy attachment specification [3]. Second, including directly payment information in UDDI in a similar way as proposed in WS-Policy attachment. Next, we explain both options.

WS-Policy attachment specification [3] proposes two alternatives to include policies (with or without payment information) in UDDI. In the first option, an UDDI entity directly references to a remotely accessible policy expressions (see Figure 6a). This model is proposed when a policy is for a unique web service. In the second approach, UDDI entities reference to policy expressions through tModels that register policy expressions (see Figure 6b). This approach is interesting because is modular and allows the reuse of policies.

Its main problem is that it does not allow distinguishing between services that requires payment and those that does not require it. To determine if a payment is required, first, we have to search services that define policies, and then, to analyze that policies to determine if payment is required or not. To cope with that problem, we propose to register each *PaymentInformation* element with a *tModel* whose category is payment information (see Figure 7a). Thus, different *businesServices* and *businessTemplates* can reuse payment information.

To include payment information in UDDI without WS-Policy define several *tModels*. First, a *tModel* to reference a *PaymentInformation* element that is remotely accessible. Second, a *tModel*, to access to *PaymentInformation* elements that are included in the

UDDI registry as a *tModel*, and finally, we need a categorization to define that a *tModel* contains payment information (see Fig. 7b).

8 Related Work

To the best of our knowledge there is not a similar proposal to the one introduced here. Despite this fact there are some proposals introduced in other environments such as ECML [9], W3C Common Markup for micropayment per-fee-links [17], UBL [4] and IOTP [5] to define information payment elements, we want to point out that the mentioned proposals have not considered the introduction of some useful elements in e-commerce models such as the requirement of presenting some credentials apart from the payment or the provision of loyalty information. Furthermore, none of the previous solutions proposes a model to incorporate them to WSDL, WS-Policy and UDDI.

From the web services specifications point of view, there are some related work such as WSLA [15], and WS-Agreement [1]. WSLA and WS-Agreement has similar goals: establish service level agreements between parties. WSLA is focused on providing a customized SLA containing such as response time, availability and throughput. This proposal is thought to negotiate the contract terms and it does not define how to utilize WSLA with UDDI directories. The WS-Agreement specification defines a language and a protocol for describing the capabilities such as advertising agreement capabilities of parties, creating agreements based on offers or monitoring agreement compliance at runtime. However, this specification is not useful to only publish information. On the other hand, this proposal, once that this previous information is discovered, if it needed or required to negotiate the terms of the agreement, then it is useful and it could be composite with our proposal. Thus, in a first step our proposal would be useful to describe the initial conditions and after that, service consumer and service provider could negotiate the agreement using the protocol proposed by WS-Agreement.

9 Conclusions and Future Work

We have presented how to describe payment information in WSDL and UDDI. In WSDL, we propose to include payment information by extending it or by means of WS-Policy. Furthermore, the extensions proposed can be complemented with other extensions such as SA-WSDL or WS-Agreement. We have also defined how to include this payment information in UDDI to facilitate web services discovery and composition. Therefore, we facilitate both the process of discovery of payment-based web services in an automatic way and other processes related to business such as prices comparison, the discovery of payment protocols supported, etc.

As future work, it is interesting to analyze how different services that are based on payment could be used in orchestration and choreographies, and how this information should be expressed, that is, if the resulting payment is the sum of the use of the individual services, if we can provide special prices, and so on.

Acknowledgments

This work has been partially funded by the PROFIT CAM4HOME (FIT350105-2007-22) and Ecospace (FP6 IST-035208) projects.

References

[1] Andrieux, A., Czajkowski, K., et al.: Web Services Agreement Specification (WSAgreement). Grid Resource Allocation and Agreement Protocol (GRAAP) WG (2005)
[2] Bajaj, S., Box, D., et al.: Web Services Policy 1.2 Framework (WS-Policy) W3C (April 2006)
[3] Bajaj, S., Box, D., et al.: Web Services Policy 1.2 Attachment (WS-PolicyAttachment). W3C (April 2006)
[4] Bosak, J., McGrath, T., Ken, G.: Universal Business Language v2.0. OASIS (2006)
[5] Burdett, D.: Internet Open Trading Protocol – IOTP Version 1.0, RFC 2801 (April 2000)
[6] Cantor, S., Kemp, J., et al.: Assertions and Protocols for the OASIS Security Assertion Markup Language (SAML) V2.0. OASIS Standard (March 2005)
[7] Chinnini, R., Moreau, J., et al.: Web Services Description Language (WSDL) Version 2.0 Part 1: Core Language. W3C Candidate Recommendation (2006)
[8] Cranor, L., et al.: The Platform for Privacy Preferences 1.1 (P3P1.1) Specification. W3C (November 2006)
[9] Eastlake 3rd D, Electronic Commerce Modeling Language (ECML). Version 2 Specification. Request for Comments (RFC) 4112 (June 2005)
[10] Farrel, J., Lausen, H.: Semantic Annotations for WSDL and XML Schema. W3C Working Draft (2007)
[11] Fensel, D.: Ontologies: Silver Bullet for Knowledge Management and Electronic Commerce. Springer, Heidelberg (2000)
[12] Glassman, S., et al.: The Millicent protocol for inexpensive electronic commerce. In: World Wide Web Journal, 4th Int. WWW Conference Proceedings, pp. 603–618 (1995)
[13] Kopecky, J., Vitvar, T., et al.: SAWSDL: Semantic Annotations for WSDL and XML Schema. IEEE Internet Comput J 6, 60–67 (2007)
[14] Kowalczyk, R., Ulieru, M., Unland, R.: Integrating Mobile and Intelligent Agents in Advanced e-Commerce: A Survey. In: Proceedings of Agent Technologies, Infrastructures, Tools, and Applications for E-Services, NODe 2002, Germany (2002)
[15] Ludwing, H., Keller, A.: Web Service Level Agreement (WSLA) Language Specification. IBM Corporation (2003)
[16] Luo, J., Montrose, B., et al.: Adding OWL-S Support to the Existing UDDI Infrastructure. In: Proceedings of IEEE International Conference on Web Services (ICWS 2006) (2006)
[17] Michel, T.: Common Markup for micropayment per-fee-links, W3C (August 1999)
[18] Ondrus, J., Pigneur, Y.: Towards a holistic analysis of mobile payments: A multiple perspectives approach. Electron. Comm. Res. and Appl. 5(3), 246–257 (2006)
[19] Verma, K., Sheth, A.: Semantically Annotating a Web Service. IEEE Internet Comput. 11(2), 83–85 (2007)
[20] Visa and Mastercad, Secure Electronic Transaction (SET) specification, v1.0 (May 1997)

An Analogical Thinking Based New Software Engineering Education Methodology

Tokuro Matsuo[1] and Takayuki Fujimoto[2]

[1] Graduate School of Science and Engineering,
Yamagata University
Jonan 4-3-16, Yonezawa,
Yamagata, 992-8510 Japan
tokuro@tokuro.net
http://www.tokuro.net

[2] Department of Future Desiggn,
Sonoda Women's University,
7-29-1, Minami-Tsukaguchi,
Amagasaki, 661–8520, Japan
me@fujimotokyo.net
http://www.fujimotokyo.com

1 Introduction

In recent years, the research of the education for software engineering is active as one area of the engineering education [1][2]. As for the features of software engineering, they not only relate to the programming but also relate business, design, requirement engineering, and technical ability. The education program based on learner's situation and curriculum design are important as well as an existing educational science and the didactics research.

In the subject of the information science of the university, many universities provide the lecture concerning software engineering. Educational issues of software engineering consist of methodologies and history of software design and life-cycle models, UML used as integrated writing method, software test and management methods. However, in software engineering education including such contents, there is a strong limitation since students do not have enough knowledge for learning software and motivations. Students do not always determine entering and selecting university based on their preferences and interests to the information science. Some students select course and university based on their knowledge level and fee. Thus, students have low motivations to the information science and software engineering. Software engineering relates not only software design and technical issues but also relates concept of business and users. Thus, students who have not considered service and business cannot easily understand the learning issues. For example, most of engineering subjects can be understood based on visual and experimental cognitive methods, such as the color reaction of chemical synthesis, experiment by electromagnetic measurement device, test of durable concrete structure. On another hand, students cannot understand by watching and hearing through an experimental method when they learn the features and qualitative phenomena of software. The student doesn't understand easily only from our teaching a field

R. Lee and H.-K. Kim (Eds.): Computer and Information Science, SCI 131, pp. 77–86, 2008.
springerlink.com

concerned frankly and straightly. In the practical lectures of programming in the university, even though students make a certain application to learn the program language and methodologies, students cannot understand essential software engineering since such applications are not developed with requirements from company and users. Even though students consider the structures and construction of systems design, students' abilities of software engineering do not always enhance. Most of students never considers about the delivery of materials to users. Actually, It is necessary to design considering the cost and the customer for the business that designs software. The business project fails when the design model with an adequate plan is not used. Recently, the purpose of university as the vocational education for the career planning is strengthening more and more [3]. A strategic educational program design is proposed by many universities to provide the good position to students and to serve employment applications from companies. However, essentially, it is necessary to give the effective education to perform such good results.

In this paper, we propose a new software engineering education model based on the analogical thinking with students' experiences and their histories. In the life-cycle model of software production process, our approach make learner understand abstraction level of concrete planning methods. Further, we give a new education program based on the role-play of business project with simple experiment and consideration since software engineering closes a business and trading with customers. One of the important study matters in software engineering is understanding of software engineering from a business administration aspect. For instance, various types of customers come to the software house and he/she receives the request of the software construction. Our proposal aims the promotion of the understanding of the advantages and the faults of multiple software making techniques for learners.

In recent years, the conception of UML is one of important learning issues in the software engineering. In learning of UML, we give a new education method where learners consider a situation of systems running. For example, when we purchase the item from a vendor machine, we unconsciously buy the commodity without any idea. We just put out money from a wallet, put in the money to the machine, push a button in which we want to purchase, and take out the item from the cage of the machine. When engineers make a system, it is important for them to consider their behavior consciously. We take an examination between straight learning using textbook and our proposed analogical thinking-based learning. As the result, our proposed methods are proven to be effective for the students.

2 Motivations

Recently, in Japan, one of problems in university education is deterioration of academic performance by university students. As a strong tendency of this problem, it is important for teachers to make efforts to education by multiple scheming. Even though there are overlapped learning issues among multiple lectures, many students forget such issues, which is taught in past. Many students cannot be conscious the relationship between them. For example, in the course of software engineering, students usually learn the word "flow charts". In the first grade in the university, all students learn in the lecture such as primary information processing. However, when we made a test to third grade of students, some students did not know the word. Further, some students answered that they are no interested in information science. Some students also answered just their level of examination about why they select the university where they entered. They searched for universities in which they can enter without their preferences and interests. Amazing reason is that some students were not good at the subjects of mother language, history, geography, and foreign language, even though scores of science-related subject were very low. Such students sometimes are not interested in the information science. However, Students who are in the course, they must take complementary lectures to graduate there. On the other hands, Teachers hope that students graduate the university with the useful knowledge and experiences. Although teachers make an effort about education and teaching, students sometimes forget and do not know the learning issues in which they had ever learn at other lecture. One of reasons why these situations occurs in Japan, students neglect and do not make an effort to learn and study in the lectures even though teachers make much efforts. Students also study in a night before an end term examination. The study issues and knowledge infiltrate in their brains. Thus, they forget without remaining in their memory.

To know the actual conditions about students' knowledge, we investigated to the students concerned with their backgrounds and knowledge of computer science and software engineering in department of engineering in the university. Our university is a public university in the rural town in Japan.

In above result of investigations, students answered that they do not have much knowledge of software and applications even though they are interested in them. Most of students do not know them without entering to the special course of Information science. In our investigations, students more than 90 percents were interested in software engineering more or less when they selected and entered to the course of information science. Namely, when most of students decided their courses, they just had the abstract knowledge about information science. Twenty percents of students had not known since they entered to the university about that software are produced and made by programming.

In this paper, we propose a new distance learning methodologies by the above learners. When there are students who learn on the web-based learning, it is very difficult for students to understand such conceptual learning issues. Thus, it is important not only to consider the e-learning environments and systems, but it is also important to make and

Table 1. Investigation of Students' Pre-Knowledge and Experience

Question 1	When did you operate a PC for the first time ?
Question 2	When did you know the word "software engineering" for the firs time ?
Question 3	When did you know that software is implemented by program languages ?
Question 4	When did you know that most of business software are produced by many engineers ?
Question 5	When were you interested in making software ?

Table 2. Result of the Investigation

	Primary school	Junior high school	High school	After entering the university	In the lecture of the university	No interest
Question 1	38	39	9	1	—	—
Question 2	10	39	22	3	0	—
Question 3	10	24	39	11	3	—
Question 4	4	12	28	39	3	—
Question 5	5	8	25	35	6	5

Table 3. Motivation Entering the Course

Question: Why did you select the course of computer science ?	Number of students.
Because I had strong preference and interest.	30
Because I had weak preference and interest.	43
Because of dependence on the score of entrance examination.	5
Because I was never interested in arts and sociology.	4
Other reason.	3

Table 4. Reasons Taking the Lecture

Question: What is the important factor to select this lecture ?	Number of students.
To take credit for graduation.	27
Learning from good teacher.	7
Learning with friends.	2
No reason.	11
Interests and preferences about contents of lecture.	25
To find good jobs.	11
For the success in the future	11

develop the learning and teaching methodologies. When they are completed, learners can learn and understand effectively.

3 Pre-discussion

Based on the above result of investigations, we thoroughly investigated about the degree of students' knowledge of software engineering. We also investigated that students have a certain concrete imagine/consideration of software engineering. Students who take the examination are same as students where we showed in the previous section. Students who take the software engineering lecture had finished taking more than twenty lectures concerned with information science. Students had also taken practical training of programming and the some languages. Table 5 shows a knowledge of design on

Table 5. Knowledge of Design on Software

	Well-known	A little known	No idea
Could you image the design process of software architecture before taking the lecture ?	3	44	38
Had you ever known the life-cycle model before taking the lecture ?	12	75	—
Did you know the flow charts before you take the lecture ?	70	15	—

Table 6. Knowledge of the UML

Question: Have you ever known the word of UML ?	Number of students.
I have already known the most of figures in UML.	2
I have already known a little figures in UML.	16
I have known only the word.	23
I have known it at the lecture.	45

software and life-cycle models. Table 6 shows a knowledge of UML. First, most of students answered that they could not imagine of design on the software. Namely, they had no experience of making software though requirements analysis and design. Most of students just made small software based on the slam-dunk model in the lecture and exercise of programming. Second, eighty percent of students did not know the lifecycle models such as waterfall model, prototype, and spiral models. There is a student who thinks about finding employment in the software house as a programmer and SE in our university. It is well-grounded to attending lecture "Software engineering" for such a student. Amazingly, twenty percent of students did not know the flowchart model and the word in itself. Seeing the result of investigations, we can forecast and consider that it is difficult for some of other students to make a flowchart with a certain concrete processing. Fifty percents of students know a word of UML because they learn it a little in the lecture and exercise of programming and forty percents in them know a little the writing methods of figures and graphs of UML. Due to limitations and constraints of lectures in the university, in exercise lecture of programming, students just make a small/prototype software and application. For example, some students make a chat system in exercise of network programming and some students make a calculator in exercise in primary programming language. Thus, students can study a basic implementation of application and can master program languages. In actual business tasks, students never study practical business implementation without taking the lecture of software engineering. However, the lecture of software engineering is just a lecture but training. Thus, we should take a deep consideration making the education and teaching program and curriculum.

To consider the above situation, there is a strong limitation that we teach based on just an abstract level. Thus, it is effective for students to learn through integrated learning both lectures and exercises. However, the software engineering is a classroom lecture in our university. It is necessary for teachers to teach and instruct based on an effective methods. To solve the problem, we propose an educational program based on analogical thinking-based learning in the software engineering education.

4 Analogical Thinking-Based Learning Method

When learners understand a new learning issue, cognitive image and figure help them promote their learning [4]. In reference, imagination is described as the chain of the image. However, software making is apart from their daily activity and life. One of effective factures is that the learners effectively use an already-known image. Thus, we develop three above new learning proves and method concretely. Our instructional approach consists of the analogical thinking, self role-play, and anthropomorphic thinking in the lectures.

4.1 Analogical Thinking-Based Learning

When a person learns a new knowledge, he/she sometimes understands it by using his/her well-known related knowledge. We propose a new learning method for students based on such behaviors. The analogical thinking-based learning method is a learning direction based on an association by similarity. Through learners' such processes, they can easily memorize and stick in their heads. On other words, a new concept in which students learn is mapped from previous behaviors. However, we should find what is the appropriate experience to promote students' learning. For example, when there is a student's experience A, we assume the mapping function F exists. And, if there is a new

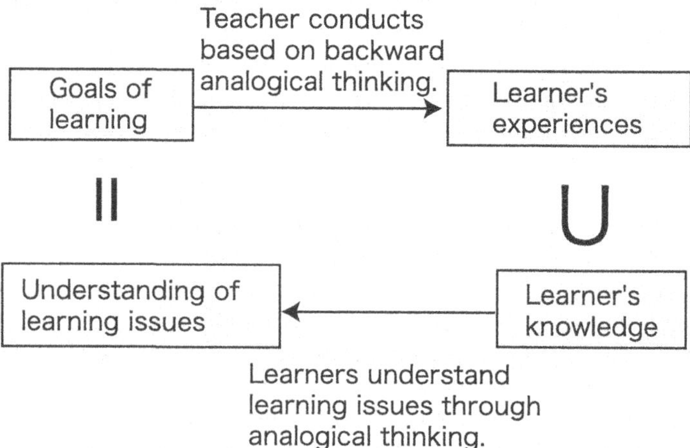

Fig. 1. Outline of Analogical Thinking

concept A' of leaning issues, $F(A)$ is essentially same A'. Thus, one of important thing is that we must find and discover a closed function with F and experience A. Figure 1 shows an outline of analogical thinking. First, teachers discover and detect the appropriate students' experiences which is analogical with learning issues. In other words, this action is like the $F^{-1}(A')$. Second, students do analogical thinking based on the experiences prepared by teachers.

4.2 Self Role-Playing-Based Learning

The Role-play-based learning is a widely well-known learning and understanding method, which is often used in psychology. For example, a learner sets a certain situation and condition. They do a role-play about characters in the situation. When there are any problems in the situation, role-players have experiences to solve the problem. Doing role-play in the learning process, they can easily find how to solve and make easily appropriate decisions if the problem occurs in actual cases. Alternatively, In our proposed learning method in this paper, students and perform role-play for multiple characters by themselves. They do it with an image of the characters' affections and situations. Further, if there are any problems in the role-play, they consider how the problems are solved. Figure 2 shows an example of learner's self role-play-based learning. Characters in learners imaginations discuss and order/serve with each other. Thus, learners can understand the situations in such buyers' and sellers' behaviors.

Fig. 2. Outline of Self role-play-based Learning

4.3 Anthropomorphic Think-Based Learning

An anthropomorphic is a method considering as if the object is like a human beings. In this paper, we apply it and propose an anthropomorphic thinking for engineering education to understand a phenomena and conditions. In an anthropomorphic think-based learning, learners consider a situation where the objects live like a human beings and communicate to users. Considering if the learner is a machine and system, the learner understands and thinks who the machine should be and service to the users.

Table 7. Result of Investigation of Analogical Think

Question: How do you think of your understanding using the analogical thinking ?	Number of students.
I think that it is easy for me to understand without examples.	4
I hope that more appropriate example with the special terms of software engineering.	18
As the lecture, I think that it is easy to understand by using experiences and example.	62

Table 8. Result of Investigation of Self Role-playing

Question: How do you think of your understanding using self role-play?	Number of students.
I think that it is easy for me to understand without examples.	7
I think that the business tasks should be shown before the lecture.	62
As the lecture, it is easy to understand by using examples of software company.	16

Table 9. Result of Investigation of Anthropomorphic Thinking

Question: Q: Did you enough understand by imagining the ATM machine ?	Number of students.
Yes.	60
No.	19
Other opinions.	5

5 Discussion

5.1 Result of Investigation After the Lecture

It is an investigation result concerning the person's of attending a lecture understanding when didactics like is used ahead as follows. Table 7 to 9 are the one that the survey content and the result were shown. Those who attended a lecture about 90 percent answered that analogized idea had helped understanding. There were of those of about 70 percent who attended a lecture by the idea that it was able to understand that it first understood the standpoint and the business of the software development company for the self role-playing, and did the self role-playing afterwards. It was answered that understanding had promoted it by having thought that the student of about 75% was oneself information system for a pseudo-human idea at the end.

6 Instruction System

In our system, the importance for learning is the thinking process rather than the learning issues. Students can generally understand viewing the textbook when they recall a

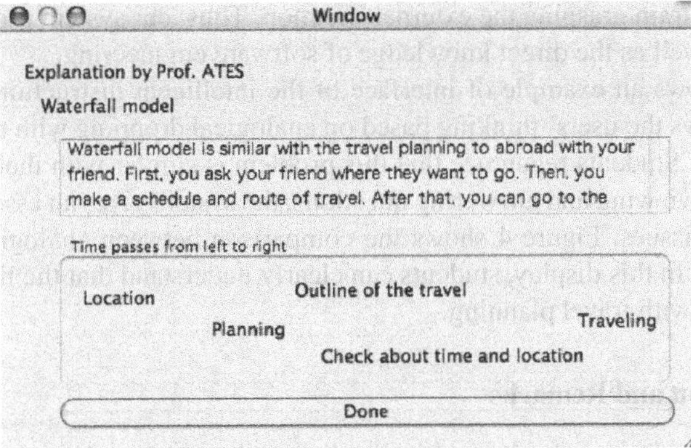

Fig. 3. Intelligent Instruction System 1

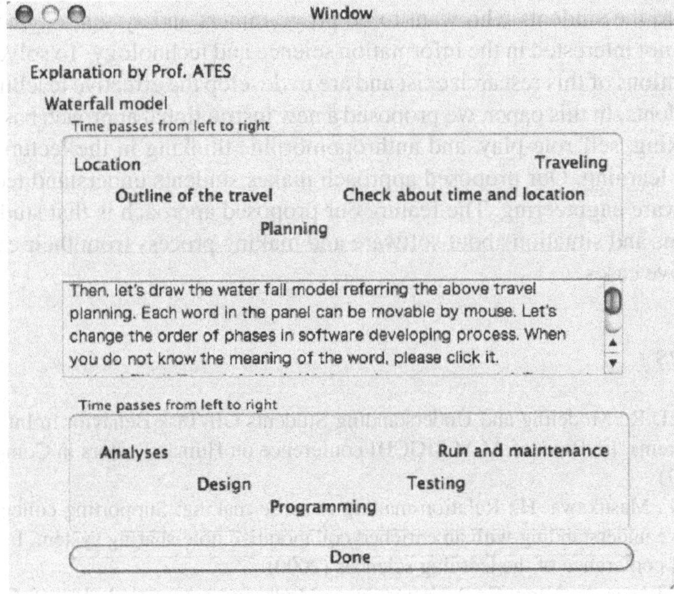

Fig. 4. Intelligent Instruction System 2

simple learning item. On the other hand, in software development, it is important for students to understand what the learning items mean and what they relate with. Even though students learn the lifecycle model and several learning issues seeing the words in the textbook, most of them fail to enhance their grasp of the learning issues. When the students learn just the system and programming, they drop off existence of actual users and their requirements. Grounds of this system are to radicate learning issues for students and to train grasping the external situation. Thus, our system provides the way of thinking as well as the direct knowledge of software engineering.

Figure 3 shows an example of interface of the intelligent instruction system. This figure also shows the users' thinking based on analogical dropping with the example of travel planning. Students recognize that this problem is similar with their experiences. After students' viewing and answering questions, the system gives an essential question about learning issues. Figure 4 shows the comparison between analogical issues and learning issues. In this display, students can clearly understand that the lifecycle model is quite similar with travel planning.

6.1 Discussion and Remarks

In this paper, we discussed a new e-learning/e-teaching methodology for students who belonging in the engineering course. Generally, it is difficult for learners who learn by WBT to understand conceptual issues on the technical methodologies. In the actual condition of students in such universities, although the course of software engineering is provided to the students who want to be programmers and system engineers, some students are not interested in the information science and technology. To solve the problems, motivations of this research exist and are to develop the effective teaching method for such students. In this paper, we proposed a new instructional approach based on analogical thinking, self role-play, and anthropomorphic thinking in the lectures used by the distance learning. Our proposed approach makes students understand technical issues in software engineering. The feature our proposed approach is that students think the conditions and situation about software and making process from their experiences and illustrative cases.

References

1. Baker, S.J.D.R.: Modeling and Understanding Students Off-Task Behavior in Intelligent Tutoring Systems. In: Proc. of ACM SIGCHI conference on Human Factors in Computing Systems (2007)
2. Miyake, N., Masukawa, H.: Relation-making to sense-making: Supporting college students' constructive understanding with an enriched collaborative note-sharing system. In: Fourth international conference of the learning sciences (2000)
3. Salton, G., Wong, A., Yang, C.: A Vector Space Model for Automatic Indexing. Communications of the ACM 18(11), 613–620 (1975)
4. Sunayama, W., Tachibana, K.: Paper Creation Support System. In: The 18th Annual Conference of Japanese Society for Artificial Intelligence (2004)

Capturing Public Knowledge to Forecast Future Events

The Forecast Accuracy of Prediction Markets and Surveys During the FIFA World Cup

Thomas Seemann and Harald Hungenberg

Departement of Business Administration, Economics and Social Sciences
Friedrich-Alexander University Erlangen-Nuremberg, Germany
{thomas.seemann,harald.hungenberg}@wiso.uni-erlangen.de

Summary. Accurate forecasts are of tremendous value for businesses and organizations. Prediction markets are a promising new forecasting mechanism that collect and aggregate information in a group of people. If prediction markets are to improve management information they need to outperform traditional forecasting methods such as surveys. However, there is little empirical evidence of their accuracy in comparison to forecasting methods such as surveys. This article compares the prediction performance of a prediction market and an independent Web-based survey for the FIFA World Cup 2006. The survey is a large scale guessing contest with roughly 220,000 participants. Our data sets include forecasts for the winner of each of the 64 World Cup matches as well as for the number of goals scored in each match. As we show, the prediction market outperformed the survey contest for both predictions, the number of goals scored in a match as well as the match winner.

1 Introduction

Organizations and businesses have always been interested in shedding light on the unknown future. In fact, it lies at the heart of all decision-making processes. In the end, predicting the future is central to doing business: What products will customers buy; what is the next strategic move of competitors; how will technology impact market shares; how will regulatory changes affect sales and margin potential? There are various methods to gathering and aggregating information in order to create a glimpse of future events or developments. Besides more traditional methods such as surveys or expert judgments, prediction markets have gained increasing attention as a promising new prediction method over the last years. Compared to a survey, where a group of participants is directly asked about their opinion, prediction markets let participants trade virtual stocks that denominate certain expectations. Prediction markets work similar to markets for other goods or securities. Markets facilitate the exchange of items between individuals in case of prediction markets participants can literally trade their expectations. The price of an item (denominated in a common currency within the market, e.g. dollars) reflects an agreement between buyer and seller and represents a quantitative measure for the prediction.

The simplest and most commonly applied contract in a prediction market is designed as "event futures" forecasting the occurrence probability of a specific event. For

R. Lee and H.-K. Kim (Eds.): Computer and Information Science, SCI 131, pp. 87–95, 2008.
springerlink.com

example, a contract will pay $10 if a certain event occurs in the future (e.g., Italy wins against France in soccer) and $0 if the event does not occur (so-called winner-takes-all contract). These contracts are then traded typically on a Web-based exchange platform. The trading price of such a contract can be interpreted as the occurrence probability of the event (in our example: the probability of Italy winning the match). Thus, the price mechanism is the aggregation method for the beliefs held by a group of people. The price reflects in a sense the consensus opinion of the traders regarding the future event. Until now there is a limited amount of empirical evidence regarding the circumstances under which prediction markets work well [1]. In particular, the forecasting accuracy of prediction markets is rarely compared to that of direct surveys despite the fact that surveys are in practice the most common alternative to aggregate the belief of a group of people. The only applications, where prediction markets have been repeatedly tested against surveys are so-called election stock markets, forecasting the result of political elections such as the Iowa Electronic Market [2]. However, election stock markets are not fully independent from election surveys. Prior to elections, representative surveys are frequently published by major news networks allowing participants of prediction markets to calibrate their expectations with the results of surveys. Insofar, the information of published surveys is incorporated by the market participants and thus is reflected in the market price of election stock markets [3].

While there is significant evidence that prediction markets are well suited to aggregate distributed information, the key question is whether they really outperform alternative methods. In this paper, we add to the existing empirical evidence by comparing a prediction market with an independent survey. Our data set is based on forecasting results of the FIFA Soccer World Cup matches in 2006. This article is structured as follows. In the subsequent paragraph we briefly review the related work on prediction markets. Paragraph 3 outlines our data sets, while paragraph 4 analyzes the prediction accuracy of the prediction market and the survey using performance metrics. This is followed by the conclusion of our analysis.

2 Related Work

Prediction markets are still in a stage of academic and corporate experimentation. Much of the prediction market research goes back to the Iowa Electronic Market (IEM) at the University of Iowa. The primary focus of their studies is related to political stock markets comparing prediction markets with the opinion polls. The results of the IEM were promising and the market outperformed the polls frequently [2]. However, as already mentioned, polls are published and are used as input by election stock market participants. Insofar, the methods are not independent. Other examples of empirical prediction market studies include Pennock et al. [9]. They show that prices on the Hollywood Stock Exchange are good indicators of actual box office returns compared to expert judgments. Wolfers and Zitzewitz [16] compare a prediction market with the forecast from a survey of around 50 economic experts. They forecast economic indicators which are frequently released by official agencies such as total non-farm payrolls data, retail trade data and business confidence measures. Both the market-based predictions and the expert survey were very close and highly correlated. A statistically meaningful difference of the two prediction models, however, was not detected. In the field of sports there is evidence that generally demonstrates the efficiency of sports betting markets. Williams [14] provides a comprehensive discussion of betting market efficiency while Sauer [11] provides empirical evidence on horse track betting. That does not mean that sports betting markets do not show any inefficiency [13] the overall functioning and predictive power of such markets is, however, undisputed. To our knowledge, the only comparison of prediction markets with a survey in the field of sports events is provided by Servan-Schreiber et al. [12]. They demonstrate that the real-money prediction market Tradesports as well as the play-money market Newsfutures outperformed the prediction of Probability Football (www.probabilityfootball.com) an online sports survey during the 2003 NFL season. Using the same dataset with an elimination of missing data points, Chen et al. [4], however, did not find a significant difference and conclude that prediction markets provide a prediction "as accurate as pooled opinions". Insofar, there is no evidence that proves the superiority of prediction markets compared to surveys in the field of sports events.

Other scholars compare prediction markets not to surveys, but to other prediction markets varying in design. These include Rosenbloom and Notz [10] who compare real-money and play-money markets in various fields and show that both markets provide reasonably accurate probability forecasts. For sports events Rosenbloom and Notz [10] did not find a significant difference between the markets. In case of non-sports events, however, the real-money markets were more accurate. In contrary, Luckner et al. [7] compare the forecast of a real-money and a play-money market related to the German soccer premier league. They found their play-money market to be even more accurate than the real-money market Bluevex. While there is evidence that prediction markets work accurately in fields such as sports, the superiority to other methods (such as surveys) remain questionable. Armstrong puts it this way: "We know little about the conditions under which prediction markets are most useful" (Armstrong, 2006, p. 17). He suggests that prediction markets should be more extensively tested against other structured group methods such as Delphi methods or surveys. Also Gruca et al. [5] conclude that prediction markets' "ability to function as an aggregator of complex information is not very well understood". In the following, we add to the empirical evidence of prediction markets research by comparing a market-based and a survey-based prediction for the FIFA World Cup 2006.

3 Data Sets

The FIFA World Cup is one of the world's most popular sports events. In 2006 the World Cup was hosted in Germany from June 9th to July 9th. The participating 32 national soccer teams first played in a preliminary round consisting of eight groups with four teams each. This resulted in 48 preliminary matches. The top two teams in each group advanced to the finals round (beginning with round of 16). The finals round applied a sudden-death system: Matches that were tie after the official 90-minute match time, were followed by a 40-minute overtime period and if necessary by a penalty shootout to determine the team qualifying for the next level. The finals round comprises 16 matches including the match for the third place. Overall, a total of 64 matches were played during the tournament. Our data set to forecast the World Cup result included the data of the real-money prediction market Tradesports as well as a survey contest hosted by the German Internet community

Knuddels. Both data sets cover all of the 64 soccer matches of the FIFA World Cup. As argued by Chen et al. [4], these games are considered suitable for such a purpose as: (1) the popularity of the FIFA World Cup represents a topic of general interest that provides liquidity for the markets and a high number of responses in the survey, and (2) intense media coverage and analyses of teams provide the general public with the information that is required for forming a knowledgeable opinion.

3.1 Tradesports Data Set

The analyzed Tradesports' contracts include two types of winner-takes-all securities that pay off 100 points if the forecast is correct and 0 points otherwise (US$ 0.10 per point): Match-win contract: pays off if the victory of a specific team or a draw is forecasted correctly. This resulted in three contracts per match: TEAM A, TEAM B, DRAW, except for the round-of-16 games and finals, where no draw was possible.

Goal contract: pays off if the sum of the goals scored in the official 90-minute playing time of a match exceeds the defined threshold. For each of the 64 matches the analysis includes the following three contracts GOALS+1.5, GOALS+2.5, GOALS+3.5 In total the data set includes 348 contracts. For determining the prediction probabilities, we considered only transactions on the actual day of the match, but prior to the start of the match. This resulted in an average of 60.7 transactions per contract which were then considered in deriving the prediction probability. For the goal contracts the average transaction price is reported as prediction probability. For the match-win contracts the average transaction price was normalized so that the probability of victory, draw and loss of each match adds up to 100%.

3.2 Knuddels Data Set

Knuddels is one of the largest Internet communities in Germany with a target audience ranging from age 16 to about age 30. During the World Cup, it hosted a guessing contest, asking its members to predict the result of each of the 64 World Cup matches. Insofar, the contest represents a typical survey, asking the participants directly to predict the result of each match. Knuddels provided the following scores for correct guesses:

- one point for predicting the right outcome
- and one point for predicting the exact result.

Various prizes were offered to the top scoring participants of the contest. In addition, participants with high scores had the privilege to add a specific icon in their personal profile an incentive that is highly valued in an active social virtual community such as Knuddels. This resulted in a high participation of roughly 220,000 participants the largest World Cup guessing game hosted. For each of the 64 games an average amount of more than 155,000 entries was recorded. Missing entries were eliminated as suggested by Chen et al. [4]. Initial guesses of a participant could be changed until the start of the specific match.

Probabilities were derived from the percentage of participants that expected a certain result (victory, draw, loss) in case of the match-win probabilities and the percentage of participants that expected a certain number of goals in case of the goal probabilities.

3.3 Summary of the Derived Probability Predictions

The above mentioned procedure results in the probability prediction of winning, losing, or draw for each of the 64 matches for Tradesports as well as the Knuddels guessing contest. Table 1 shows an excerpt of the resulting probability predictions. For match 1, for example, Tradesports predicted a probability of 78.2% while Knuddels predicted a probability of 90.7% for a victory of team 1. Table 2 shows the predicted probabilities for the number of goals scored. For example, in match 1 Tradesports forecasted a 79.0% probability that more than 1 goal is scored (GOALS+1.5) and a probability of 35.3% that more than 3 goals are scored (GOALS+3.5).

Table 1. Example of predicted match-win probability

	Tradesports			Knuddels		
Match No.	Victory Team 1	Victory Team 2	Draw	Victory Team 1	Victory Team 2	Draw
1	78.2%	6.2%	15.6%	90.7%	5.3%	4.0%
2	46.6%	22.9%	30.5%	56.5%	20.8%	22.6%
3	60.9%	12.9%	26.2%	82.8%	8.0%	9.2%
4	4.7%	81.1%	14.2%	12.3%	72.3%	15.4%
5	56.2%	15.7%	28.1%	76.9%	10.6%	12.5%
...						

Table 2. Example of predicted goal probabilities

	Tradesports			Knuddels		
Match No.	GOALS+1.5	GOALS+2.5	GOALS+3.5	GOALS+1.5	GOALS+2.5	GOALS+3.5
1	79.0%	54.6%	35.3%	95.5%	78.0%	37.1%
2	69.9%	39.7%	20.0%	85.5%	59.0%	26.4%
3	71.7%	41.5%	20.9%	93.9%	75.8%	46.4%
4	82.6%	56.6%	36.4%	83.4%	57.4%	28.3%
5	71.8%	43.5%	26.0%	92.0%	73.9%	43.3%
...						

4 Results and Interpretation

While we identified a systematic difference in the information aggregation of the two mechanisms, we will evaluate the accuracy of Tradesports' and Knuddels' predictions using the performance metrics.

4.1 Performance Metrics

We apply common performance measures used by Servan-Schreiber et al. [12] to evaluate the forecasting accuracy of prediction markets as well as of survey predictions. The same approach has also been applied by Chen et al. [4] and by Rosenbloom and Notz [10]. All measures are based on the deviation of the prediction probability and the

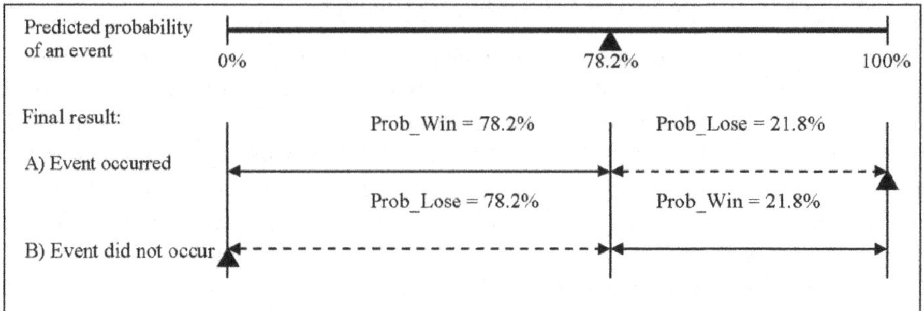

Fig. 1. Determining Prob_Win and Prob_Lose

final result that has been observed. Prob_Win denotes the probability that has been assigned to the event that has occurred, while Prob_Lose is the probability that has been assigned to option that has not occurred. For example, Tradesports predicted a 78.2% chance for team 1 winning match 1. If this event occurs, Prob_Win equals 78.2%. If, however, the event does not occur Prob_Lose (or the prediction error) equals 78.2% (see Figure 1).

In order to determine the quality of the two prediction mechanisms, the Prob_Win and Prob_Lose terms need to be aggregated across all observations. To do so, we apply the following metrics:

(1) Root Mean Square Error (RMSE) = $\sqrt{\text{average}(\text{Prob\_Lose}^2)}$. Based on its quadratic nature, the Root Mean Square Error penalizes predictions that significantly diverge from the final result. The metric is particularly useful when larger prediction errors have an over-proportional impact. A lower metric value is more accurate.

(2) Average Quadratic Score (AQS) = $\text{average}(100-400\cdot\text{Prob\_Lose}^2)$. The Average Quadratic Score is a scoring function that goes back to the "ProbabilityFootball" contest used by Servan-Schreiber et al. [12]. The Average Quadratic Score rewards confident predictions more when they are correct, and penalizes confident predictions more when they are incorrect. For example, a prediction of 90% earns 96 points if the chosen team wins and loses 224 points if the chosen team loses. A prediction of 50% earns no points. A higher (or less negative) Average Quadratic Score is more accurate.

(3) Average Logarithmic Score (ALS) = $\text{average}(\log(\text{Prob\_Win}))$. The underlying idea of the Average Logarithmic Score is to calculate the likelihood of the observation, given the probability prediction. This likelihood is logarithmized to obtain the so-called log-likelihood. Dividing the log-likelihood by the number of items, results in the Average Logarithmic Score as defined above. A higher (or less negative) metric is more accurate. We will not apply the Mean Absolute Error (MAE) = average(Prob_Lose) which is used by both Servan-Schreiber et al. [12] and Rosenbloom and Notz [10]. This metric is not, as Winkler and Murphy [15] calls it, a "good" assessor of a probability forecast. The metric provides an incentive for the forecaster to "hedge" his prediction. This means, the forecaster can minimize the expected score, if he does not reveal his true assessment, but if he overestimates the probability in case he expects the event

Table 3. Performance metrics for match-win predictions

Metric	Tradesports	Knuddels
Root Mean Square Error	**0.389**	0.395
Average Quadratic Score	**-93.308**	-112.579
Average Logarithmic Score	**-0.201**	-0.206

Table 4. Performance metrics of goal predictions

Metric	Tradesports	Knuddels
Root Mean Square Error	**0.461**	0.509
Average Quadratic Score	**-49.198**	-55.696
Average Logarithmic Score	**-0.267**	-0.324

occurring with more than 50% probability and if he underestimates the probability in case he regards the change of occurrence lower than 50%. All other introduced scores are referred to as proper scoring rules. These scoring rules are defined in a way that the forecaster can improve his expected score only if he reveals an "honest" assessment of this event forecast [8].

We calculated the performance metrics based on Tradesports' and Knuddels' predictions. Table 3 shows the performance metrics for the prediction of victory and defeat, while Table 4 shows the metrics for the prediction of the number of goals scored in each match (values in bold font indicate a higher accuracy).

Tradesports' metrics dominate those of Knuddels' regarding the predictions of the number of goals as well as regarding the match winners. This result will be further verified with a randomization test that measures the statistical significance of the metric differences.

4.2 Randomization Test

To assess the significance of the scoring rules results, we follow the approach of Servan-Schreiber et al. [12] and Chen et al. [4] and employ a randomization test. The randomization test attempts to evaluate, whether the difference of a performance metric is statistically significant given the final results of the matches. The procedure is as follows: we randomly swap Knuddels' and Tradesports' predictions, creating two new groups of reshuffled predictions. We then calculate the difference of the performance metric for these reshuffled predictions. This step is repeated $10,000$ times, which results in a probability distribution for the metric difference. By recording the percentile of the originally observed metric difference compared to the metric difference of the $10,000$ randomly re-shuffled predictions, an indication of the statistical likelihood of a given metric difference is provided. For example, the difference of the Root Mean Square Error for the match-win prediction is $0.389 - 0.395 = -0.006$. By creating $10,000$ pairs of reshuffled predictions (one half randomly taken from Tradesports' predictions, the remaining from Knuddels' prediction) we obtain an empirical distribution of the difference in Root Mean Square Error. By comparing the difference of -0.006 with this

Table 5. Match-win predictions: confidence levels of randomization test ($n = 10,000$)

Metric	Difference: Tradesports - Knuddels	Superior	Confidence level randomization test (n=10,000)
Root Mean Square Error	-0.006	Tradesports	65.8%
Average Quadratic Score	19.271	Tradesports	**99.9%**
Average Logarithmic Score	0.005	Tradesports	66.7%

Table 6. Goal predictions: confidence levels of randomization test ($n = 10,000$)

Metric	Difference: Tradesports - Knuddels	Superior	Confidence level randomization test (n=10,000)
Root Mean Square Error	-0.048	Tradesports	**99.9%**
Average Quadratic Score	6.498	Tradesports	86.7%
Average Logarithmic Score	0.057	Tradesports	**99.9%**

distribution, we find that in 65.8% of the (10,000) cases, the difference is smaller than -0.006. Insofar, the difference of Root Mean Square Error is not significant at a 95% confidence level. The full results of the randomization test are outlined in Table 5 and Table 6 (statistically significant results at a 95% confidence level are printed in bold).

For the match-win predictions the Average Quadratic Score shows a statistically significant result in favor of Tradesports while the other two metrics fail to achieve the 95% confidence level. For the goal predictions Tradesports demonstrates a significant result for two out of the three performance metrics the Root Mean Square Error and the Average Logarithmic Score. Insofar, the scoring rules demonstrate a superiority of Tradesports' prediction compared to the survey-based Knuddels' prediction.

5 Conclusion

In our analysis we compared the predications of the real-money prediction market Tradesports and the survey-based contest Knuddels. For both mechanisms we derived probability predictions for the winner of the 64 World Cup matches as well as for the number of goals scored in each match. We compared the forecasting accuracy of Tradesports and Knuddels using performance metrics. All three performance metrics: Root Mean Square Error, Average Quadratic Score and Average Logarithmic Score analyzed

- qualify as a proper scoring rule,
- show a superiority of Tradesports' prediction compared to Knuddels' for both the match-win as well as the goal predictions, and
- show statistically significant results in the randomization test for at least one of the metrics.

Therefore, we conclude that the real-money prediction market Tradesports outperformed the predictions of the survey-based contest Knuddels in forecasting the FIFA World Cup 2006 matches for both the match winner and the number of goals scored. This shows that the market mechanism is efficient in aggregating beliefs and predicting sports events. In fact, the market outperforms the large-scale survey in predicting the World Cup results. This strengthens again the evidence that prediction markets provide the potential to improve forecasting processes. Whether prediction markets are capable of revolutionizing corporate forecasting and decision making, as some advocates promote [6], is still questionable. However, there are good reasons to wonder if betting in prediction markets can at least help to uncover untapped knowledge distributed in their organizations and improve forecasting. A more comprehensive understanding of the conditions under which prediction markets produce results superior to other forecasting methods is an essential prerequisite for a more widespread application of such methods.

References

1. Armstrong, J.S.: Findings from evidence-based forecasting: Methods for reducing forecast error. International Journal of Forecasting 22, 583–598 (2006)
2. Berg, J., Forsythe, R., Nelson, F., Rietz, T.: Results from a Dozen Years of Election Markets Research. Working Draft, University of Iowa, College of Business Administration (for eventual publication in: Plott, C.R., Smith, V.L., (eds.): The Handbook of Experimental Economics Results) (2003)
3. Berg, J., Nelson, F., Rietz, T.: Accuracy and Forecast Standard Error of Prediction Markets, Working Draft, University of Iowa, College of Business Administration (2003)
4. Chen, Y., Chu, C.-H., Mullen, T., Pennock, D.M.: Information Markets vs. Opinion Pools: An Empirical Comparison. In: ACM Conference on Electronic Commerce (EC 2005), Vancouver, BC, Canada (2005)
5. Gruca, T.S., Berg, J.E., Cipriano, M.: Consensus and Differences of Opinion in Electronic Prediction Markets. Electronic Markets 15(1), 13–22 (2005)
6. Kiviat, B.: The End of Management, TIME Magazine, Inside Business (July 2004), `http://www.time.com`
7. Luckner, S., Weinhardt, C., Studer, R.: Predictive Power of Markets: A Comparison of Two Sports Forecasting Exchanges. In: Dreier, T., Studer, R., Weinhardt, C. (eds.) Information Management and Market Engineering, Universittsverlag Karlsruhe (2006)
8. Murphy, A.H., Winkler, R.L.: Forecasters and Probability Forecasts: Some Current Problems. Bulletin of the American Meteorological Society 52, 239–248 (1971)
9. Pennock, D.M., Lawrence, S., Nielsen, F.A., Giles, C.L.: Extracting Collective Probabilistic Forecasts from Web Games. In: 7th ACM SIGKDD International Conference on Knowledge Discovery and Data Mining (KDD-2001), New York (2001)
10. Rosenbloom, E.S., Notz, W.: Statistical Tests of Real-Money versus Play-Money Prediction Markets. Electronic Markets 16, 63–69 (2006)
11. Sauer, R.D.: The Economics of Wagering Markets. Journal of Economic Literature 36, 2021–2064 (1998)
12. Servan-Schreiber, E., Wolfers, J., Pennock, D.M., Galebach, B.: Prediction Markets: Does Money Matter? Electronic Markets 14(3) (2004)
13. Tetlock, P.C.: How Efficient Are Information Markets? Evidence from an Online Exchange, Working Draft, Department of Economics, Harvard University (2004)
14. Williams, L.V.: Information Efficiency in Financial and Betting Markets. Cambridge University Press, Cambridge (2005)
15. Winkler, R.L., Murphy, A.H.: 'Good' Probability Assessors. Journal of Applied Meteorology 7, 751–758 (1968)
16. Wolfers, J., Zitzewitz, E.: Prediction Markets. Journal of Economic Perspectives 18, 107–126 (2004)

An Agent Based Semantic OSGi Service Architecture

Pablo Cabezas, Saioa Arrizabalaga, Antonio Salterain, and Jon Legarda

CEIT and Tecnun (University of Navarra) Manuel de Lardizábal 15,
20018 San Sebastián, Spain
{pcabezas,sarrizabalaga,asalterain,jlegarda}@ceit.es

Summary. The OSGi Framework offers a cooperative environment for the deployment and management of services for multidisciplinary software applications, achieving interoperability between systems. But even so, its service registry lack of non-syntactic information prohibits agents making this framework available for a wider range of applications. This paper proposes an agent-based novel architecture with a semantic in-memory OSGi service registry based on OWL. It enhances the potential of OSGi with semantic data extracted from services deployed in the framework, using software agents in conjunction with Java Annotations and Java Reflection API to dynamically obtain and invoke all required information. As a result, service development and deployment in the OSGi framework will get another view, avoiding the commonly used Interfaces pattern. This Architecture has been successfully applied to a domotic environment managed by a Service Residential Gateway (SRG).

Keywords: semantic architecture, OSGi, service agents, residential gateway, domotic environment.

1 Introduction

Information Science studies the application and usage of knowledge in organizations, the interaction between different actors, focuses on understanding problems, applying information and other technologies as needed.

In cooperation with the Information Society, the need for new paradigms related to the management of services in various environments provides new challenges in this field. Frameworks like the Open Services Gateway Initiative (OSGi) have been created to contribute with a standard service container in which services can be easily managed.

In a producer-consumer system environment several actors take part: the Client (end-user of the application), the Application Manager (usually a client, but with more access level than normal users) and the Service Provider (that takes care of lowlevel application management adding or removing functionalities, debugging and solving errors, etc.).

Using the OSGi framework in application development, the solution is divided into several modules that provide services for each concerning part. This way, if a module wants to be upgraded, the system does not have to be restarted and that new module can be connected hot-plugged. OSGi provides a powerful way to install and uninstall new services and functionalities, simply by following implementation patterns via Interfaces. In this specific circumstance, the Service Provider wants to add more support

R. Lee and H.-K. Kim (Eds.): Computer and Information Science, SCI 131, pp. 97–106, 2008.
springerlink.com

and functionalities to a certain module, but do not want to take care about the API nor know its interfaces, programming according to them. A Semantic OSGi Service Registry (SOR) [1] is proposed in this work, replacing the previous functional version of the application in order to allow the Service Provider to upload contents and services, publishing them via semantic terms in a known ontology in the environment, rather than following an interface with strict methods that have to be implemented.

In the following sections OSGi and Semantic Web Services are reviewed, and how they are applied to SOR. In the fourth section, SOR Architecture is analyzed indepthly, showing its modules and explaining how it enhances the existing OSGi Service Registry, adding semantic indexable information, and getting results through invocation of services via software agents.

Finally, in the fifth section a case study is presented, in which a Service Residential Gateway (SRG) [2] with a domotic network installation raises a new scenario, and how SOR helps adding functionalities with its semantic solution.

2 OSGi

OSGi [3, 4] is an independent, non-profit organization that defines and promotes an open specification for the delivery of managed services to networked environments.

The original market of OSGi was the home service gateways [5-6]. This area has now been extended to include any networked environment, such as the automotive industry, software development environments like Eclipse, mobile devices and application servers. All these devices can use the distributed computing ability to connect the outer network environment through the OSGi-based Service Gateway [7].

The primary goal of the OSGi service framework is the use of the independence of the Java programming language platform and the dynamic code-loading capacity in order to develop and dynamically deploy applications for small-memory devices. It also provides a life cycle management functionality (see Fig. 1) that permits application developers to partition applications into small self-installable components. These components are called bundles [8], and they can be downloaded on demand and removed when they are no longer needed. When a bundle is installed and activated in the framework, it can register a number of services that can be used by other bundles.

A bundle is a physical and logical unit that allocates and delivers service implementations in the OSGi service registry. From the physical perspective, a bundle is distributed in a Java archive file format (JAR) that includes code, resources and manifest files. The manifest file informs the framework of the bundle class execution path and declares Java packages that will be shared with other bundles. It also has information regarding the activator class of the bundle. From the logical perspective, it is a service provider for a certain service or a service requester that intends to use a particular service within the framework during execution time. The OSGi framework provides the mechanism for managing the bundle lifecycle. If a bundle is installed and executed in a framework, it can provide its services, find and use other services available on the framework and can be grouped with other bundle services through the framework's service registry.

The service registry contains all registered services and provides ways to access, get them or track their execution in the framework. Each service is registered under a

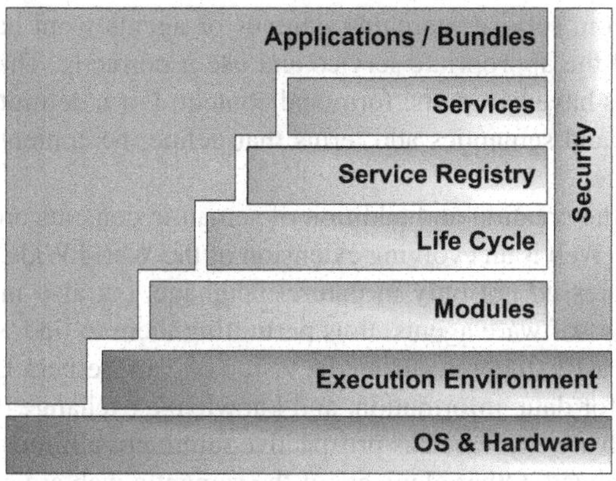

Fig. 1. OSGi Architecture Overview

specified class name, and defines its own properties at registration time, in the form of attribute-value pairs. They are there to differentiate it from other services in the framework, as well as to provide additional information.

The class name string is one of the ways to retrieve a service from the service registry, casting it to the desired class that will provide the services for the client application. It is also possible to apply filters with the properties and their values in this stage, if the string under which the service was registered is not known previously. If the search is successful, the client application will get a ServiceRegistration object, which finally contains the invokable services. But these services will not be able to be invoked if the client does not know the object class and, therefore, will not be able to cast the object into the needed class (or interface). This is the reason why interfaces are so powerful in OSGi. If a class providing services implements an interface known by the client, there will be no problems when getting recently implemented and deployed services, because this will be absolutely transparent for the client application. The client will get the object through the interface, because all the available services will be those contained in that interface.

Even using interface patterns and property filters, server and client bundles are highly coupled through simple text expressions and interfaces that cannot be enriched with the addition of new methods and services. When a service provider entity wants to extend the capabilities of a existing functional system, those additional services will have to be implemented strictly using the interfaces and properties defined for that environment.

The SOR Architecture aims to solve this problematic situation by adding semantic information to the OSGi service registry, avoiding the trouble with interfaces and properties by means of using Intelligent SOR Agents for service deployment and invocation in the framework with additional and consultable information.

3 Semantic Web Services

Semantics refers to aspects of meaning, and contrasts with syntax, which is the study of the structure of sign systems, that focuses on the form, not the meaning. With mere syntactic information, service searching engines or agents wont have any or enough information to find the appropriate service and use it correctly. Therefore, declarative service descriptions have two parts: form and content. Form delimits what can be said and how to say it, and semantics add terms that define the content and other aspects related to the service.

Most efforts in the creation and addition of semantic contents are made in the Web area. The Semantic Web is an evolving extension of the World Wide Web in which web content can be expressed not only in natural language, but also in a format that can be read and used by software agents, thus permitting them to find, share and integrate information more easily. It derives from the vision of Tim Berners-Lee of the Web as a universal medium for data, information, and knowledge exchange [9]. Some elements of the semantic web are expressed as prospective future possibilities that have yet to be implemented or realized. Other elements of the semantic web are expressed in formal specifications, some of which include Resource Description Framework (RDF), a variety of data interchange formats (e.g. RDF/XML, N3, Turtle, N-Triples), and notations such as RDF Schema (RDFS) and the Web Ontology Language (OWL), all of which are intended to provide a formal description of concepts, terms, and relationships within a given knowledge domain.

Semantic Web Services are self-contained, self-describing, semantically markedup software resources that can be published, discovered, composed and executed across the Web in a task driven semi-automatic way. Semantic Web Services can be defined as the dynamic part of the semantic web [10-12]. The mainstream XML standards for interoperation of web services specify only syntactic interoperability, not the semantic meaning of messages. For example, WSDL can specify the operations available through a web service and the structure of data sent and received but cannot specify semantic meaning of the data or semantic constraints on the data. This requires programmers to reach specific agreements on the. interaction of web services and makes automatic web service composition difficult. Semantic web services solve these problems by providing another layer on top of the web service infrastructure to supply semantic meaning for web services.

Semantic Web Services can enable the dynamic discovery, composition and execution of functionality with the aim of providing a higher order level of value-added services. There are many ontologies developed to support the deployment and interoperability of Semantic Web Services, among which it is worth highlighting OWL-S [13].

3.1 OWL-S

OWL-S is an ontology built on top of Web Ontology Language (OWL) by the DARPA DAML program. It replaces the former DAML-S ontology. OWL-S is an ontology within the OWL-based framework of the Semantic Web for describing Semantic Web Services. It enables users and software agents to automatically discover, invoke, compose, and monitor Web resources offering services, under specified constraints. Development of OWL-S aims to enable the following tasks: Automatic Web service discovery, Automatic Web service invocation and Automatic Web service composition and interoperation. OWL-S is an accepted and worldwide used technology based on

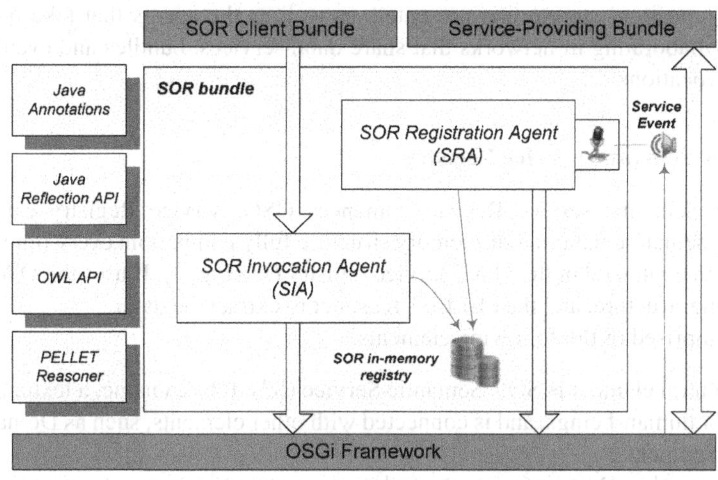

Fig. 2. SOR Architecture Overview

widely spread technologies such as OWL and RDF. The OWL-S ontology consists of four main classes:

- Service, with concepts tying the parts of an OWL-S service description together, holds a textual description of the service.
- Profile, presents the inputs, outputs, preconditions and effects of a service.
- Model, which has properties used to describe how the service works.
- Grounding, with properties used to specify how the service is activated, including details on communication protocols, message formats, port numbers, etc.

In this paper we present an architecture with its core composed by the OSGi mechanisms for publication, discovery and invocation of services, and with information based on the semantic description capabilities proposed by the OWL-S ontology. SOR will provide semantic (and syntactic) content about the services deployed in it, modifying the existing service registry with the paradigm of Semantic Web Services.

4 SOR Architecture

The OSGi service registry enables a bundle to publish objects to a shared registry, advertised via a given set of Java interfaces. Published services also have service properties associated with them in the registry. But all this information is merely syntactic and cannot be easily parsed by a software agent, in order to automate searches, invocations or other tasks. Therefore, if a bundle wants to be deployed, it must inexorably follow that interface publishing pattern in the OSGi framework. The following figure displays an overview of the architecture:

SOR provides an easy way to publish additional information about services, without modifying the default behavior of the OSGi framework. In this chapter SOR internal semantic repository structure is presented, as well as the agents that take part in the system, collaborating in networks that share data, services, bundles and even produce remote invocations.

4.1 SOR Semantic Service Registry

The SOR Semantic service Registry enhances OSGi service Registry capabilities, by adding semantic data in a in-memory structure fully compatible every time with the information contained in the OSGi service registry (see Fig. 3). It uses the OWLAPI to conform the structure, and the PELLET reasoner to extract the data.

It is comprised of the following elements:

- The central element is SOR Semantic Service (S3). It has a name, a textual description for human-beings, and is connected with other elements, such as Domain, Type and Arguments.
- A Domain has DomainConcepts, and contains terms in external ontologies for the S3s to be described. It has a valid identifier and a name, which is an indexable term in an upper level ontology.
- A DomainConcept is a term in a low-level ontology, and has a valid identifier, a name and a textual description.
- A Type has a name, and is related to an S3 object (through its hasS3Return Type connection) and with Arguments (through their hasType relationship).
- An Argument has a valid identifier, a textual description, a name, a Type and a DomainConcept which it is related to.

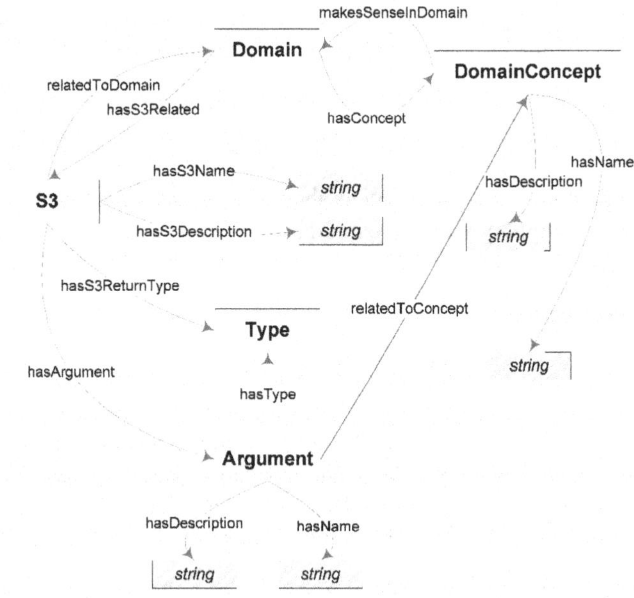

Fig. 3. Diagram of the SOR Architecture Semantic Registry Data and Internal Relations

Besides the basic relationships, inverse relations have been created in order to establish searches from one point of the relation to another and vice versa, where required.

Initially, when a bundle is registered into the framework, the agent will look for the property "SOR.PUBLISH" which, if set to "true", will indicate the agent to add the semantic information provided to the SOR registry.

The information will be extracted from each valid method, defining a valid one as a method that supplies data in the form of Java Annotations. A service providing bundle will not have to know anything about the SOR bundle, only having to:

- Apply the correct property to the initial set at deployment time.
- Specify the correct Java Annotation for each method that is wanted to be deployed. The Java Annotations name is S3Method, and has the following fields:
- Domain, corresponding to the Domain object in the SOR registry data structure. Defines a term in a high-level ontology, and can be a key in S3 searching.
- Concept, corresponding to the DomainConcept object in the SOR registry data structure, defines a more specific term in a low-level ontology. It can be defined as a technical specification or detail in a specific area.
- Description, which is a textual description for human-beings about the service.
- Arguments, which are an array of new Java Annotations, each one called S3MethodArgument.

Each S3MethodArgument has the following fields:

- Order of the argument in the method declaration, as an integer number.
- Concept, corresponding to the DomainConcept object in the SOR registry data structure, defines a more specific term for this method argument.
- Description.

Note that the information relative to name, argument types, return type, etc. has been avoided, only asking for the necessary and more crucial information in order not to overload the code with interminable annotations.

Once a bundle has been deployed and started within a framework, the SOR Registration Agent begins its work.

4.2 SOR Registration Agent

The SOR Registration Agent (SRA) implements the OSGi ServiceListener pattern, looking for events in the framework concerning valid or SOR-compliant serviceproviding bundles.

When an event of the type STARTED or STOPPED is released, the SRA will parse and get the information if and only if the bundle is SOR-compliant. As it has been commented in the previous section, the property "SOR.PUBLISH" will be the key.

At this time, the SRA will get the ServiceReference from the ServiceEvent object, and from another property called "objectClass" the names of classes deployed in this bundle will be obtained. Now, for each valid object class string a generic class will be obtained, from which the SRA will start getting method information.

For the first case, in a STARTED event, SRA will only get information from method annotations (and from the methods themselves) if the Annotation is valid; it is, if information can only partially be retrieved, in the end it will not allow the method to be invoked, so this method parsing process is discarded. If a valid Annotation is found, then SRA will make use of the Java Reflection API to get all required data from its name, arguments, types, etc. In the second case, when a STOPPED event is fired, SRA will have to find if a SOR-compliant bundle has been involved in that event, in order to leave the SOR registry information in a consistent state with respect to the information published in the OSGi registry.

Using Java Annotations and the Java Reflection API, service-providing bundles will not have previous SOR bundle knowledge, nor will they have to include a reference to that bundle in their manifest. This way, the deployment of services in the framework is absolutely dynamic and independent of the interface pattern commented before.

4.3 SOR Invocation Agent

When a S3 is registered into the SOR architecture, it then can be invoked by a client bundle. This time the client will do have to get a reference to SOR, by adding an external reference to it. It will then have to specify to the SOR Invocation Agent (SIA) details relative to the DomainConcept which the method can be related to, and a set of Arguments to execute the method with.

Then, the SIA will look for the Domain related to the main DomainConcept term provided by the client, to get a top-level view of services in the system. When finding services related to that Domain and that specific DomainConcept term, then the process will take a step forward and look for compatibilities in the number and type of arguments provided. If a valid S3 is found, then it is executed and the result is returned to the client. In case no S3 can be found, the client will get a SORNoS3CompatibleFoundException, indicating this eventuality. The client won't have to note if the service is static or not, because the SIA will get the class object instance if required (or not) and make the invocation in an absolutely transparent way for the end-user.

5 Case Study

It has been developed a gateway that controls a number of interconnected Homes and provides services for them, such as Internet connection management, firewall, QoS, video door entry configuration and usage, file sharing, parental control and domotic installation management.

In the SRG project, a Service Provider wanted to add more bundles to the systems in order to cover and support more domotic devices. Even in the domotic network modification stage, if a new domotic sensor or actuator is introduced to the system, the SRG must know its type and get the services from the required bundle to get it to work. Furthermore, the Service Provider didnt want to follow the interfaces provided by the system development team, and wanted to reach a higher abstraction level in the installation of bundles in the system.

A laboratory has been used as a certain room in a Home, and it has been equipped with several KONNEX sensors and actuators, managed through a KNX-to-IP gateway that will send data and different commands, as well as serve as entry point to the installation for the SRG. In the following picture the place in which the installation took place can be seen:

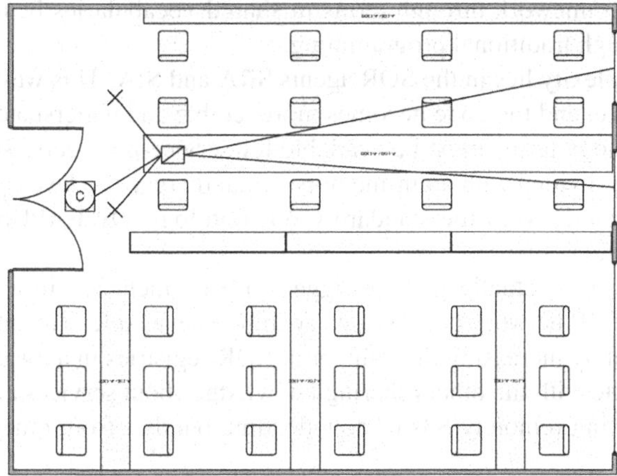

Fig. 4. Map of the domotic installation

There are installed and running elements like motion sensors and actuators to lower and raise the blinds. Now the Service Provider wants to add Lighting sensors and actuators. In the old OSGi interface-style way, the Service Provider would have had to adapt its implementation for this kind of element to the Interface which the SRG is working with, situation that can lead to trouble if the code has been developed by another unknown company (for example, if the code has been extracted from a freeopen project in the Internet).

With SOR, the Service Provider will look for the public methods that will be accessible to the client applications (for the SRG, in this case), and will add the property "SOR.PUBLISH" in the Activator class, besides the Java Annotations S3Method and S3MethodArgument, containing and providing valid terms in the SRG Domotic Ontology, for each method wanting to be published in the semantic registry. The Service Provider will not have to get the SOR bundle to finish the new bundle, what gives him more flexibility and makes the framework to reach a higher level of dynamic code-loading and invocation. More importantly, the client does not have to know technical details about the methods, only accessing them through terms in shared ontologies and getting the desired results. This low-coupling solution makes the system favorable for an easier addition of new functionalities into the SRG, reducing the development time and, consequently, reducing the time that the client will have to wait until the new functionality for his/her domotic network will be available.

6 Conclusions and Future Work

The SOR Architecture has been created for a Multi-Service Multi-Residential Gateway, but its nature has caused to finally get to be a Horizontal architecture with potential applications in many projects. The addition of semantic information to the OSGi registry was a need in order to facilitate to the client the access to the Semantic Services deployed in the framework through terms in shared vocabularies better than through interfaces and rough traditional programming.

Also, the complexity lies in the SOR agents SRA and SIA. This way, the client bundle reduces its size, and the code becomes more legible and understandable for human beings. The ontology terms must be available for service-providers, SOR and clients, but this is easily provided with an in-memory semantic registry developed, created, accessed and maintained with the standard OWL ontology, OWLAPI libraries and the PELLET reasoner.

The SOR scope is actually getting bigger with its inclusion in a sensor network project, with TCP/IP network connection capabilities available. The aim is to enhance the SOR potential by increasing the number of SOR registries in a local network, making them cooperate with the others sharing knowledge about services and ontologies in the system, invoking remotely S3s or downloading bundles from other frameworks if needed.

In addition, the development team will make a SOR system compliant with an OWL-S external system, as OWL-S is seen as the future standard de-facto in semantic web services throughout the World Wide Web. This will make SOR capable of getting all kinds of information from the biggest service database available in the World.

References

1. Cabezas, P., Arrizabalaga, S., Salterain, A., Legarda, J.: SOR: Semantic OSGi service Registry. In: UCAmI 2007, Zaragoza, Spain (2007)
2. Bull, P.M., Benyon, P.R., Limb, P.R.: Residential gateways. Bt Technology Journal 20, 73–81 (2002)
3. Alliance, T.O.: OSGi Service Platform Core Specification, 4 edn. (August 2005)
4. Alliance, T.O.: OSGi Service Platform Service Compendium (July 2006)
5. Hwang, T., Park, H., Chung, J.W.: Design and implementation of the home service delivery and management system based on OSGi service platform. In: Consumer Electronics, 2006. ICCE 2006. 2006 Digest of Technical Papers. International Conference (2006)
6. Zhang, H., Wang, F.-Y., Ai, Y.: An OSGi and agent based control system architecture for smart home. In: Networking, Sensing and Control, Proceedings. IEEE, Los Alamitos (2005)
7. Lee, C., Nordstedt, D., Helal, S.: Enabling smart spaces with OSGi. In: Pervasive Computing, vol. 2, pp. 89–94. IEEE, Los Alamitos (2003)
8. Hall, R.S., Cervantes, H.: Challenges in building service-oriented applications for OSGi. In: Communications Magazine, vol. 42, pp. 144–149. IEEE, Los Alamitos (2004)
9. Tim Berners-Lee, J. H. a. O. L.: The Semantic Web (2001)
10. Hepp, M.: Semantic Web and semantic Web services: father and son or indivisible twins? IEEE Internet Computing 10, 85–88 (2006)
11. Gu, T., Pung, H.K., Zhang, D.Q.: Toward an OSGi-based infrastructure for contextaware applications. In: Pervasive Computing, vol. 3, pp. 66–74. IEEE, Los Alamitos (2004)
12. Dieter Fensel, H.L., Polleres, A., de Bruijn, J., Stollberg, M., Roman, D., Domingue, J.: Enabling Semantic Web Services. In: The Web Service Modeling Ontology. Springer, Heidelberg (2007)
13. Steffen Balzer, T.L., Wagner, M.: Pitfalls of OWLS – A Practical Semantic Web Use Case, p. 10 (2004)

A Framework for SOA-Based Application on Agile of Small and Medium Enterprise

Seung Woo Shin[1] and Haeng Kon Kim[2]

[1] Department of Computer information & Communication Engineering, Catholic Univ. of Daegu, Korea
selab@cu.ac.kr

[2] Department of Computer information & Communication Engineering, Catholic Univ. of Daegu, Korea
hangkon@cu.ac.kr

Summary. SOA (Service Oriented Architecture) is a very promising solution that many small and medium sized enterprises may use for developing web service environments. SOA, however, has an inherent limitation: there is no specific method associated with it. Thus, there are no definitive answers to support those enterprises.

In this paper, we propose Xplus with the use of Agile methodologies as a framework for SOA-based applications. The employment of these methodologies will provide small to medium sized organizations with successful results for a wide variety of cases. The use of an Xplus framework on SOA systems will surpass methodologies as a niche for these organizations, allowing for the creation of productive, high-quality software.

Keywords: Software Development Methodology, Service Oriented Architecture, Agile Methodology.

1 Introduction

Most large enterprises are already employing web service environments. For most small and medium sized organizations, though, the move into this new field is still being made. Many of these businesses are strongly considering the use of an SOA-based system for the import of their web service environment of the reconstruction of existing web services.

In recent years, numerous studies have attempted to build web-based business transaction environments, or to reconstruct existing web applications, on the basis of a Service-Oriented Architecture. SOA's lack of a specific methodology, though, has meant that there are no definitive answers for the small and medium sized enterprises seeking to implement these applications, and thus making it very difficult to construct them on SOA-based systems.

In this paper, we describe an attempt to implement Xplus as a framework for Service-Oriented Architecture by use of Agile methodologies. These methodologies are suitable for use by small and medium sized enterprises, to increase the agility required by the

R. Lee and H.-K. Kim (Eds.): Computer and Information Science, SCI 131, pp. 107–120, 2008.
springerlink.com

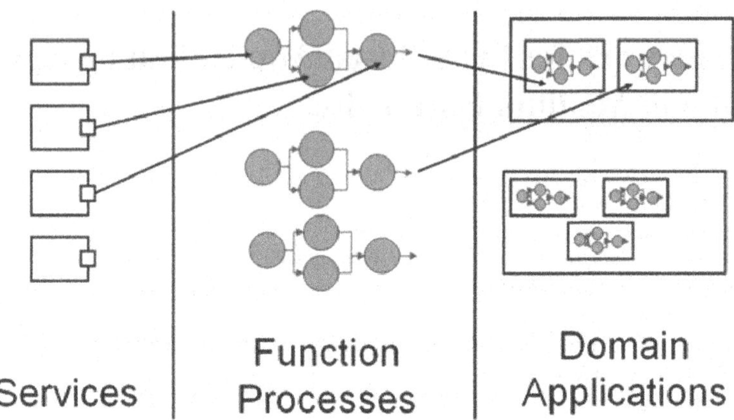

Fig. 1. SOA basic concepts

specific business. Through the introduction of SOA, it is possible to effectively provide the customer with the required services.

2 Related Works

2.1 SOA(Service Oriented Architecture)

Service-Oriented Architecture (SOA) is an architectural framework that supports integrating business tasks as linked services that can be accessed when needed over a network. Figure 1 presents the relationship between service, processes and the applications in SOA. Within SOA architecture, all functions, such as check service inventory, software distribution and payment services, etc., are defined as a kinds of the services [1]. For SOA there are three important architectural perspectives as shown in Figure 2.

The Application Architecture

This is the most promised business solution which consumes services from one or more providers and integrates them into the business processes.

The Service Architecture

This provides a bridge between the implementations and the consuming applications, creating a logical view of sets of services which are available for use, invoked by a common interface and management architecture.

The Component Architecture

This describes the various environments supporting the implemented applications, the business objects and their implementations [2].

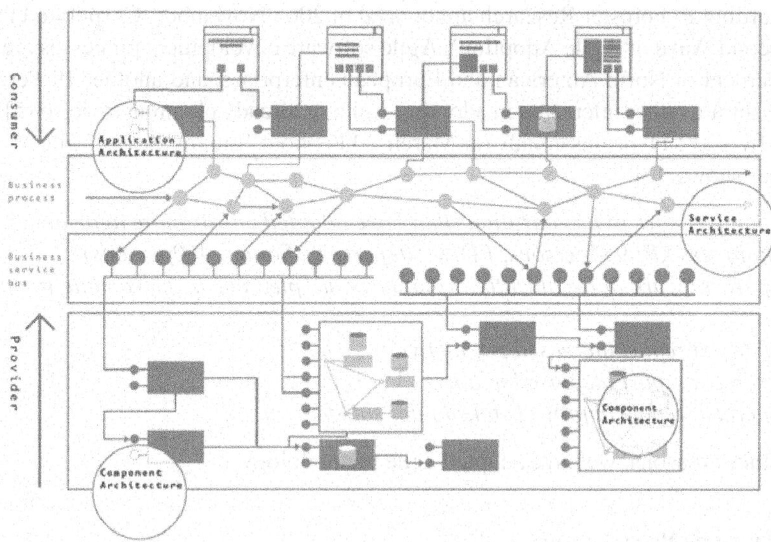

Fig. 2. Three architectural perspectives

2.2 Agile Methodology

Agile software development started to take off during the late 1990s when Kent Beck promoted Extreme Programming (XP) with a set of values, principles, and practices for planning, coding, designing, and testing software. All agile software development approaches have several values in common, such as:

- *Frequent inspection and adaptation*
- *Frequent delivery*
- *Collaboration and close communication*
- *Reflective improvement*
- *Emergence of requirements (incremental), technology, and team capabilities*
- *Empowerment and self-organization*
- *Dealing with reality, not artifacts*
- *Courage and respect*

From these values, the various agile methods in use today emphasize different practices.

In February 2001, the Agile Manifesto was defined, valuing individuals and interactions over processes and tools, working software over comprehensive documentation, customer collaboration over contract negotiation, and responding to change over following a plan. It is the foundation of all agile methods in use today [21].

Methodologies similar to Agile created prior to 2000 include Scrum (1986), Crystal Clear, Extreme Programming (1996), Adaptive Software Development, Feature Driven Development, and DSDM (1995).

According to Forester Research announced in 2005 November 'Corporate IT Leads The Second Wave of Agile Adoption', Agile software development processes are in use at 14 Percent of North American and European enterprises, and another 19 Percent of enterprises are either interested in adopting Agile or already planning to do so [5]. Also, survey by 4232 IT professionals on March 2006 about 'Survey Says: Agile Works in Practice' shows: [6]

- *65 percent work in organizations that have adopted one or more agile development techniques (XP: 954persons, FDD: 502persons, Scrum: 460persons)*
- *41 percent work in organizations that have adopted one or more agile methodologies*
- *60 percent report increased productivity*
- *66 percent report increased quality*
- *58 percent report improved stakeholder satisfaction*

For these reasons, we have adopted agile methodology.

2.3 Extreme Programming (XP)

eXtreme Programming(XP) is a discipline of software development based on values of simplicity, communication, feedback, and courage. It works by bringing the whole team together in the presence of simple practices, with enough feedback to enable the team to see where they are and to tune the practices to their unique situation. Figure 3

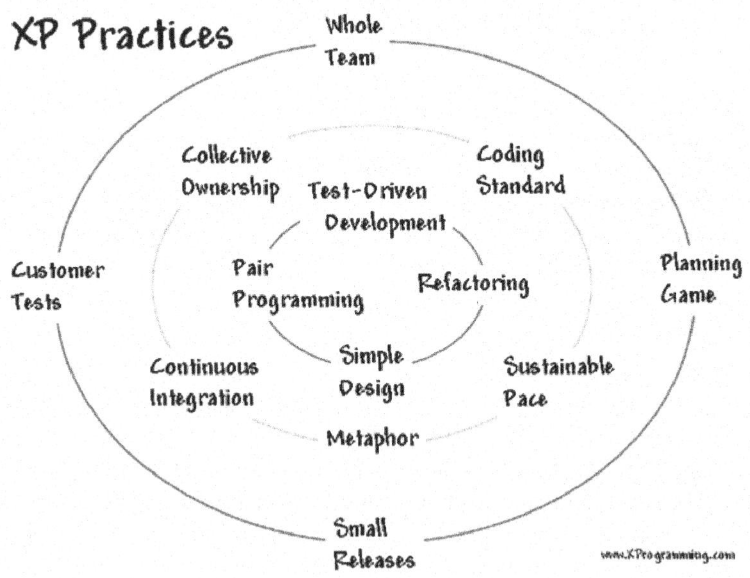

Fig. 3. Circles of 13 XP practices

presents circles of 13 XP practices [7]. Grouped into four areas, derived from the best practices of software engineering: [8]

- ***Fine scale feedback:*** *Pair programming, Planning Game, Test driven development, Whole team*
- ***Continuous process:*** *Continuous Integration, Refactoring, Small Releases*
- ***Shared understanding:*** *Coding Standards, Collective Code Ownership, Simple Design, System Metaphor, Customer Tests*
- ***Programmer welfare:*** *Sustainable Pace*

Figure 4 shows process of XP project [9]. Start out collecting user stories and conducting spike solutions for things that seem risky. Do this for few weeks. Then schedule a release planning meeting. Invite customers, developers, and managers to create a schedule that everyone agrees on. Begin your iterative development with an iteration planning meeting [9].

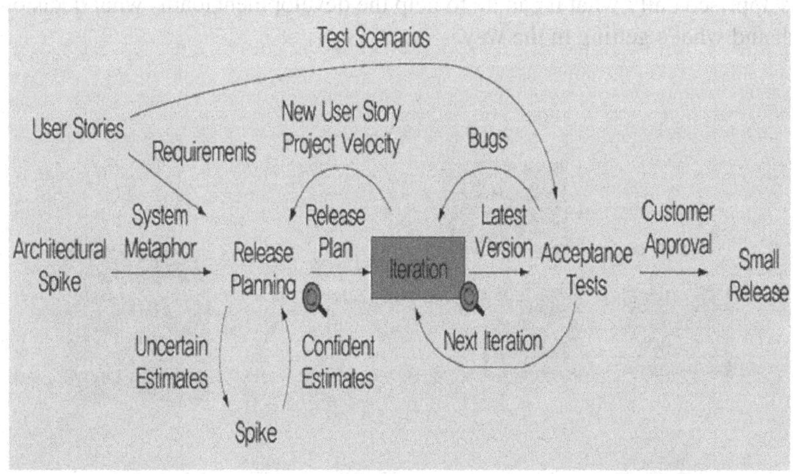

Fig. 4. Process of XP project

2.4 Scrum

Scrum is an iterative, incremental process for developing any product or managing any work. It produces a potentially shippable set of functionality at the end of every iteration. It's attributes are: [10]

- *Scrum is an agile process to manage and control development work.*
- *Scrum is a wrapper for existing engineering practices.*
- *Scrum is a team-based approach to iteratively, incrementally develop systems and products when requirements are rapidly changing.*

- *Scrum is a process that controls the chaos of conflicting interests and needs.*
- *Scrum is a way to improve communications and maximize co-operation.*
- *Scrum is a way to detect and cause the removal of anything that gets in the way of developing and delivering products.*
- *Scrum is a way to maximize productivity.*
- *Scrum is scalable from single projects to entire organizations. Scrum has controlled and organized development and implementation for multiple interrelated products and projects with over a thousand developers and implementers.*
- *Scrum is a way for everyone to feel good about their job, their contributions, and that they have done the very best they possibly could [10].*

Figure 5 [11] illustrate diagram of Scrum flow. Scrum provides daily status on team progress, and iterative (every 30 days) reviews of product progress. Everything is visible what's to be worked on, how work is progressing, and what has been built supporting management decisions regarding cost, time, quality, and functionality. Plus, management is apprised daily what it can do to help the development teams what decisions are needed, and what's getting in the way.

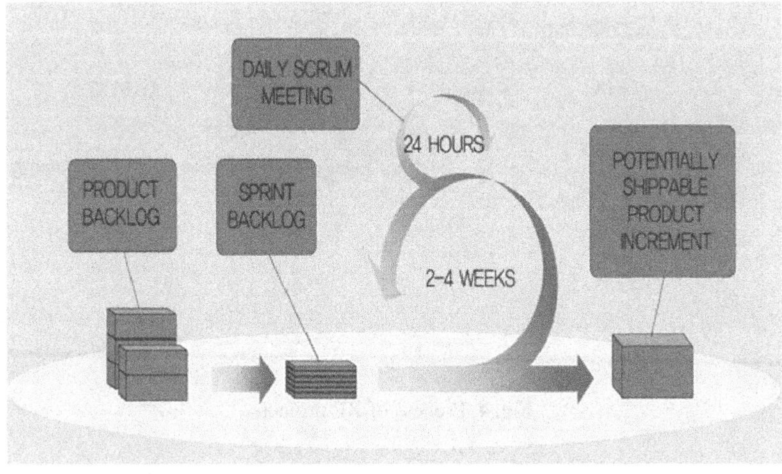

Fig. 5. Diagram of scrum flow

3 Framework for SOA Application Using Agile Methodology

This section defines the described framework for SOA systems employing Agile methodology, for use by small and medium sized enterprises.

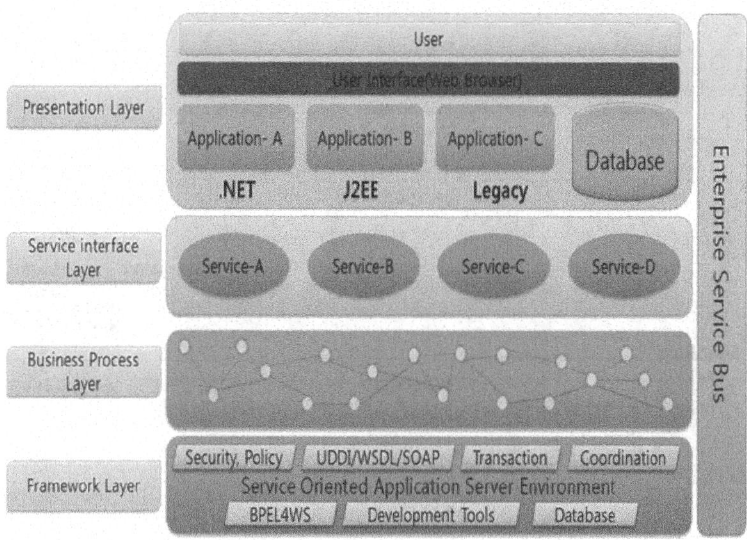

Fig. 6. Architecture of SOA-based application development

3.1 Architecture

A primary goal of deploying SOAs is to simplify development and implementation of new applications and capabilities through aggregating many low-level tasks into higher-level services. To use a human analogy, the heart is a discrete system comprised of smaller, specialized systems, such as valves, arteries, and veins. An SOA is similar. All electronic business processes are implemented at the lowest levels as specialized functions designed to perform specific tasks.

Low-level services are designed to behave like text messaging as short, to-the-point interactions. These low-level services are created and maintained by people who have the expertise required to weave them together to produce a desired business result. A service doesn't necessarily know about other services or how they operate. For example, when a financial reporting application needs data, it simply asks for it from a software service without knowing or caring from where or how the data is supplied. And at higher levels of aggregation, neither does a Web application, portal, or application system worry about what other applications do. In this paper, we have proposed the use of Xplus Architecture as a framework for SOA-based application development, as shown in Figure 6. This architecture essentially refers to the Service-Oriented Architecture reference model [16, 17].

This architecture employs four separate layers: the framework layer, the business process layer, the service interface layer and the presentation layer. The applications use the independent designed platforms for legacy applications through wrapping techniques, and each layer uses standardized technologies. In top-down order, the

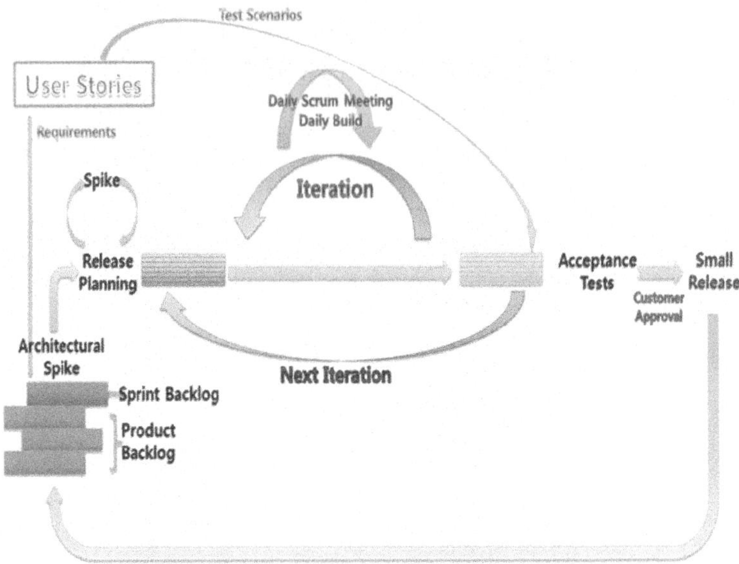

Fig. 7. Xplus Framework Process Model

presentation layer makes use of HTML, XML, SQL, Ajax, JSP, and ASP technologies. The service interface layer provides an API for heterogeneous applications. The business process layer works to combine business process. Finally, the framework layer provides a platform for the Service Oriented Application server environment.

3.2 Process Model

For SOA applications, we propose a process methodology that combines two separate Agile development process methodologies. Figure 7 presents Xplus on Scrum-based flow unities and XP practice elements. In this paper, we dubbed it the 'Xplus process model'. As shown in Figure 7, this process model centers on the diagram of Scrum flow for project management. For these management tasks, Scrum provides a powerful model, but this model is not detailed enough to use for developing specific elements. The description of proposed models follows below.

1. ***Planning Game:*** *Project teams create user stories and test scenarios. If the team wishes to improve clarity by estimating the whole planning process, they can use a spike solution (Architectural Spike, Spike) and exploration (technical testing of new adopting technique).*
2. ***Release Planning:*** *Project teams estimate suitable unit iterations from product backlog.*

3. ***Sprint Backlog:*** *According to release planning, project teams program and create the test unit code, refactoring and iteration. Each iteration should be about a 10 30 day period, and should include the Daily Build and Daily Scrum Meeting.*
4. ***Acceptance test:*** *Development teams use the customer to test the test scenarios.*
5. ***Small Release:*** *If the customer approves of the product during acceptance testing, a small release of the product may be implemented.*
6. ***Next Product Backlog Sprint:*** *Project teams sprint for performance of next iteration, gathering additional a previous requirements (includes product backlog) after the small release. The entire process-flow is based on Xplus practices.*

Our Xplus has the following several advantages:

- Effectively supports both developer and product manager by integrating two methodologies which provide specificity for each practice.
- The framework is easily adapted by deleting ambiguous elements through use of XP practices.
- The strictness of Scrum time boxing is reduced through the use of pair programming and sustainable pace practices, which support a pleasurable development cycle and long-term work sustainability.
- Manageability is increased by applying XP's incremental practices.
- Flexibility is maximized by the use of the Service-Oriented Architecture.

4 Case Study Using Xplus Framework

Here we describe a sample project that was performed with the Xplus framework. Figures 8 and 9 display the first activity in the process: the creation of user stories. In Figure 8 we see the initial user role identification; and if Figure 9 we have the initial modeling of that user role.

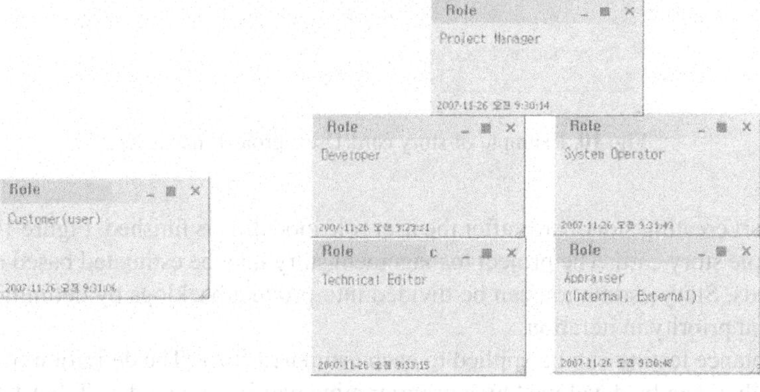

Fig. 8. User role identify

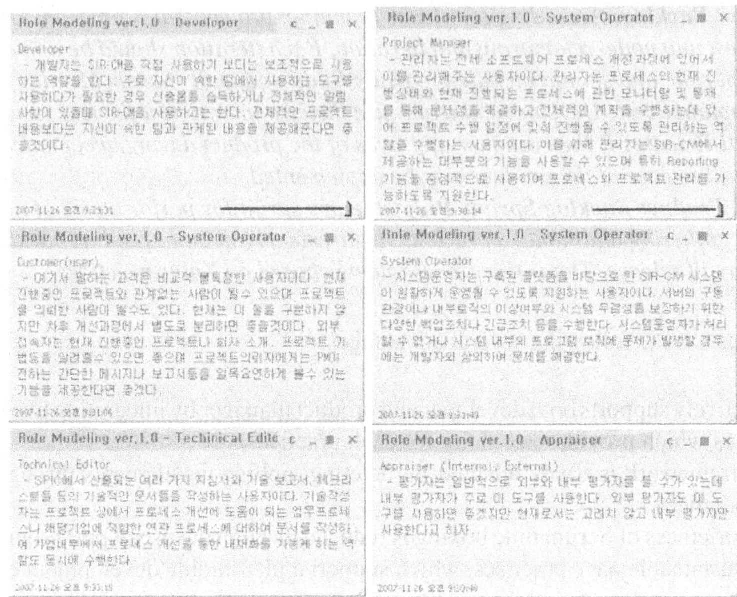

Fig. 9. User role modeling

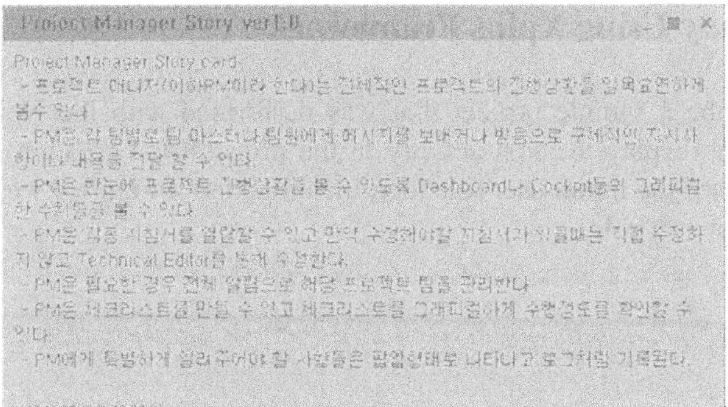

Fig. 10. Example of story card(User: project manager)

We start creating story cards after the user role modeling is finished. Figure 10 shows an example story card. The project management story may be estimated based on these story cards. Story estimation can be divided into product backlogs by defining the development priority in iteration.

Acceptance testing can be applied to each estimated story. The definition of the 'estimate value' can be developed by a programming pair in about a day. Table 1 displays a story card's resulting story estimate.

Table 1. Story estimate result

Story	Estimate	Priority
Project Manager (PM) shows whole project progress	7	1
PM using graphical view (dashboard or project cockpit) for whole project elements	14	4
PM uses guideline and document. If need modify it, Demand messaging to 'Technical Editor'	4	2
If PM wants to notice to all project stakeholder, PM can Messaging to all project stakeholder	6	3
PM is able to create any checklist. PM can watch graphical view.	10	5
If specially inform to PM, show the pop up window and write log file.	4	6

Table 2. Release planning

Iteration 1(10)	Iteration 2(14)
Project Manager (PM) shows whole project progress. (7)	PM using graphical view (dashboard or project cockpit) for whole project elements (14)
PM uses guideline and document. If need modify it, Demand messaging to 'Technical Editor'. (4)	PM is able to create any checklist. PM can watch graphical view. (10)
If PM wants to notice to all project stakeholder, PM can Messaging to all project stakeholder. (6)	If specially inform to PM, show the pop up window and write log file. (4)

Release point planning was established based on estimate values and determined priorities. The list of release planning is shown in Table 2. An iteration also may have a story, where suitable.

Figure 11 shows the dashboard user interface for communication to customers. The UI illustrates an example of a story titled 'PM using graphical view (dashboard or project cockpit) for whole project element'. After finishing release planning, the Xplus process moves to the first 'sprint' and 'Daily Scrum Meeting'. At this point, CRC (Class, Responsibilities and Collaboration) cards must be created.

Table 3 summarizes the advantages of the Xplus model for the development of SOA applications. The results of our sample project have made us confident the use of the Xplus model can guarantee successful projects.

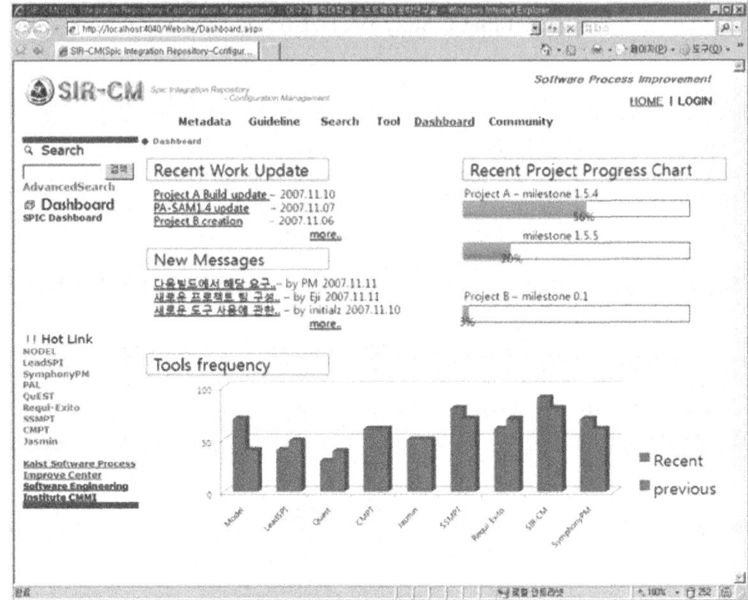

Fig. 11. Dashboard user interface

Table 3. Advantages of Xplus model for developing SOA application

Xplus Practices	SOA Principles	Advantage
Small Release	Loose Coupling, Service Granularity	Small release practice is designed one stand-alone component, in SOA applications assurance of loose coupling and service granularity
TDD	Autonomous	Test-Driven Development practice core is automation testing.
Coding Standard and Simple Design	SOA Simplifies Development	Standardized coding practice to guidance simplify development
Refactoring	Designed for reuse, well defined interface	From refactoring code or design to guidance reusable and has well defined interface

5 Conclusions and Future Works

In this paper, it was proposed that Service-Oriented Architecture using an Xplus framework is ideal for small and medium enterprises. A case study was performed and documented in order to verify our proposal. An SOA based system will be easier for small and medium sized organizations through the use of agility principals on an Xplus framework.

This system will surpass other methodologies in its niche for these small and medium sized businesses. Xplus using agile methodology can be applied to produce very high quality software.

Future studies will be performed using a project approach based on this process model, and will report the weaknesses and advantages of this system.

Acknowledgments

"This research was supported by the MIC(Ministry of Information and Communication), Korea, under the ITRC(Information Technology Research Center) support program supervised by the IITA(Institute of Information Technology Advancement)" (IITA-2008-(C1090-0801-0032)).

References

1. Chen, I.-Y., Huang, C.-C.: An SOA-based software deployment management system. In: Proceedings of the 2006 IEEE/WIC/ACM International Conference on Web Intelligence, pp. 617–620. IEEE, Los Alamitos (2006)
2. Sprott, D., Wilkes, L.: Understanding Service Oriented Architecture. Microsoft Architecture Journal, Microsoft 1, 10–17 (2004)
3. Craig, L., Basili, V.R.: Iterative and Incremental Development: A Brief History. Computer 36(6), 47–56 (2003)
4. Beck, K., Beedle, M., van Bennekum, A., et al.: Manifesto for Agile Software Development (2001), http://www.agilemanifesto.org/
5. Schwaber, C., Fichera, R.: Corporate IT Leads The Second Wave Of Agile Adoption. Forester Research (2005)
6. Ambler, S.: Survey Says: Agile Works in Practice. Dr. Dobbs Portal (2006), http://www.ddj.com/architect/191800169
7. Jeffries, R.: What is Extreme Programming? XProgramming.com: an agile software development resource (2001), http://www.xprogramming.com/xpmag/whatisxp.htm
8. Wikipedia, Extreme Programming. Wikipedia (2007), http://en.wikipedia.org/wiki/Extreme_programming/
9. Don, J.: Wells, Extreme Programming Project, Extreme Programming: a gentle introduction (2000), http://www.extremeprogramming.org/
10. Schwaber, K.: What is Scrum? Scrum: Its about common sense (2007), http://www.controlchaos.com/about/
11. Schwaber, K., Beedle, M., Martin, R.C.: Agile Development with Scrum. Prentice-Hall, Englewood Cliffs (2001)
12. Subramaniam, V., Hunt, A.: Practices of an Agile Developer. The Pragmatic Programmers, LLC (2006)
13. Martin, R.C.: Agile Software Development: Principles, Patterns, and Practices. Prentice-Hall, Englewood Cliffs (2003)
14. Jeffries, R.E., Anderson, A., Hendrickson, C.: Extreme Programming Installed. Pearson Education, London (2003)
15. Korea eXtreme Programming Users' Group, Korea eXtreme Programming Users' Group (2007), http://www.xper.org/
16. Earl, T.: Service-Oriented Architecture: A field guide to integrating XML and Web Service. Pearson Education, London (2004)

17. Earl, T.: Service-Oriented Architecture: Concepts, Technology, and Design. Prentice-Hall, Englewood Cliffs (2005)
18. Cohn, M.: User Stories Applied: For agile software development. Pearson Education, London (2004)
19. OASIS Open, Reference Model for Service Oriented Architecture 1.0 (2006), http://docs.oasis-open.org/soa-rm/v1.0/
20. Pietri, W.: An XP Team Room (2004), http://www.scissor.com/resources/teamroom/
21. Krogdahl, P., Luef, G., Steindl, C.: Service-oriented agility: Methods for successful Service-Oriented Architecture (SOA) development, Part 1: Basics of SOA and agile methods (2005), http://www.ibm.com/developerworks/webservices/library/ws-agile1
22. Sidky, A., Arthur, J.: A Disciplined Approach to Adopting Agile Practices: The Agile Adoption Framework. Agile Journal (2007), http://www.agilejournal.com/articles/articles/a-disciplined-approach-to-adopting-agile-practices:-the-agile-adoption-framework,-part-1.html
23. Gannod, G.C., Burge, J.E., Urban, S.D.: Issues in the Design of Flexivle and Dynamic Service Oriented Systems. In: International Workshop on Systems Development in SOA Environments. IEEE, Los Alamitos (2007)
24. Alegria, J.A.H., Bastarrica, M.C.: Implementing CMMI using a Combination of Agile Methods. CLEI electric journal. Paper. 7 9(1) (2006), http://www.clei.cl/cleiej/paper.php?id=119/
25. Tilkov, S.: 10 Principles of SOA. Stefan Tilkovs Weblog (2006), http://www.innoq.com/blog/st/2006/12/13/10_principles_of_soa.html
26. Kajko-Mattsson, M., Lewis, G., Smith, D.B.: A Framework for Roles for Development, Evolution and Maintenance of SOA-Based Systems. In: International Workshop on Systems Development in SOA Environments. IEEE, Los Alamitos (2007)
27. Beck, K.: Extreme Programming Explained: Embrace change 2/E. Pearson Education, London (2005)

Optimizing Text Summarization Based on Fuzzy Logic

Hamid Khosravi[1], Esfandiar Eslami[2], Farshad Kyoomarsi[3], and Pooya Khosravyan Dehkordy[4]

[1] International Center for Science & High Technology & Environmental Sciences, University of Shahid Bahonar Kerman, Kerman, Iran
Hkhosravi@mail.uk.ac.ir

[2] The center of Excellence for Fuzzy system and applications, University of Shahid Bahonar Kerman, Kerman, Iran
Eeslami@mail.uk.ac.ir

[3] Department of Mathematics and Computer, University of Shahid Bahonar Kerman, Kerman, Iran
Kumarci_farshad@graduate.uk.ac.ir

[4] Department of Computer, Islamic Azad University (Arak branch), Arak, Iran
Pooya_khd@iaun.ac.ir

Summary. In this paper we first analyze some state of the art methods for text summarization. We discuss what the main disadvantages of these methods are and then propose a new method using fuzzy logic. Comparisons of results show that our method beats most methods which use machine learning as their core.

1 Introduction

Nowadays, people need much more information in work and life, especially the use of internet makes information more easily gained. Extensive use of internet is one of the reasons why automatic text summarization draws substantial interest. It provides a solution to the information overload problem people face in this digital era.

Text Summarization is not a new idea. Research on automatic text summarization has a very long history, which can date back at least 40 years ago, from the first system built at IBM. Several researchers continued investigating various approaches to this problem through the seventies and eighties. Many innovative approaches began to be explored such as statistical and information-centric approaches, linguistic approaches and the combination of them. In next section, we will deal with the main approaches, which have been used or proposed.

Summary construction is, in general, a complex task which ideally would involve deep natural language processing capacities. In order to simplify the problem, current research is focused on extractivesummary generation [3]. An extractive summary is simply a subset of the sentences of the original text. These summaries do not guarantee a good narrative coherence, but they can conveniently represent an approximate content of the text for relevance judgment.

A summary can be employed in an indicative way – as a pointer to some parts of the original document, or in an informative way – to cover all relevant information of

R. Lee and H.-K. Kim (Eds.): Computer and Information Science, SCI 131, pp. 121–130, 2008.
springerlink.com

the text [5,7]. In both cases the most important advantage of using a summary is its reduced reading time. Summary generation by an automatic procedure has also other advantages: (i) the size of the summary can be controlled; (ii) its content is determinist; and (iii) the link between a text element in the summary and its position in the original text can be easily established[10,13].

2 Summarization Approaches

The main steps of text summarization are identifying the essential content, understanding it clearly and generating a short text. Understanding the major emphasis of a text is a very hard problem of NLP[15]. This process involves many techniques including semantic analysis, discourse processing and inferential interpretation and so on. Most of the research in automatic summarization has Been focused on extraction. But as in [5,8] the author described, when humans produce summaries of documents, they do not simply extract sentences and concatenate them, rather they create new sentences that are grammatical, that cohere with one another, and that capture the most salient pieces of information in the original document. So, the most pressing need is to develop some new techniques that do more than surface sentence extraction, without depending tightly on the source type. These need intermediated techniques including passage extraction and linking; deep phrase selection and ordering; entity identification and relating, rhetorical structure building and so on. Here we discuss some main approaches which have been used and proposed.

Summarization methods can be roughly grouped into three categories: Statistical approach, general linguistic approach, and hybrid methods that use a combination of these two approaches.

2.1 Statistical Approach

The statistical approach summarizes without understanding; it relies on the statistical distribution of certain features [9]. Recent representative works include the classification-based method, the IR-based method [4] and the position-based method among others. For the classification-based method, it is to classify sentences as either summary-worthy or not, depending on the training data. A classifier can be used to perform this task. A training procedure, that is an algorithm is used to calculate the parameters of the classifier. For the IR based methods, usually we make full use of techniques form Information Retrieval. But normally it is hard to give a satisfactory result, because the output summary is not very coherent. For the position-based method, we should notice that, important sentences are located at positions that are genre-dependent. Briefly speaking the statistical approach is domain-independent and fast, but has an inherent upper bound of performance.

2.2 Linguistic Approach

Summarization based on these method needs linguistic knowledge so that the computer can analyze the sentences semantically and then decides which sentences to choose considering the position of the verb, subject, noun and etc. these methods are more difficult than statistical methods.

3 A Review of Text Summarization Based on Machine Learning

The methods of summarization can be classified, in terms of the level in the linguistic space, in two broad groups: (a) shallow approaches, which are restricted to the syntactic level of representation and try to extract salient parts of the text in a convenient way; and (b) deeper approaches, which assume a semantics level of representation of the original text and involve linguistic processing at some level.

In the first approach the aim of the preprocessing step is to reduce the dimensionality of the representation space, and it normally includes: (i) stop-word elimination – common words with no semantics and which do not aggregate relevant information to the task (e.g., "the", "a") are eliminated; (ii) case folding: consists of converting all the characters to the same kind of letter case – either upper case or lower case; (iii) stemming: syntactically-similar words, such as plurals, verbal variations, etc. are considered similar; the purpose of this procedure is to obtain the stem or radix of each word, which emphasize its semantics.

A frequently employed text model is the vectorial model. After the preprocessing step each text element – a sentence in the case of text summarization – is considered as a N-dimensional vector. So it is possible to use some metric in this space to measure similarity between text elements. The most employed metric is the cosine measure, defined as $\cos q = (\langle x.y \rangle) / (|x| \, . \, |y|)$ for vectors x and y, where $(\langle , \rangle)$ indicates the scalar product, and $|x|$ indicates the module of x. Therefore maximum similarity corresponds to $\cos q = 1$, whereas $\cos q = 0$ indicates total discrepancy between the text elements.

4 The Used Attribute in Text Summarization

We concentrate our presentation in two main points: (1) the set of employed features; and (2) the framework defined for the trainable summarizer, including the employed classifiers.

A large variety of features can be found in the text-summarization literature. In our proposal we employ the following set of features:

(a) Sentence Length. This feature is employed to penalize sentences that are too short, since these sentences are not expected to belong to the summary [11,14]. We use the normalized length of the sentence, which is the ratio of the number of words occurring in the sentence over the number of words occurring in the longest sentence of the document.
(b) Sentence Position. This feature can involve several items, such as the position of a sentence in the document as a whole, its the position in a section, in a paragraph, etc., and has presented good results in several research projects.
(c) Similarity to Title. According to the vectorial model, this feature is obtained by using the title of the document as a query against all the sentences of the document; then the similarity of the documents title and each sentence is computed by the cosine similarity measure [2,6].
(d) Similarity to Keywords. This feature is obtained analogously to the previous one, considering the similarity between the set of keywords of the document and each sentence which compose the document, according to the cosine similarity. For the next two

features we employ the concept of text cohesion. Its basic principle is that sentences with higher degree of cohesion are more relevant and should be selected to be included in the summary.
(e) Sentence-to-Sentence Cohesion. This feature is obtained as follows: for each sentence s we first compute the similarity between s and each other sentence s of the document; then we add up those similarity values, obtaining the raw value of this feature for s; the process is repeated for all sentences. The normalized value (in the range [0, 1]) of this feature for a sentence s is obtained by computing the ratio of the raw feature value for s over the largest raw feature value among all sentences in the document. Values closer to 1.0 indicate sentences with larger cohesion.
(f) Occurrence of proper names. The motivation for this feature is that the occurrence of proper names, referring to people and places, are clues that a sentence is relevant for the summary. This is considered here as a binary feature, indicating whether a sentence s contains (value "true") at least one proper name or not (value "false"). Proper names were detected by a part-of-speech tagger.

We summarized this text(Based on this method) in which the results of performing this method is shown in table No. 1(after this text).

The Race of Man

If you stood in a busy place in big cosmopolitan city, like Times Square in New York City or Piccadilly Circus in London, and watched people go by, you would soon realize how the people of the modern world intermixed are Anthropologists speak of three major races of man. These races are identified as there distinct group of people. Each group has certain physical characteristics that are inherited. These three groups belong to one human family, and all may have been the same originally.

However, as they moved to different parts of the earth, they developed different features adapted to the conditions of climate and food in the places where they lived for a long period of time. In more modern times, these groups of people have been intermixing. Some groups have been conquered; other groups have intermarried.

Nevertheless, if you watched the passers-by in Times Square carefully, you would probably recognize several major types of people. A man with yellowish skin, straight black hair, high cheekbones, and almond-shaped eyes probably belongs to the people of eastern Asia called the mongoloid race. American Indians, who live in America and have reddish-yellowish skin, also belong to this group, If a man is from Africa south of the Sahara desert, he is likely to have a long head with black or dark brown skin, a broad, flat nose; thick, protruding lips; and tightly curled hair. He belongs to the negroid race. Other men like him can be found in the south pacific islands. a third group of men had their original home on the content of Europe. This group is known as the Caucasoid race, because the earliest skull of this type of human being was found in the Caucasus Mountains region in southeastern Europe. The Caucasians are called the white race because their skin color is generally lighter than the yellow, brown, or black skin tones of the other races. People of the white race have a variety of head shapes. And their hair varies from silky straight to curly. People of the race living near the Mediterranean Sea are usually darker-haired and darker-eyes than people in areas farther north. Nowadays people of the Caucasoid people live ia all parts of the world.

Some scientists speak of a fourth group, the Australia, who are darkskinned aborigines living on the content of Australia. There are also some people, like the ainu of Japan, who do not seem to belong to any one of the major races.

If you wanted to make a map showing the races of mankind, it could not have only three or four colors for the main racial group; it would have to show many tints and shades. As men mingle more and more in the modern world of easy travel, the races become more intermixed. It is sometimes difficult to label a man as belonging to one race or another. Is it even desirable to emphasize the differences between races? Mans great problem is to learn how to live peacefully with people different from himself. As members of one family, men must "live like brothers or die like beasts."

The summarized text based on this method is described as below:

If you stood in a busy place in big cosmopolitan city, like Times Square in New York City or Piccadilly Circus in London, and watched people go by, you would soon realize how the people of the modern world intermixed are. If you wanted to make a map showing the races of mankind, it could not have only three or four colors for the main racial group; it would have to show many tints and shades. If a man is from Africa south of

Table 1. The results of summarization based on vector method

Sentence number	Sentence position in paragraph	Sentence length	Similarity to the Title	Similarity to the keyword	Similarity to the text concept	Prorer noun	Sentence cohesion	Final similarity	Rank
1	1	1	0	1	1	1	0.383275261	13.35	1
23	1	0.85366	0	0.333333	0.5	0	0.244897959	12.098	2
12	1	0.85366	0	0	0.25	1	0.12244898	12.06	3
9	1	0.46341	0	1	0.75	0	0.827067669	11.633	4
21	1	0.4878	0	0	0.25	0.5	0.142857143	11.521	5
16	0.84	0.68293	0	0	0.25	0.5	0.051020408	11.391	6
15	1	0.34146	0	0.333333	0.25	0.5	0	11.175	7
22	0.75	0.56098	0	0	0.25	0.5	0.248447205	10.995	8
10	0.75	0.60976	0	0	0.5	1	0.171428571	10.968	9
24	0.8	0.43902	0	0	0.5	0	0.238095238	10.558	10
3	0.66	0.2439	0	0.333333	0.25	0	1	10.24	11
17	0.5	0.63415	0	0	0	0.5	0	10.199	12
18	0.34	0.4878	0	0.333333	0.25	0	0.5	9.711	13
11	0.5	0.39024	0	0	0.5	0.5	0.267857143	9.547	14
6	0.36	0.82927	0	0	0.5	0	0	9.446	15
25	0.6	0.36585	0	0	0.25	0	0.476190476	9.378	16
26	0.4	0.60976	0	0.333333	0.25	0	0.514285714	8.978	17
2	1	0.19512	1	0.333333	0.25	0	0	8.884	18
5	0.86	0.39024	0	0	0	0	0	8.046	19
13	0.75	0.14634	0	0	0	0	0.238095238	8.02	20
4	0.61	0.21951	0	0	0	0	0	7.93	21
7	0.47	0.26829	0	0.333333	0.25	0	0.909090909	7.327	22
14	0.5	0.29268	0	0	0	0	0	7.151	23
19	0.34	0.53659	1	0.666667	0.5	0	0.454545455	7.049	24
20	0.18	0.31707	0	0.666667	0.25	0	0.879120879	6.758	25
27	0.2	0.34146	0	0	0.25	0	0.102040816	6.103	26
8	0.06	0.21951	0	0	0	0	0	3.993	27

the Sahara desert, he is likely to have a long head with black or dark brown skin, a broad, flat nose; thick, protruding lips; and tightly curled hair.

Nevertheless, if you watched the passers-by in Times Square carefully, you would probably recognize several major types of people. Some scientists speak of a fourth group, the Australia, who are dark-skinned aborigines living on the content of Australia. This group is known as the Caucasoid race, because the earliest skull of this type of human being was found in the Caucasus Mountains region in southeastern Europe.

5 The Problems of Machine Learning Method and Their Solution

Some of the features used in this method such as main concepts, the occurrence of proper nouns and non essential information have binary attributes such as zero and one which sometimes are not exact. For example, one of these features is the "main concepts" attributes; that is, if one sentence contain at least one of the given nouns, the value of that sentence is one and otherwise it is zero.

What is obvious is that the sentence containing one noun has less value than the sentence containing two nouns. But there is no deference between these two sentences in the ordinary methods. To solve this problem, we try to define these attributes as fuzzy quantities; that is each sentence, depending on the presence of each attribute, has the value ranging from zero to one. Also, to compare different sentences, we use COS formula which depends on cross product. Since all of the vector dimension (sentence attributes) are the same in cross product, each of these attributes are the same in the final result. What is clear is that some of the attributes have more importance and some have less and so they should have balance weight in computations and we use fuzzy logic to solve this problem.

6 Fuzzy Logic

As the classic logic is the basic of ordinary expert logic, fuzzy logic is also the basic of fuzzy expert system. fuzzy expert systems, in addition to dealing with uncertainty, are able to model common sense reasoning which is very difficult for general systems. One of the basic limitation of classic logic is that it is restricted to two values, true or false and its advantage is that it is easy to model the two-value logic systems and also we can have a precise deduction.

The major shortcoming of this logic is that, the number of the twovalue subjects in the real world is few. The real world is an analogical world not a numerical one.

We can consider fuzzy logic as an extension of a multi-value logic, but the goals and application of fuzzy logic is different from multivalue logic since fuzzy logic is a relative reasoning logic not a precise multi-value logic. In general, approximation or fuzzy reasoning is the deduction of a possible and imprecise conclusion out of a possible and imprecise initial set [1].

7 Text Summarization Based on Fuzzy Logic

In order to implement text summarization based on fuzzy logic, we used MATLAB since it is possible to simulate fuzzy logic in this software. To do so; first, we consider

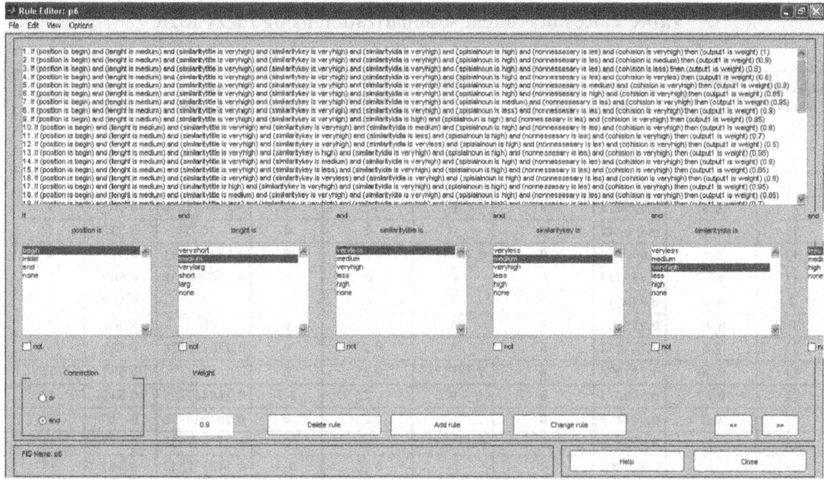

Fig. 1. Rules definition of goal function

each characteristic of a text such as sentence length, similarity to little, similarity to key word and etc, which was mentioned in the previous part, as the input of fuzzy system. Then, we enter all the rules needed for summarization, in the knowledge base of this system (All those rules are formulated by several expends in this field like figure 1).

After ward, a value from zero to one is obtained for each sentence in the output based on sentence characteristics and the available rules in the knowledge base. The obtained value in the output determines the degree of the importance of the sentence in the final summary.

To do these steps, we summarize the same text using fuzzy logic. The obtained results are presented in table 2.

The summarized text based on this method is described as below:

If you stood in a busy place in big cosmopolitan city, like Times Square in New York City or Piccadilly Circus in London, and watched people go by, you would soon realize how the people of the modern world intermixed are. Anthropologists speak of three major races of man. These races are identified as there distinct group of people. Each group has certain physical characteristics that are inherited. These three groups belong to one human family, and all may have been the same originally.

However, as they moved to different parts of the earth, they developed different features adapted to the Conditions of climate and food in the places where they lived for a long period of time. A man with yellowish skin, straight black hair, high cheekbones, and almond-shaped eyes probably belongs to the people of eastern Asia called the mongoloid race. If a man is from Africa south of the Sahara desert, he is likely to have a long head with black or dark brown skin, a broad, flat nose; thick, protruding lips; and tightly curled hair. He belongs to the negroid race. Other men like him can be found in the south pacific islands.

Table 2. Results of text summarization using fuzzy logic

Sentence number	Sentence position in paragraph	Sentence length	Similarity to the Title	Similarity to the keyword	Similarity to the text concept	Proper noun	Sentence cohesion	Final similarity	Rnank
1	0.2	0.829268	0	0	0.5	0	0	0.643	1
2	0.5	0.390244	0	0	0.5	0.5	0.267857143	0.636	2
3	0.5	0.292683	0	0	0	0	0	0.614	3
6	0.8	0.439024	0	0	0.5	0	0.238095238	0.602	4
10	0.66	0.390244	0	0	0	0	0	0.587	5
12	0.4	0.609756	0	0.3333333	0.25	0	0.514285714	0.573	6
13	0.75	0.146341	0	0	0	0	0.238095238	0.568	7
4	0.61	0.219512	0	0	0	0	0	0.568	8
8	0.47	0.219512	0	0	0	0	0	0.564	9
25	0.6	0.365854	0	0	0.25	0	0.476190476	0.561	10
14	0.46	0.243902	0	0.3333333	0.25	0	1	0.553	11
22	0.75	0.560976	0	0	0.25	0.5	0.248447205	0.546	12
27	0.2	0.341463	0	0	0.25	0	0.102040816	0.546	13
23	1	0.853659	0	0.3333333	0.5	0	0.244897959	0.545	14
21	1	0.487805	0	0	0.25	0.5	0.142857143	0.544	15
18	0.84	0.487805	0	0.3333333	0.25	0	0.5	0.535	16
7	0.7	0.268293	0	0.3333333	0.25	0	0.909090909	0.535	17
16	0.54	0.682927	0	0	0.25	0.5	0.051020408	0.531	18
20	0.38	0.317073	0	0.6666667	0.25	0	0.879120879	0.53	19
17	0.22	0.634146	0	0	0	0.5	0	0.522	20
19	0.06	0.536585	1	0.6666667	0.5	0	0.454545455	0.51	21
24	1	0.853659	0	0	0.25	1	0.12244898	0.509	22
9	1	0.463415	0	1	0.75	0	0.827067669	0.509	23
15	1	0.341463	0	0.3333333	0.25	0.5	0	0.509	24
26	0.75	0.609756	0	0	0.5	1	0.171428571	0.509	25
11	1	0.195122	1	0.3333333	0.25	0	0	0.507	26
5	1	1	0	1	1	1	0.383275261	0.505	27

8 Comparison and Conclusion

We chose 10 general text out of TOEFL text to compare the result of vectorial method with fuzzy method. We gave these texts and the summaries produced by both vectorial and fuzzy methods to 5 judges who had an M.A. in TEFL(teaching English as a foreign language). We asked the judge to read the main texts and to score the summaries

Table 3. The results of comparing vector and fuzzy logic method presented by judges

Judges / Methods	The First judge	The second judge	The third judge	The four judge	The five judge
Score of vector method	%65	%68	%72	%66	%64
Score of fuzzy method	%75	%80	%79	%78	%75

produced by the two methods considering the degree to which they represent the main concepts. This means that if a user has to read one of these summaries instead of reading the main text, which summary conveys concept of the main text. The given score by the judges using the two methods are shown in table No.3.

The results show that all the judge gave a better score to the summaries produced by fuzzy method. This indicates that fuzzy method worked better in parts of the sentence which contained uncertainty due to the use of fuzzy quantities. Therefore by using fuzzy approach in text summarization, we can improve the effect of available quantities for choosing sentences used in the final summaries. In Other word, we can make the summaries more intelligent.

References

[1] Buckley, J.J., Eslami, E.: An introduction to fuzzy logic and fuzzy sets. In: Advances in Soft Computing. Physica-Verlag, Germany (2002)

[2] Fisher, S., Roark, B.: Query-focused summarization by supervised sentences ranking and skewed word distribution. In: Proceedings of DUC (2006)

[3] Hirst, G., DiMarco, C., Hovy, E., Parsons, K.: Authoring and generating healtheducation documents that are tailored to the needs of the individual patient (1997)

[4] Hand, T.F.: A proposal for task-based evaluation of text summarization systems. In: Mani, I., Maybury, M. (eds.) Proceedings of the ACL/EACL 1997 Workshop on Intelligent Scalable Text Summarization, Madrid, Spain July 11 (1997)

[5] Inderjcet Main, the MITRE corporation 11493 Sanset Hills noad, USA (2003)

[6] Ferrier, L.: A Maximum Entropy Approach to Text Summarization, School of Artificial Intelligence, Division of Informatics, University of Edinburgh (2001)

[7] Mani, I.: Automatic Summarization. John Benjamins Publishing Company, Amsterdam (2001)

[8] Mani, I., Maybury, M.T.: Advances in Automatic Text Summarization. MIT Press, Cambridge (1999)

[9] Passonneau, R., KuKich, K., Robin, J., Hatzivassiloglou, V., Lefkowitz, L., Jing, H.: Generating summaries of work flow giagrams. In: Proceedings of the International Conference on Natural Language Processing and Industrial Applications, New Brunswick, Canada, University of Moncton (1996)

[10] Passonneau, R., Kukich, K., McKeown, K., Radev, D., Jing, H.: summarizing web traffic: A portability exercise, Department of Computer Science, Columbia University (1997)

[11] Radev, D.R.: Language Reuse and Regeneration: Generating Natural Language summaries from Multiple On-Line Sources. PhD thesis, Department of Computer Science, Columbia University, New York (1998)

[12] Stairmond, M.A.: A Computational Analysis of Lexical Cohesion with Applications in Information Retrieval. Ph. D. thesis, Center for Computational Linguistics, UMIST, Manchester (1999)

[13] Strzalkowski, T., Stein, G., Wang, J., Wise, B.: A Robust Practical Text summarizer. In: Nani, I., Maybury, M. (eds.) Adv. in Autom. Text Summarization. The MIT Press, Cambridge (1999)

[14] Teufel, S., Moens, M.: Argumentayive classification of extracted sentences as a first step towards flexible abstracting. In: Mani, I., Maybury, M. (eds.) Advances in automatic text summarization, The MIT Press, Cambridge (1999)

[15] Hatzivassiloglou, V., Klavans, J., Eskin, E.: Detecting text similarity over short passages: Exploring linguistic feature combinations via machine learning. In: Proceedings of the Joint SIGDAT Conference on Empirical Methods in Natural Language Processing and Very Large Corpora (1999)

Abbreviations

NLP. Natural language processing
IR. Information Retrieval
ML. Machine Learning

Reusability Enhancement by Using Flexible Topology Architecture for Network Management System

Hee Won Lee[1], Chan Kyou Hwang[1], Jae-Hyoung Yoo[1], Ho-Jin Choi[2], Sungwon Kang[2], and Dan Hyung Lee[2]

[1] KT Network Technology Lab., Daejeon, Republic of Korea
{hotwing,ckhwang,styoo}@kt.com
[2] Information and Communications Univ., Daejeon, Rebulic of Korea
{hjchoi,kangsw,danlee}@icu.ac.kr

Summary. With the expansion of IP networks, a diversity of network management systems have been developed in order to support network operations. Based on the common functionalities of these network management systems, a platform for network management systems was developed by KT, on which future network management systems could be developed on the platform. However, in the process of the development, it was discovered that the reusability of platform modules was stagnating at around 30%. In order to further increase reusability in developing network management systems, this paper proposes a proactive reuse method to build a flexible network management topology architecture. The goal of the method is to overcome the limitation of the reusability approach that utilizes a platform for network management systems developed based on only common functionalities.

1 Introduction

As many new services are emerging with the growth of IP networks, many network management systems have been developed to manage these networks. For example, Nespot NMS was developed for the Nespot service, and IP Access NMS for Megapass ADSL. As each service has its own network management system, the number of network management systems has increased rapidly with the growth of services. A new project for developing a new network management system has been created with the advent of a new service. In the initial stage when a project was formed with a service, there were no links among those projects, even though network management systems had common functionalities.

In KT (formerly called Korea Telecom), the biggest telecommunications operator in Korea, there has been an effort to develop a platform for network management systems, because network management systems have similar architecture and common functionalities. Since then, many new network management systems have been developed based on the platform. However, there is a limitation of this platform from the viewpoint of reusability, because the reuse rate of the platform is below 35%. Since most of the network management systems have many common features and the number of network management systems is growing, the main idea of software product lines can be utilized to overcome this limitation.

R. Lee and H.-K. Kim (Eds.): Computer and Information Science, SCI 131, pp. 131–145, 2008.
springerlink.com

2 Background

In KT, the number of network management systems has been increasing with the growth of IP-based services. In 2001 and 2002, there were only three network management systems, including IP Core NMS, NAS NMS, and Nespot NMS. However, the number of network management systems increased rapidly in 2003. For instance, VPN NMS, IRIMS, ICAS, IP Access NMS, KSMS and so forth were additionally developed. The number of network management systems continued to grow and by 2006 it had risen to a total of 19. Fig. 1 shows the growth in the number of network management systems over the years.

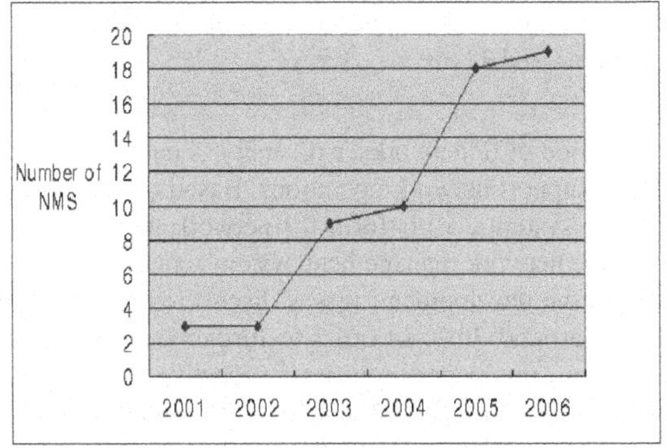

Fig. 1. Growth of the number of network management systems

As a way to reduce development costs and shorten the development period, a platform for network management systems was developed based on IP Core NMS in 2003. However, the modules of the platform could be reused with a reusability rate of approximately 30%. For reusability enhancement, strategic reuse is required. Strategic reuse, which is the key in software product lines, can be adopted in developing network management systems in order to boost reusability, because network management systems have common functionalities and architecture.

3 Related Work

Strategic reuse, which is the basic idea of software product lines, can be an initiative for increasing reusability in the development of network management systems.

A product line approach to software involves three essential activities, as described below.

> *... fielding of a product line involves core asset development and product development using the core assets, both under the aegis of technical and organizational management [2].*

In developing a software system that has a number of functionalities analogous to the previous systems, the reactive approach is generally taken [7]. However, in core asset development, which is one of the three essential activities of software product lines, a proactive approach is required [4] [6].

Central to the concept of software product lines is the architecture of products in an application domain. The product line architecture is the foundation for developing an application based on the software product line approach. The reason why software architecture is important is that (i) software architecture represents the earliest design decisions, because it is hardest aspect to change, most critical aspect to get right, and a communication vehicle amongst stakeholders, (ii) it is also the first design artifact addressing performance, modifiability, reliability, and security, and (iii) it is a key to systematic reuse, because it is a transferable and reusable abstraction [9].

The product line approach makes it possible to enhance reusability by adopting proactive reuse based on the architecture of products in an application domain. Reusability is analyzed in a platform for network management systems, and a method to increase reusability is studied in this paper, which is based on the architecture of network management systems.

4 Product Line Scoping of Network Management Systems

Product line scoping is an integral part of every product family endeavor [3]. In this section, a product line scoping for network management systems is performed, which can contribute to enhancing reusability of the platform for network management systems.

4.1 Product Line Scoping of Network Management Systems

Scoping is an activity that bounds a set of systems by defining those behaviors or aspects that are "in" and those behaviors or aspects that are "out" [2]. It is recognized as an important part of product line technology, and methods are emerging to help with scoping [3].

In these methods, there are examinations of existing products and context diagramming. First, let us examine existing products. In telecommunications companies,

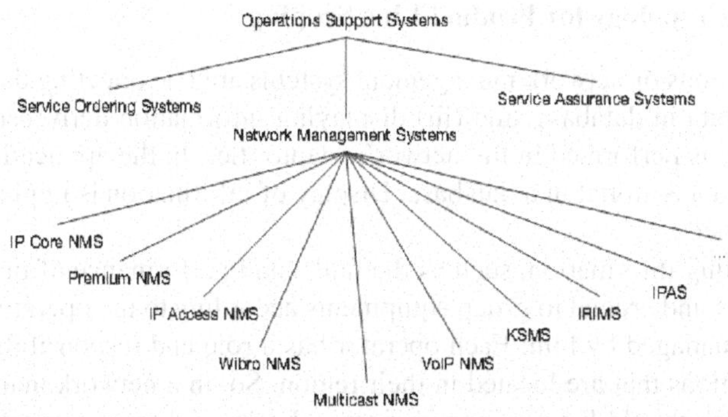

Fig. 2. Operations support systems of KT

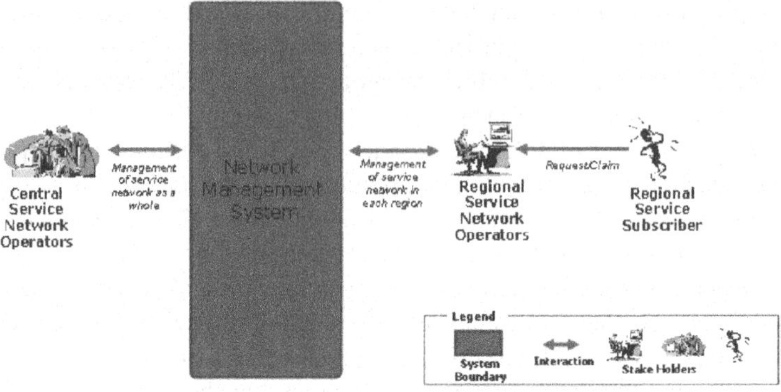

Fig. 3. Context diagram of network management system

operations support systems have been developed in order to support the services that are operated on their network. Customer's orders are received through Service Ordering Systems. Service Assurance Systems can assure the operation of customers' service network without the blackout of their service. Network Management Systems support the operations of customers' service networks, which are managed by network operators. In the branch of Network Management Systems, various network management systems have been developed to support network operations. Fig. 2 shows some of the network management systems that have been developed by KT. Since all network management systems have similar architecture and functionalities, Network Management Systems is a good candidate for product line development.

Second, let us draw context diagram, which is shown in Fig. 3. A network management system is administered and managed by central service network operators. The equipment located in each region is managed by each regional service network operator. Regional service network operators receive requests or claims from their customers who subscribe their service. From this context diagram, it is recognized that a network management system can be operated by multiple operators who take charge of equipments that are located in their own region.

4.2 Utilizing Topology for Product Line Scoping

The main functions of network management systems are (i) collecting data, (ii) storing the collected data in database, and (iii) displaying information derived from the data. Data collection is performed in the network adaptor tier. In the application server tier, the collected data is stored in a database. Display of information is implemented in the user interface tier.

In representing information, such as the faults and performance of managed equipment, it is basic and crucial to group equipments according to the operator, because the equipment is managed by him. Each operator has a role and responsibility for service network operations that are located in their region. So, in a network management system, an operator should have an appropriate authority and management scope within which he can manage his equipment.

Therefore, an operator's equipment and management scope should be appropriately connected. Generally, this has been achieved by connecting an operator's equipment to his organization in his company. This can be easily expressed by network management topology. That is, equipment is grouped into a scope within which an operator can manage his equipment and a connection is composed between equipment, between a piece of equipment and a scope, or between scopes.

Thus, since a network management system should be managed by a number of operators and each operator should have their own authority and management scope, network management topology is a prerequisite for developing a network management system. If network management systems are set to a product line scope, it is important to design flexible network management topology architecture, because each network management system has their own network management topology, which makes it hard to reuse the modules related to it.

5 Building a Flexible Topology Architecture for Network Management System

As a way to increase the reusability of the platform for network management systems, a flexible topology architecture is used. The flexible topology architecture is introduced in this section.

5.1 Importance of Topology in Developing Network Management System

Topology is defined as a representation of how a node and its connection are arranged and it can be represented by a map or tree. In building a network management system, topology is generally used because it is an effective way of representing the relationship between a management unit and a network element. An operator can manage network elements under his responsibility through his part in the entire topology shown in the network management system. However, once topology is designed and implemented, it is very costly to change, because many functions are built based on the topology on which the location of displayed information is represented. Therefore, if a topology is designed to provide flexibility, modifications to a network management system can be minimized and the network management system can adapt itself to the rapidly-changing service environment.

5.2 Flexible Topology Architecture for Network Management System

In conventional topology architecture, topology is static, which means that if topology is modified, such as through the change of an organization name, for example, then the change should be conducted at the database level. A conventional topology architecture is shown in Fig. 4(a)

A primitive way to give flexibility to topology in response to an addition or modification of topology is to use a database. However, for flexible addition or modification of topology, it is not enough to change the name of a management unit, for example, an

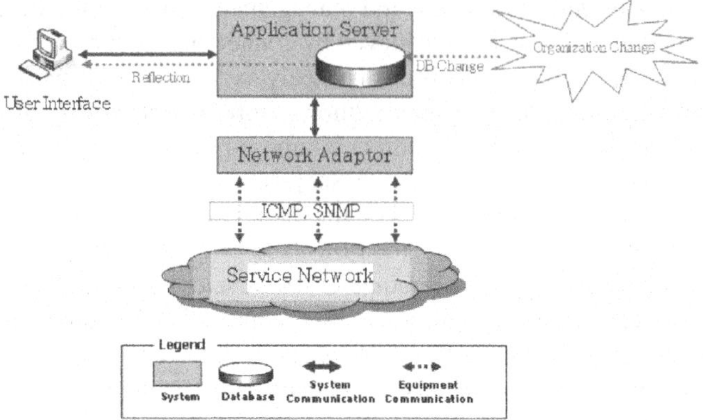

(a) Conventional topology architecture

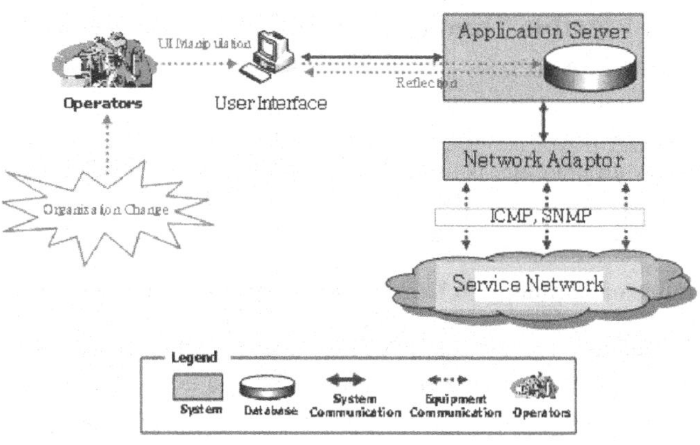

(b) Flexible topology architecture

Fig. 4. Network management topology architecture

organization, using a database. If a flexible topology architecture is designed and implemented in a system, modules that are related to topology can be reused when developing a subsequent network management system.

Fig. 4(b) shows a flexible topology architecture. In a flexible topology architecture, if topology is changed, through an action such as the removal of an organization, for example, an operator performs an operation to remove the organization at the user interface level, so the change is reflected on the database of the application server tier. Application Server distributes the changed information to every client application on the user interface tier, and the topology change is completed.

For a flexible topology architecture that can be fit to any network management system, the topology should be able to be composed by its operator at the user interface level. For this, models for composing topology are required. This paper uses a model called *Equip and Scope* [17] for a flexible topology architecture.

5.3 Conceptual Model for Flexible Topology Architecture: Equip and Scope

In the concept of *Equip and Scope* [17], *Scope* is defined as a management unit of equipment, which can be composed by an operator according to its configuration. A piece of equipment can be included in more than one scope. It is possible to duplicate a piece of equipment in several scopes. Since more than one piece of equipment can be located in several scopes, it should be defined which equipment is master, and which is slave. The concept of master/slave is required in the case that a fault occurs in a piece equipment and it is necessary to find out which equipment of which part of topology has priority for that fault.

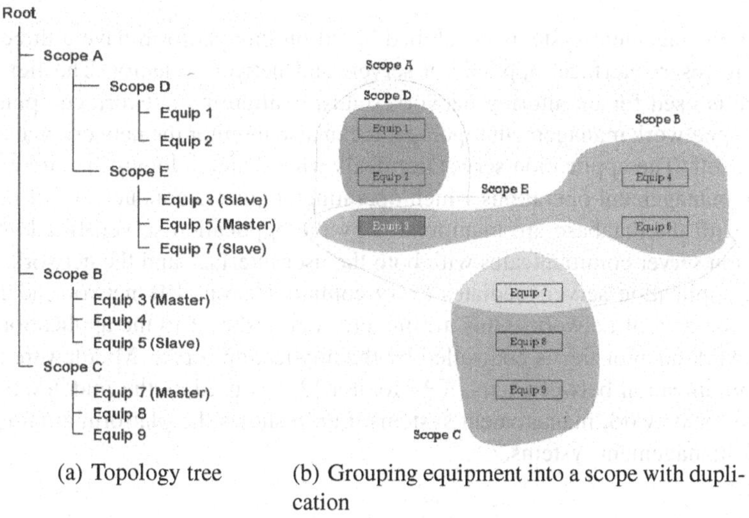

(a) Topology tree (b) Grouping equipment into a scope with duplication

Fig. 5. Flexibility of *Equip and Scope*

The flexibility of *Equip and Scope* for flexible network topology architecture is represented in Fig. 5. Pieces of equipment can be grouped into a scope and several scopes can also be grouped into a scope. Equipment can be duplicated in more than one scope, but only one piece of equipment in a scope is marked 'Master' and the other identical equipment is marked 'Slave'. The flexibility of this concept is high, because equipment can be grouped into any form of scope regardless of whether the equipment is already within another scope or not.

6 Reusability Enhancement

By applying the flexible network management topology architecture we proposed in Section 5 to the development of a network management system, the reusability of a network management system can be increased. A platform that is designed based on the consolidation of common functionalities of network management systems has a limitation in terms of reusing the platform modules. Reusability can be improved by using a flexible network management topology architecture at the user interface tier that has a low reuse rate.

Since the platform for network management systems was applied in developing a network management system, reusability has become around 30%. Design based on common functionalities of network management systems raised the reusability of the platform to the level of 30 percentages. In order to increase the usefulness of the platform, reusability should be increased further. To this end, reusability is analyzed based on the three-tier architecture of the network management systems. If each tier has a different reuse strategy, reusability can be effectively enhanced.

6.1 The Platform Architecture of Network Management Systems

Network management systems developed based on this platform have a three-tier architecture: user interface, application server, and network adaptor. The tier of user interface is used for monitoring network status, configuring network equipment, and managing network management topology. It can also monitor the network management system itself. The application server tier deals with business logics that are related to network management operations which operating engineers conduct. All of the transactions with the database are manipulated by the application server. In addition, the application server communicates with both the user interface and the network adaptor. That is, application server mediates every communication. The network adaptor tier collects the data of network status, traffic, etc. and pushes it to the application server. The network adaptor tier is controlled by the application server. Middleware controls the communication between tiers. TP-Monitor [21] is used as the middleware in the platform for network management systems. Fig. 6 shows the platform architecture of network management systems.

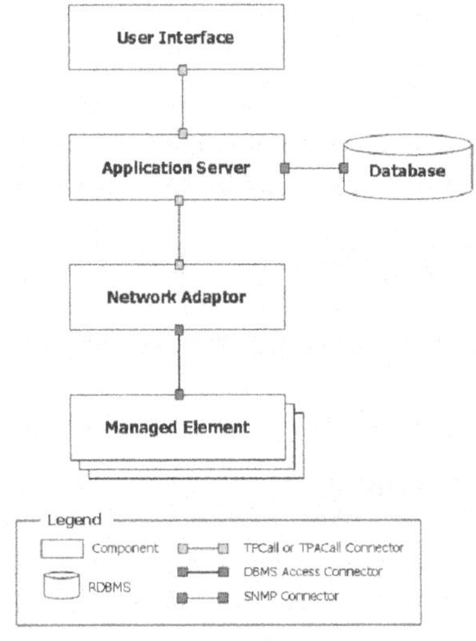

Fig. 6. The platform architecture of network management systems

6.2 Reusability Analysis of the NMS Platform

In VoIP NMS, reuse ratio [14] was evaluated according to three criteria; no modification, 10% modification, and 20% modification. Reuse ratio with no modification is equivalent to reuse percent [11]. If the parameters of a function that takes charge of the communication between the framework of middleware and each module are customized, their module is counted as reused with 10% modification. If a module is changed according to the modification of database columns that comes from the change of business logics, it is regarded as reused with 20% modification.

In VoIP NMS that applied the platform for network management systems, reusability was 33.3% with tolerance of 20% modification as a whole. Reusability can be analyzed based on the three-tier architecture of the network management systems. Fig. 7 shows that the network adaptor has a reuse level of 69.4%, the application server 57.7%, and the user interface 2.6%. In the reuse metric, strictly referring to the reuse ratio, in terms

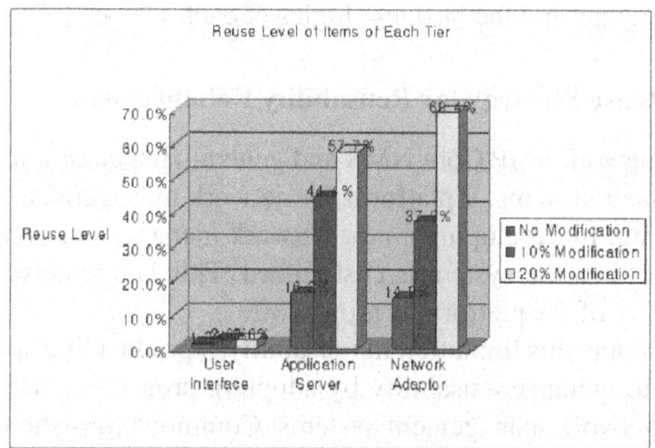

(a) Reuse level of items of user interface, application server, and network adaptor

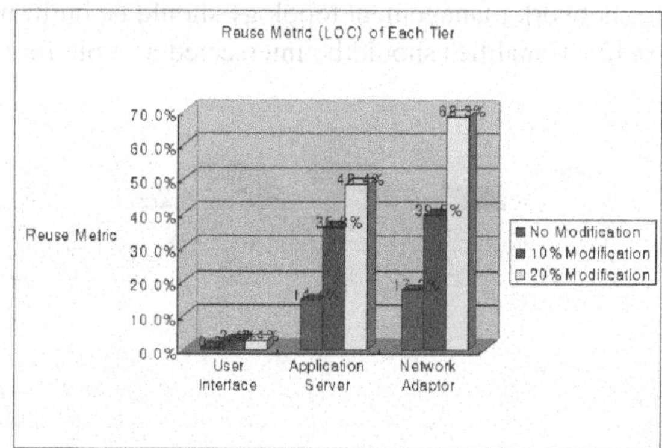

(b) Reuse metric (LOC) of user interface, application server, and network adaptor

Fig. 7. Analysis of reusability based on the three-tier architecture

of LOC, the network adaptor has 68.3%, the application server 48.4%, and the user interface 2.4%.

The network adaptor has high reusability and the application server has medium-high reusability, but the user interface has very low reusability. That is, most parts of the user interface tier are not utilizing the platform for network management systems. This shows that a platform for network management systems designed based on the common functionalities of network management systems has a limitation in enhancing reusability. In terms of functionality, each network management system has different functions in their user interface tiers.

The reason why tiers are arranged in order of amount of reusability as network adaptor, application server, and user interface is that (i) the network adaptor has many functions that every network management system requires, such as ping fault management, SNMP resource management, SNMP Opersatus [1], CPU usage rate, etc., (ii) the application server has not only many common functionalities, but also has system-specific functions related to business logics, and (iii) the user interface is easy to change according to user requirements and the business logics of each system.

6.3 Proactive Reuse Strategy for Reusability Enhancement

By taking the framework of IP Core NMS and generalizing common functionalities of network management systems, a platform for network management systems has been built and is evolving. In developing a new network management system, the platform is first applied and then the system is customized. This is a reactive approach, under which the reusability of the platform is around 30%.

In order to overcome this limitation in reusability, a product line approach is needed. This is a method to enhance reusability by adopting proactive reuse based on the architecture of the network management systems. Commonalities should be proactively considered according to the architecture. It is necessary to employ different strategies on each tier, according to the three-tier architecture of the network management systems.

This paper suggests that in the user interface tier which has a low reuse rate a conceptual model (CM) for network management topology should be built; in the application server tier, common functionalities should be intersected as a platform module; and in

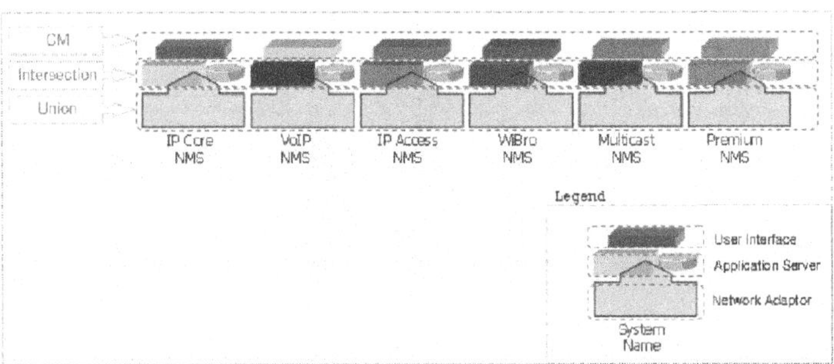

Fig. 8. Reusability enhancement strategy according to the three-tier architecture of network management systems

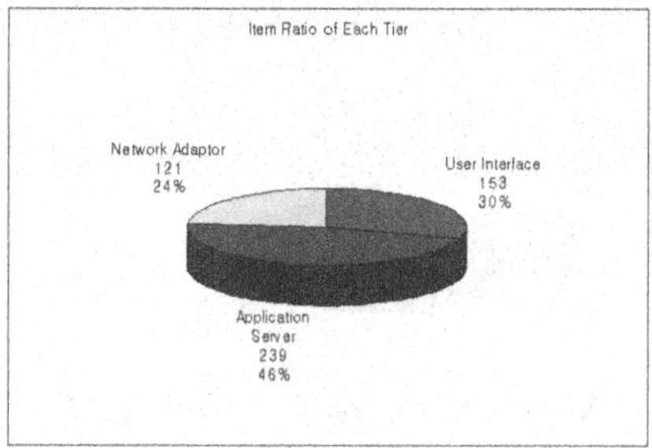

(a) Item ratio of user interface, application server, and network adaptor

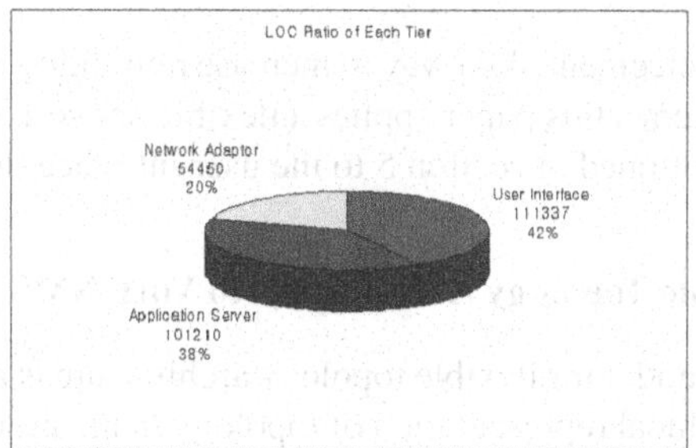

(b) LOC ratio of user interface, application server, and network adaptor

Fig. 9. Ratio of each tier: (a) By item (b) By LOC

the network adaptor tier, common functionalities should be unionized. Fig. 8 shows a reusability enhancement strategy, according to the three-tier architecture of network management systems.

As shown in Fig. 9, in VoIP NMS, the user interface, the network adaptor, and the application server respectively take up 30%, 46%, and 24% in terms of the number of items. In this paper, an item [12] is defined as a class in the user interface tier, or a service that is a module designed for a database transaction in the application server tier or network adaptor tier. In terms of LOC, the sizes of the user interface, the network adaptor, and the application server are 42%, 38%, and 20% respectively. Even though the user interface takes the biggest piece of the pie in terms of LOC, its reusability is only 2.4%, as shown in Fig. 7.

As shown in the above reusability analysis, which is the result of adopting the platform for network management systems to the development of VoIP NMS, the reusability of the user interface needs to be raised for true reusability enhancement. On the other hand, Fig 10 shows that in the main functions of the user interface tier, LOC for implementing network management topology takes up 72%. This shows that it requires a large costs to implement a network management topology.

Since network management topology is the biggest slice of the pie in the user interface tier, if it is reused, then reusability would be enhanced in the user interface, resulting in

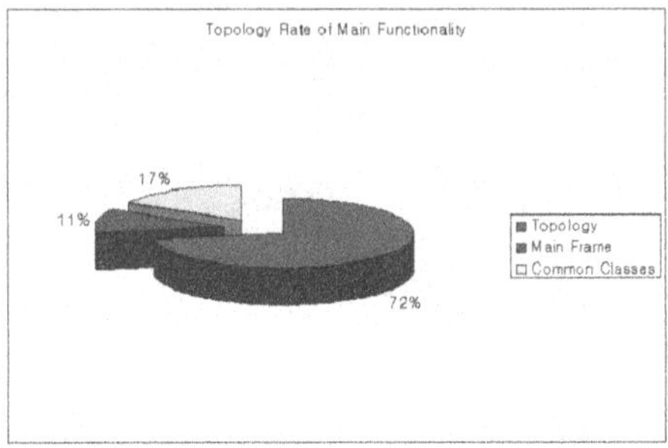

Fig. 10. LOC ratio of the main functionalities of the user interface tier

overall reusability enhancement. As a way to increase reusability of the platform for network management systems, this paper applies a flexible network management topology architecture that is mentioned in section 5 to the user interface tier.

6.4 Applying Flexible Topology Architecture to VoIP NMS

The *Equip and Scope* model for flexible topology architecture is adopted in VoIP NMS. Fig. 11 shows the graphical representation of topology in the user interface tier, and the classes corresponding to each representation.

In *Equip and Scope*, 'CNMSScope' class represents scope and 'CNMSEquip' class represents equipment. Fig. 12 shows the inheritance tree of classes for flexible topology in the user interface tier. The 'CDiagramTopologyView' class inherits CNMSEquip, CNMSScope, CNMSLink, etc. Through this class, multi-selection and navigation is possible. VoIP NMS inherits 'CDiagramTopologyView', producing 'CVoIPNMSView' class. This means that in the other network management system topology can be reused by inheriting 'CDiagramTopologyView' and customizing 'CVoIPNMSSView'.

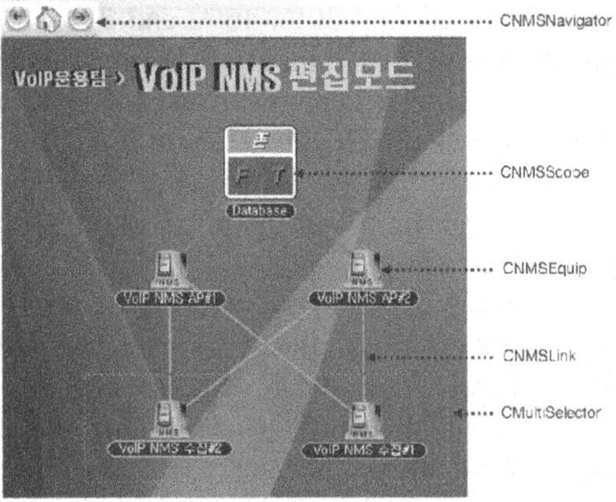

Fig. 11. Graphical representation of topology and corresponding classes in VoIP NMS

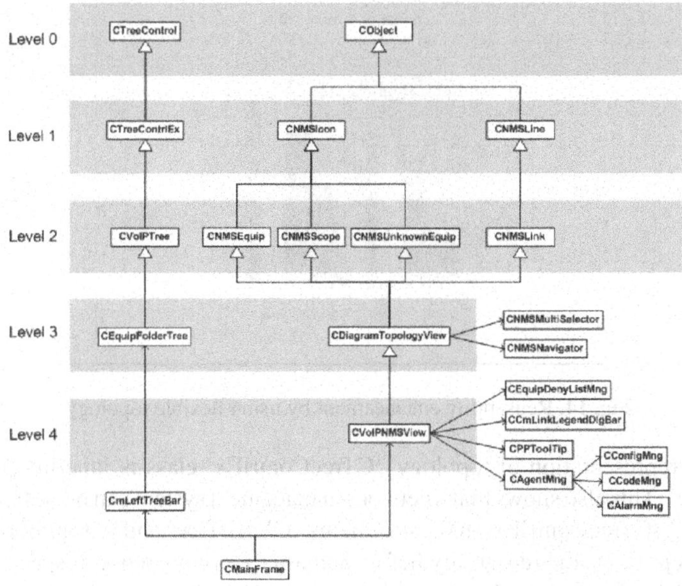

Fig. 12. Inheritance tree of classes for flexible topology in the user interface tier

Metric Type	Classification	User Interface	Application Server	Network Adaptor	Total
Number of Items	Intrinsic	149	101	37	287
	10% Modification	2	67	27	96
	20% Modification	0	32	39	71
	Reuse Without Modification	2	39	18	59
	Total	153	239	121	513
LOC	Intrinsic	108631	52186	17266	178083
	10% Modification	2325	21634	12087	36046
	20% Modification	0	12820	15693	28513
	Reuse Without Modification	381	14570	9404	24355
	Total	111337	101210	54450	266997
Reuse Metric (Percentage)		0,3%	14,4%	17,3%	9,1%
Reuse Level		1,3%	16,3%	14,9%	11,5%
Reuse Ratio	Modification less than 10% by LOC	2,4%	35,8%	39,5%	22,6%
	Modification less than 10% by Item	2,6%	44,4%	37,2%	30,2%
	Modification less than 20% by LOC	2,4%	48,4%	68,3%	33,3%
	Modification less than 20% by Item	2,6%	57,7%	69,4%	44,1%

(a) Reusability before topology reuse

Metric Type	Classification	User Interface	Application Server	Network Adaptor	Total
Number of Items	Intrinsic	127	101	37	265
	10% Modification	2	67	27	96
	20% Modification	0	92	39	71
	Reuse Without Modification	24	39	18	81
	Total	153	239	121	513
LOC	Intrinsic	78402	52188	17266	147854
	10% Modification	2325	21634	12087	36046
	20% Modification	0	12620	15693	28513
	Reuse Without Modification	30610	14570	9404	54584
	Total	111337	101210	54450	266997
Reuse Metric (Percentage)		27,5%	14,4%	17,3%	20,4%
Reuse Level		15,7%	16,3%	14,9%	15,8%
Reuse Ratio	Modification less than 10% by LOC	29,6%	35,8%	39,5%	33,9%
	Modification less than 10% by Item	17,0%	44,4%	37,2%	34,5%
	Modification less than 20% by LOC	29,6%	48,4%	68,3%	44,6%
	Modification less than 20% by Item	17,0%	57,7%	69,4%	46,3%

(b) Reusability after topology reuse

Fig. 13. Comparison of reusability before and after topology reuse

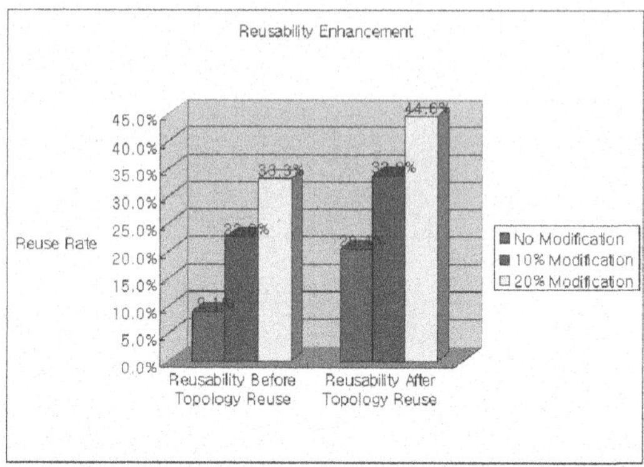

Fig. 14. Reusability enhancement by using flexible topology

For tree representation of topology, 'CTreeContrlEx' class is inherited, producing 'CVoIPTree'. This also shows that a network management system can reuse tree topology by inheriting 'CTreeContrlEx' and customizing 'CVoIPTree' and 'CEquipFolderTree'.

Fig. 13 represents the reusability before and after topology reuse, respectively.

In VoIP NMS, the network management topology of the user interface tier was designed and implemented based on the flexible topology architecture using the concept of *Equip and Scope*. If the topology module having the concept of *Equip and Scope* becomes a module of the platform for network management systems, reusability of the user interface of VoIP NMS is increased from 0.3% to 27.5% in terms of LOC (i.e. reuse metric). The reusability of the platform for network management systems is enhanced as a whole from 33.3% to 44.6% in terms of LOC with a tolerance of 20% modification, as shown in Fig. 14. That is, through the use of a flexible topology architecture, reusability is enhanced by 10%.

7 Conclusion

By applying the flexible network management topology architecture we proposes to the development of a network management system, the reusability of the platform for network management systems can be enhanced. There is a limitation in reusing the platform based on the consolidation of common functionalities of network management systems. However, reusability can be improved by using the flexible topology architecture, which is based on a proactive reuse strategy. The conceptual model of *Equip and Scope* was used to build the flexible network management topology. By applying the flexible topology architecture using *Equip and Scope* to VoIP NMS, it was proven that the reusability of the platform for network management systems could be enhanced.

References

1. Stallings, W.: SNMP, SNMPv2, SNMPv3, and RMON 1 and 2. Addison-Wesley, Reading (1999)
2. Clements, P., Northrop, L.: Software Product Lines: Practices and Patterns. Addison-Wesley, Reading (2002)

3. Clements, P.: On the Importance of Product Line Scope. In: van der Linden, F.J. (ed.) PFE 2002. LNCS, vol. 2290. Springer, Heidelberg (2002)
4. Clements, P.C., Northrop, L.M.: Salion, Inc.: A Software Product Line Case Study, Technical Report, CMU/SEI-2002-TR-038 (November 2002)
5. Krueger, C.W.: Easing the Transition to Software Mass Customization. In: Proceeding of the 4th International Workshop on Product Family Engineering, Bilbao, Spain (2001)
6. Krueger, C.W.: Software Mass Customization (May 2005), Available: `http://www.biglever.com/papers/BigLeverMassCustomization.pdf`
7. Buhrdorf, R., Churchett, D., Krueger, C.: Salion's Experience with a Reactive Software Product Line Approach. In: Proceedings of the 5th International Workshop on Product Family Engineering, Siena, Italy, pp. 317–322. Springer, Heidelberg (2003)
8. Bosch, J.: Design and Use of Software Architectures: Adopting and Evolving a Product-Line Approach, May 2000, Addison Wesley Professional. ACM Press (2000)
9. Northrop, L.: Software Product Lines: Reuse That Makes Business Sense. In: 2006 Presentations, Australian Software Engineering Conference (ASWEC 2006), Sydney, Australia (April 2006)
10. van Ommering, R., Bosch, J.: Widening the Scope of Software Product Lines – From Variation to Composition. In: Chastek, G.J. (ed.) SPLC 2002. LNCS, vol. 2379, pp. 328–347. Springer, Heidelberg (2002)
11. Poulin, J., Caruso, J.: A Reuse Metrics and Return on Investment Model. In: Proceedings of the 2nd workshop on Software Reuse: Advances in Software Reusability. IEEE, Los Alamitos (1993)
12. Frakes, W., Terry, C.: Reuse Level Metrics. In: Proceedings of the 3rd International Conference on Software Reuse: Advances in Software Reusability, IEEE, Los Alamitos (1994)
13. Frakes, W., Terry, C.: Software Reuse: Metrics and Models. ACM Computing Surveys 28(2) (June 1996)
14. Devanbu, P., et al.: Analytical and Empirical Evaluation of Software Reuse Metrics. In: Proceedings of the 18th International Conference on Software Engineering, Berlin, Germany (1996)
15. Curry, W., et al.: Empirical Analysis of the Correlation between Amount-of-Reuse Metrics in the C Programming Language. In: SSR 1999, pp. 135–140. ACM, New York (1999)
16. Mascena, J.C.C.P., de Almeida, E.S., de Lemoss Meira, S.R.: A Comparative Study on Software Reuse Metrics and Economic Models from a Traceability Perspective. IEEE, Los Alamitos (2005)
17. Lee, H.W., Kim, Y.D., Hwang, C.K., Yoo, J.-H.: Flexible Topology Architecture for Network Management System. In: Ata, S., Hong, C.S. (eds.) APNOMS 2007. LNCS, vol. 4773. Springer, Heidelberg (2007)
18. Hirata, T., Mimura, I.: Flexible Service Creation Node Architecture and its Implementation. In: Computer Communications, CCW 2003. Proceedings. 2003 IEEE 18th Annual Workshop, October 2003, pp. 166–171 (2003)
19. KT Homepage, Available: `http://www.kt.co.kr`
20. KT VoIP Service Homepage, Available: `http://www.allup.co.kr`
21. TMaxSoft Homepage, Available: `http://www.tmax.co.kr`

Process Deployment in a Multi-site CMMI Level 3 Organization: A Case Study

Bayona Luz Sussy[1], Calvo-Manzano Jose Antonio[1], Cuevas Gonzalo[1], San Feliu Tomás[1], and Sánchez Angel[2]

[1] Languages and Informatics Systems and Software Engineering Department
Faculty of Computer Science, Polytechnic University of Madrid
Campus Montegancedo, 28660 Boadilla del Monte Madrid, Spain
sbayona@zipi.fi.upm.es, jacalvo@fi.upm.es, gcuevas@fi.upm.es, tsanfe@fi.upm.es

[2] *everis*, Madrid, Spain
Angel.Sanchez@everis.com

Summary. This paper presents a case study about deployment process and shows the results of the deployment processes on five software development and maintenance sites of one organization. This paper focuses on the deployment process elements, critical success factors and impact on the process. It highlights the importance of having an effective deployment process strategy for the organizational processes to be used, adopted and institutionalized. It also presents the level of acceptance and use of these software processes. Finally, the lessons learned during the deployment process are presented.

Keywords: CMMI, Defined Process, Process Deployment, Change management, Human factors.

1 Introduction

Methodologies and process models developed have reached maturity in recent years; one of the widely used is the "Capability Maturity Model" (CMM) that has just evolved into CMMI, "Capability Maturity Model Integration" (CMMI) [1]. The Software Engineering Institute (SEI) developed the CMMI models, now composed of three constellations: acquisition, development and services.

CMMI development constellation has two representations: staged and continuous. In this paper, we have considered the staged representation. The development constellation promotes the continuous improvement of organizations with the aim of producing software and quality services within the schedules and costs. The current CMMI for the Development model has 22 process areas.

Each process area is defined by a set of generic and specific goals and practices. In order to develop the activities of each process, each specific practice says "what should be done" and what the products to be obtained are, but it does not say "how it should be done".

R. Lee and H.-K. Kim (Eds.): Computer and Information Science, SCI 131, pp. 147–156, 2008.
springerlink.com

CMMI has five maturity levels: initial process, managed process, defined process, quantitatively managed process and optimizing process. For each maturity level, a set of process areas represent the critical issues that must be under control to achieve that level. CMMI level 3, called *Defined*, requires that the organization has a defined and measured process. The process is described in terms of standards, procedures, tools and methods. That is, the process is established, documented and measured [2].

Many companies decide to follow this model motivated by the need to be certified at a certain level, to improve the quality of the processes and productivity. In recent years, the number of organizations certified in CMMI level 3 has increased. According to the SEI of 1712 enterprises, 36.1% are at the "defined" level [3]. An organization at CMMI Maturity Level 3 has established a process library denominated Process Asset Library (PAL) [1].

Today, organizations recognize the importance of using standard processes and establishing the Process Asset Library (PAL) for software development projects. A competitive organization must have standard processes and software development methods, particularly when it has software factories or software development sites distributed in different locations.

Once the organization's standard processes are defined, they must be deployed and institutionalized. However, it is a difficult task to ensure that the new processes are adopted and institutionalized because it is conditioned by the relationship that exists between people (work teams, users and the manager) [4] and artefacts (methods, tools, processes and paradigms) to be deployed.

A change management approach is necessary to ensure that the staff work as expected. Additionally, it is important to have a deployment strategy composed of training aspects, integrated teams, communication and metrics.

In the previous paragraphs we referred to the institutionalization that CMMI defines as "The ingrained way of doing business that an organization follows routinely as part of its corporate culture" [1].

Moreover, it is necessary to recognize the critical difference between process *implementation* and process *deployment*. The concept of deployment goes beyond the simple instantiation of an implemented process to address the effective deployment process specification needed to achieve multiple implementations process across an organization. Each deployment must be tailored to the culture of each site of the same organization [5].

The International Process Research Consortium (IPRC) has included the topic of Process Deployment in a list of research items for the next 10 years because intensive research into the human factor and change management is needed [6].

This paper presents the results of a research work to know aspects that influence process deployment in five software development and maintenance sites. This paper focuses on the organization, deployment critical success factors and lessons learned. It focuses on identifying the impact of a formal methodology in the process of the deployment process. This paper is organized as follows: section 2 has a brief introduction to the process deployment elements. Section 3 presents the methodology used and the survey results. Section 4 shows the process deployment lessons learned. Finally, section 5 indicates the conclusions and future research work.

2 Process Deployment

We define process deployment as "the process that allows the implementation, adoption, management and institutionalization of the processes generated in Process Engineering, allowing multiple implementations of the process across the organization".

The process of process deployment has focused on people, incorporated aspects such as training to develop collective skills, the motivation to use processes and effective communication. During the process deployment, some problems may arise such as a lack of support and commitment from senior management, lack of leadership, processes defined inadequately, lack of methodology for the process deployment, resistance to change and organizational culture not ready for change. The main elements for the deployment of processes are: (1) the organization in which the deployment of the processes will be carried out, (2) the processes library to be deployed, (3) the human resources with the necessary knowledge and skills, (4) the process of process deployment, (5) the deployment methodology and (6) change management.

2.1 The Organization

This research work was conducted in everis *consultancy* firm. everis is a leading Spanish consulting firm whose corporate headquarters are in Madrid (Spain) and several subsidiaries in Latin America and other countries in Europe. *everis* develops business and information technology consultancy services, as well as system integration delivery and outsourcing. The firm has more than 5000 professionals in different areas of expertise and forecast revenue of 300 million euros for 2007. One of the most vibrant divisions is "*everis* sites", which is in charge of the set-up and operation of system applications development and maintenance, providing a first rate service to several customers worldwide. *everis* considers CMMI as a basic prerequisite to operate in a global market where quality is a must. At the sites, the process deployment has the commitment of senior management. The sites have similar functional organizations.

2.2 The Process Library

A generic objective of CMMI maturity level 3 is to institutionalize the processes that constitute the process library. At this level the processes are characterized, understood and described by means of standards, procedures, tools and methods. The set of standard processes of the organization are supported by the process areas Organizational Process Definition (OPD) and Organizational Process Focus (OPF) [1]. The goal of the process area OPD is to establish and maintain a usable set of organizational process assets (standards), to establish life-cycle model descriptions, thus defining the tailoring criteria and guidelines to define and establish the organizations' measurement repository and process asset library.

It is important that the processes are well defined because they are the input to the process deployment. Therefore, if the processes have been poorly defined (are unclear, not flexible), there is a serious deployment risk.

Unformalized processes are abandoned in times of crisis, and people are unable to repeat the same process [8]. The organization is using CMMI model to define the processes to be deployed at the sites.

2.3 Human Resources

People play a major role in the definition and deployment of processes. Their knowledge and skills will enable them to perform effectively the process activities. Everyone in the organization has an individual culture that cannot be changed easily. So, a strategy is required to encourage them to accept the changes to be produced. In addition, people must be convinced that processes are used to improve and achieve benefits and have to be monitored [9].

Each site has a site manager, project managers, product lines coordinators and those in charge of them, analysts, encoders, supplier managers and those in charge of quality and audit.

2.4 The Deployment Process

Processes deployment is focused on people at all levels of analysis: individuals, teams, groups, organizations, countries and cultures. At the organizational level, it is the character of persons monitoring and putting processes into practice in their daily work that has to be controlled [5].

The results of the process deployment enable the organization to obtain usage patterns that provide feedback on process improvement. As in every activity related to process improvement, it will be necessary to establish a measurement process to control the process of process deployment. This provides feedback for future deployment processes through lessons learned.

2.5 The Methodology

It is important for the organization to have a process deployment methodology that supports and promotes efficient process deployment. This methodology should include the model; in this case, it is CMMI.

The methodology contains the procedures, guidelines and rules for carrying out the deployment of processes, the activities to be developed for the training in the new processes, including the motivation and communication strategies to be established. Its application will be more or less complex depending on the complexity of the projects, the knowledge and skills of people, the need to work in teams, the work environment and the amount of information that is collected for process improvement [7].

2.6 Change Management

Nowadays, change is constant in the business environment [8]. A change is any alteration that occurs in the work environment. The deployment of new processes is considered a change and should be managed. In a situation like this, employees are conditioned by the feelings, values, motivations, behaviour and needs that are originated by the change.

The fear of losing skills and the lack of efficiency generate frustration in the staff. The change also implies new learning and acquiring new skills. For this reason communication plays an important role in achieving acceptance and reducing rejection.

In many cases, one forgets that it is people who have to change. If the members of the organization do not change their behaviour, the success of the project can hardly be achieved.

3 Methodology and Results

The research was carried out at five software development and maintenance sites. The main objective of the research was to know how the deployment process had been carried out at each site, which methods or procedures were followed, the content of the processes library, the use of processes and the lessons learned. It was also interesting for the research to identify the critical success factors for the deployment based on past experience and the factors that will influence future deployment.

Critical success factor is a business term for an element, which is necessary for an organization or project in order to achieve its mission [9].

The following activities were carried out to achieve the main objective:

1. Identify the issues to research and develop the work plan.
2. Identify those in charge of the implementation of processes at the sites.
3. Identify the organizational processes.
4. Identify in advance some of the critical success factors for the implementation of processes.
5. Develop a survey with open and closed questions.
6. Carry out the survey with the personnel in charge of implementing the processes.
7. Analyze the results.

This paper focuses on the activities 5, 6 and 7.

3.1 The Survey

The survey is divided into three main modules:

- Module I. Related to the organization, the organizational structure and the staff.
- Module II. Related to the deployed processes, its use and complexity.
- Module III. Related to the methodology used in deployment process and, the critical success factors of deployment.

An example of survey questions used to get the results of this paper were:

- *Related to the organization.* The questions in the survey refer to a description of the organization, its structure and staff.
- *Related to the critical success factors.* The question was to list, in order of importance, the critical success factors of the deployment process.
- *Related to the methodology used for the deployment process.* The question was if the site had a deployment strategy for new procedures.
- *Related to the situation of implemented processes.* The question was to indicate what has been the level of acceptance by the members of the project.
- *Related to the lessons learned.* The question was to list the positive and negative lessons learned.

The survey was carried out at five sites: Three sites were maintenance sites and two sites were software factories. Each site established its own procedures to carry out the implementation of the processes.

3.2 Results Obtained through the Surveys

Some of the results of the survey are presented below.

3.2.1 Related to the Organization

Each site consists of four basic areas:

1. the layer of Coordination and the Quality Committee which are in charge of the monitoring and control of the projects, coordination and tracking of indicators, quality control and definition, and procedures implementation and methodology
2. those in charge of functional applications that are partners and receive and estimate requests
3. the production lines in charge of resolving the requests and carrying out tests and
4. the management of requirements is the main mechanism for the relationship between the functional part and the production lines.

The software factories are virtual software development sites (with several physical sites). They have three fundamental areas: (1) production that is structured in various technological lines customized to each client, (2) technical support and (3) quality area, under a light and agile structure management. The demand management is separate from the production area. The production area is divided into areas of knowledge which constitute the production lines.

3.2.2 Related to the Critical Success Factors

Table 1 shows the critical success factors according to the managers experience in order of importance.

It also shows the order of importance given to the factors for a future deployment of processes.

Each site was numbered from 1 to 5. Every site has two columns, the "B" (before) column indicates the order of importance assigned to the factor when the implementation took place. The "A" (after) column indicates the order of importance for a future process implementation.

At site 1, located in Madrid, all the factors listed in the table are important and, in a future implementation of processes, the factors would be taken into account in the same order. The process implementation would be in charge of the site staff. They have experience in implementing processes at other sites. The staff that used the processes participated in their definitions.

At site 2, located in Buenos Aires, the order of importance of the implementation factors in the future has changed. They consider the motivation factor for the use of procedures important. At this site the procedures implemented had to be customized taking into account the needs of the site. The processes were defined at site 1 with the participation of staff from site 2, which facilitated the process implementation. At the

Table 1. Success factors by order of criteria

SITES / FACTORS	Operation Sites						Software Factory			
	Site 1		Site 2		Site 3		Site 4		Site 5	
	B	A	B	A	B	A	B	A	B	A
Vision shared	1	1	7	3	1	1	1	1	1	1
Organizational Culture	2	2	6	6	1	1	3	1	2	2
Leadership and management	3	3	5	7	3	1	1	1	1	1
Motivation and use procedures	5	5	2	1	2	2	1	1	1	1
Training	6	6	1	2	2	2	1	1	1	1
Communication	4	4	3	4	1	1	1	1	1	1
Change Management	7	7	4	5	3	3	1	1	3	3
Audit and Quality	8	8	-	-	3	3	2	2	2	2
External Consultancy	N	N	N	N	N	N	S*	S*	N	N

B : Before A: After.

* External consultancy was only required for CMMI implementation and appraisal process.

beginning resistance to change was detected because they had to change their know-how. Their own personnel participated in the process implementation.

At site 3, located in Barcelona, the vision of the organization, organizational culture and communication are factors that came first. They have projects with a heavy workload. During the change process there was resistance to change motivated by the extra workload.

At site 4, located in Sao Paolo, most of the factors are in first place. These factors were critical during the process implementation. They initiated the process implementation using their own personnel, but then they had problems managing the change because their staff was moved to productive tasks. During the process implementation external support was provided by a consultancy firm. The staff is still highly motivated. They believe that external support will be provided in a future process implementation.

At site 5, located in Seville, the change management factor is in third place, because at this site the resistance to using new procedures was managed.

3.2.3 Related to the Methodology for the Deployment of Processes

The survey also found that every site defined its own procedures to implement the new processes. All had the same objective: to adopt and use the new process.

The communication and training strategies were different for each site. All the strategies shared the same objective, which was to use and adopt the processes.

At site 4, a specialized external consultancy firm was chosen to manage the change with positive results.

3.2.4 Related to the Situation of Implemented Processes

A brief summary of the process status acceptance is showed in Figure 1. It shows the level of process acceptance at each of the sites by the staff.

The acceptance levels of the procedures were categorized in the survey using the following scale: *Very High, High, Medium and Low.*

At site 1, in the *Very high* category the process acceptance level was 86.5%. The staff participated actively throughout the development of the procedures. The processes were customized to their needs. All implemented processes are being used.

At site 2, in the *High* category the process acceptance level was 39%. For the *Medium* and *Low* categories, the acceptance levels were 21.7% and 34.8% respectively. At this site, the definitions of implemented processes were provided by site 1.

At site 3, a similar behavior is shown. At site 4 in the *High* category, the acceptance level was 73.7%. At this site, the change resistance was managed successfully. The staff was motivated to use the process. All the processes are being used.

At site 5, 38.5% of the processes have a *High* acceptance level and the 53.8% a *Medium* acceptance level.

In order to ensure the success of the process deployment, the organization must be prepared and mature like sites 1 and 4 that are at CMMI maturity level 3.

For organizations with maturity levels lower than 3, there is a tendency to improve, although there are difficulties during process implementation.

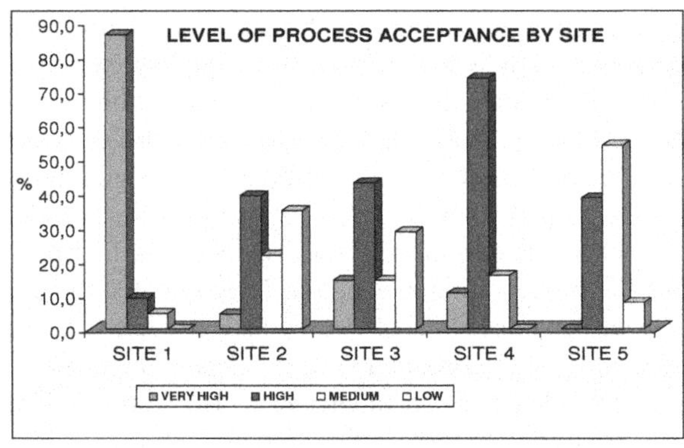

Fig. 1. Level of process acceptance by site

4 Organizational Lessons Learned

A number of lessons could be learned from these process deployments, but we present only the obtained from surveys:

- To perform an assessment before starting a change because the organization has to be ready to deal with change. The organization must have organizational infrastructure, with well-trained human resources ready to carry out the new activities related to change.
- To involve the staff during the process definition and implementation in order to facilitate their adoption. In this way the staff will identify with the process defined.
- To customize the procedures in accordance with the environmental aspects. When there are unresolved issues such as linguistic, cultural and geographical aspects, it will not be easy to implement a process.
- To have an established method for the process deployment that includes communication, training and management of change factors. An initial comprehensive plan is necessary. This plan must be followed up taking deadlines into account and corrective actions can be taken when the deployment project deviates significantly from the plan.
- To train the staff continuously to improve their skills and increase their knowledge in order to minimize the resistance to change.

5 Conclusions and Future Work

This paper presents some issues that were observed during the process deployment in five software development and maintenance sites. It can be concluded that:

- It is important to recognize the difference between process implementation and process deployment. Organizations normally use the term implementation. The deployment concept goes beyond the single instantiation of a process implemented.
- The process deployment is focused on the human factor and change management. The objective of process deployment is to put the process into practice, minimizing the resistance to change and achieving its institutionalization. The organization should establish a formal method focused on the human factor to avoid the potential risks.
- Good practices for deployment process must be collected, written and stored in a repository. It is necessary to learn from the mistakes and to share the acquired knowledge.

A future research work that arises is to identify the relationship between the critical success factors and their impact on the process of deployment. A strategy for the process deployment based on previous critical success factors could be developed.

Acknowledgments

This paper is sponsored by ENDESA, everis Foundation and Sun Microsystems through *"Research Group of Software Process Improvement for Spain and Latin America"*.

References

[1] Carnegie Mellon Software Engineering Institute, CMMI® for Development, Version 1.2. CMU/SEI- 2006-TR-008 (2006) (Accessed October 10, 2007), http://www.cmu.edu/pub/documents/06.reports/pdf/06tr008.pdf
[2] Kulpa, M., Johnson, K.A.: Interpreting the CMMI: A process Improvement Approach. Auerbach Publications (2003)
[3] Software Engineering Institute. Carnegie Mellon University, CMMI Process Maturity Profile CMMI 2007: Appraisal results 2006 End-year Update (2007) (Accessed September 26, 2007), http://www.sei.cmu.edu/appraisalprogram/profile/pdf/CMMI/2007marCMMI.pdf
[4] Humphrey, W.: TSP: Leading a Development Team. Addison-Wesley Publishing Company Inc, Reading (2005)
[5] Forrester, E. (ed.): The International Process Research Consortium, A Process Research Framework. Software Engineering Institute (2006)
[6] Quinn, Cameron: Diagnosing and Changing Organizational Culture. Jossey-Bass (1999)
[7] Software Guru, Conocimiento en práctica 10-13 (2006) (Accessed September 30, 2007), http://www.softwareguru.com.mx/downloads/SG-200606.pdf
[8] Kotter John, P.: Leading Change. Harvard Business School Press (1996)
[9] Daniel, D.R.: Management Information Crisis. Harvard Business Review (1961)

Experiments with Content-Based Image Retrieval for Medical Images

Gongzhu Hu and Xiaohui Huang

Department of Computer Science
Central Michigan University
Mount Pleasant, Michigan, USA
hu1g@cmich.edu

Summary. Content-Based Image Retrieval (CBIR) is to retrieve digital images from an image data repository by the contents in the image, such as shape, texture, color, and other information that can be extracted from the image. CBIR is also referred to as query by image contents. As large image collections being created and become available, more and more applications rely on CBIR techniques to retrieve images from the collections. One important application area is the medical field. Many medical and health care institutions have started using various CBIR systems to assist and improve diagnosis and treatment of diseases. In this paper, we introduce a method for image retrieval and classification using low-level image features. This method is based on selection of prominent features in the high dimension feature space and the parameter of the k-NN algorithm. We also combine non-image features (patient records) and image features to improve the accuracy of the results. Both the patient data and images are from a clinical trail studying aloe in treating the side effects due to radiation on oral cancer patients at Mid-Michigan Medical Center. A *MatLab* image engine is used for image feature retrieval, and principal component analysis is applied to reduce the feature space for optimizing the performance.

Keywords: content-based image retrieval, principle component analysis, medical images.

1 Introduction

In early days, searching for an image from a collection of images was commonly done through the description of the image, such as image id, caption, category type, and other text-based information associated with the images. As the number of image collections (databases, repositories, archives, etc.) and the size of each collection grow dramatically in recent years, there is also a growing needs for searching for images based on the information that can be extracted from the images themselves rather than their text descriptions. *Content-based image retrieval* (CBIR) is an approach for meeting this need. CBIR is to retrieve digital images by the actual contents in the image. The contents are the *features* of the image such as color, texture, shape, salient point, and other information about the image including some statistic measures of the image. Using CBIR-based systems, a user can provide (or click on) an existing image to find images in the repository that are similar or related to the given image.

As a technique that uses visual contents to search images from large scale image databases according to users' interests, CBIR has been an active and fast advancing

R. Lee and H.-K. Kim (Eds.): Computer and Information Science, SCI 131, pp. 157–168, 2008.
springerlink.com

research area since the 1990s. During the past decade, remarkable progress has been made in both theoretical research and system development. However, there remain many challenging research problems that continue to attract researchers from multiple disciplines. One of the challenging areas is the medical field. Many medical and health care institutions, particularly in rural areas, do not have modern systems that provide support for image management and usage.

A regional medical institution, the Mid-Michigan Medical Center (MMC), has conducted cancer experiments and research for years. One of the clinical experiments was to investigate the effectiveness of the drug *aloe vera distillate* in reducing the side effect on cancer patients who receive radiation treatments. A group of N ($N = 25$ in our experiments) patients with head and neck cancer who received radiation were enrolled in a clinical trial for two weeks. They were divided into two groups: the Aloe Vera group that includes those using *aloe vera distillate*, and the Placebo group for the remaining patients who went through normal treatments without using the *aloe vera distillate* drug. Photo images of patient's tongue and throat were taken weekly for each patient for clinical evaluation. Other measurements (such as patient's height, weight, serum albumin, stage of cancer, food intake, and symptom distress, etc.) were also taken during the trial period. At the end of the period, the images and text-based measurements are available for the doctors to evaluate the effectiveness of the drug.

To help and enhance the evaluation process, an image retrieval system was built to support managing medical images and patient data [12]. The purpose of such a system is to assist doctors to identify images of patients with similar patterns and compare them with related cases. For researchers, it can be used to build predictive models for pattern recognition and prediction. The image retrieval system was developed in *MatLab* based on a naïve image processing module. Although the system can generally do its job, it has its weaknesses, primarily the poor run-time performance due to high dimensions of image features.

In this paper, we present a method based on the *principle component analysis* and the selection of k in the k-NN (*nearest neighbor* method for classification and retrieval. Principle component analysis was applied to the images to find an optimal set of image features so that the dimensionality of the feature space can be significantly reduced to enhance the performance while retaining good accuracy for images classification. In addition, the improved system also used some patients records (measurements) as "added features" to the image features to form a slightly larger feature space to improve the classification accuracy.

2 Related Work

As database management techniques were widely used in the late 1970's, image retrieval systems based on the database technologies also attracted a lot of attentions of researchers [2, 3]. Early techniques were mostly based on the textual annotation of images using traditional database management systems. A survey of text-based image retrieval can be found in [10].

As the volume of digital images increased dramatically since the 1990's, text-based image retrieval methods were no longer be able to handle the task simply because manual annotation of the large volumes of images became prohibitively expensive or impossible. There was an increasing need for image retrieval methods based on the content of images, rather than the text descriptions. Various approaches were proposed, from edge-based [13], semantic-based [9], color and orientation histograms [1], to query by image content [5]. In recent years, content-based image retrieval has evolved from low-level image representation to semantic concept models to higher-level semantic inferences [11]. The fundamentals of CBIR can be found in [8]; and [4] provided an good survey and discussed new trends in this area.

Research on content-based image retrieval applying to medical images have been conducted to assist diagnosis and patient treatment. It is mostly an interdisciplinary task, mostly combination of medical, computing, and statistics. Complex applications normally consists of several layers of information models, from raw data representation to high level knowledge [7].

In this paper, we present the results of the experiments using the improved content-based image retrieval system we developed for the Mid-Michigan Medical Center.

3 Methodology

One of the basic modules of the content-based image retrieval system at MMC is an image feature generating engine. It generates a feature vector of 520 features for each input image. For most of the images, the feature vector generated contains a lot of missing values and zero values, which makes a collection of feature vectors a sparse matrix. And, the processing time is quite long for engine to generate the features, especially when we deal with a large set of images.

Our aim was to find the most significant ones among all the generated features and only use these significant features for classification. This will dramatically speed up the processing time while retaining the discrimination power so that the accuracy of classification remains pretty much at the same level as using the original feature space.

To describe this task a bit more formally, let $\mathscr{F}_N$ be the the feature space of N features generated by the image engine, and Θ be the average accuracy of the classification resulted using $\mathscr{F}_N$. The task we are targeting to is to find a *minimum* n ($n \ll N$) such that $\theta \simeq \Theta$, where θ is the classification accuracy using $\mathscr{F}_n$ ($\mathscr{F}_n \subset \mathscr{F}_N$) and $\simeq$ means "approximately equal."

In addition, we use some non-image features (medical measurements, such as cancer condition, weight, drug dosage, etc.) to slightly extend the reduced image feature space when we believe these non-image features may play an important role in classification.

3.1 Image Database and Feature Space

Data about the patients, images, and other related information are stored in a relational database. The `Patient` table contains many attributes like this

id	*name*	*address*	*entering_date*	...

where *id* is a unique identification number (an integer) automatically generated. For each patient, multiple images were taken each week during the trial period. The images are stored in a file structure and the information about the images are stored in the `Image` database table:

patient_id	*week*	*image_#*	*file_path*

The *patient_id–week–image_#* code is used to identify an image. For example 21-03-05 means the image is the 5th image of patient 35 taken on week 3.

The image features are extracted for each digitized image by running the image processing engine. The engine generates $N = 520$ image features including colors and edges. For M images, the resulting features are stored in an $M \times 520$ matrix.

Six non-image features are selected to add to the feature space. They are measurements related to cancer condition, weight, oral condition, clinical evaluation, pain evaluation and dosage, which were taken each week for eight weeks. All the features are normalized.

3.2 Similarity Measure and Classification

Distance measures and probability measures are commonly used methods to measure the similarity of two cases. Since probability measures require a large number of cases as the training set (at least larger than the number of variables, or features in our discussion) and the sizes of the training sets in our experiments are not large enough, we choose to use simple Euclidean distance measure. Since each image is an N (520 features in our case) dimensional vector, the distance of two images $\mathbf{X}$ and $\mathbf{Y}$ is simply

$$D_{\mathbf{XY}} = |\mathbf{X} - \mathbf{Y}| = \sqrt{\sum_{i=1}^{N} (x_i - y_i)^2}$$

There are quite a few algorithms for classification, including Bayesian, k-nearest neighbor (k-NN), and support vector machine (SVM). In the work presented in this paper, we used the k-NN classifier. It is a method for classifying objects based on closest training samples in the feature space. The basic idea of the k-NN algorithm is like this: Let $C = \{c_i, i = 1, \ldots, l\}$ be a set of l categories and $S = \{s_i\}$ be a set of training samples where each sample s_i has a value v_i and a label $c_j \in C$. Let x be a new sample to be classified. The k-NN algorithm finds the k samples in S with the smallest distances from x (k nearest neighbors of x). If the majority of the k samples are labeled c_t, x is classified as belonging to category t.

For the patient cases we dealt with, we randomly choose m cases as the training base, in which each case is known to belong to a group (Aloe or Placebo). Given a testing case x, we apply the k-NN algorithm to calculate the retrieval function $R(x,k)$ defined as

$$R(x,k) = \begin{cases} 1 & \text{if the majority of the } k \text{ cases belong to the Aloe group} \\ 0 & \text{otherwise} \end{cases}$$

The case x is said to belong to the Aloe group if $R(x,k) = 1$, to the Placebo group if $R(x,k) = 0$.

For m testing cases whose true groups are known, we can calculate the accuracy of the classification method that is the ratio of the correctly classified cases among all the m cases. That is,

$$accuracy = \frac{1}{m}\sum_{i=1}^{m} T_i$$

where

$$T_i = \begin{cases} 1 & \text{if case } i \text{ was correctly classified} \\ 0 & \text{otherwise} \end{cases}$$

A classification model is built by determining the optimal value of k for the given m training set.

3.3 Principle Component Analysis

Since the feature space has a large dimensionality (520 features in our medical images), the feature extraction and classification process take quite a long time to finish. This is the problem common to all applications involving data of high dimensions. It is desirable to find a way to express the data in a multivariate (feature) space of a much lower dimension without lose much information in the data.

There are many dimensionality reduction approaches, including both linear and non-linear methods. For the experiments we did on the medical image, the commonly used *principle component analysis* method [6] was used. Basically PCA is to find the *principle components* from the data with each principle component (a vector) pointing to a direction in which the "spread" (or differences) of the data is explicitly expressed.

A set of M data points of N dimensions can be expressed as a $M \times N$ mean-adjusted matrix $\mathbf{X}_{M\times N}$. Each row of the matrix is a data point in the N-D space. In our medical application, a row represents a trial case in the 520-D feature space.

The basic algorithm of PCA is to calculate the covariant matrix, $\mathbf{C}_X$, of $\mathbf{X}_{M\times N}$, and the eigenvalues and eigenvectors of $\mathbf{C}_X$. Then, select the eigenvectors that correspond to the n ($n \ll N$) largest eigenvalues. And finally, transform the original data to the much reduced n-D feature space by applying the eigenvector matrix transposed to the original data $\mathbf{X}_{M\times N}$ transposed. The PCA transformation is given in Algorithm 1.

Algorithm 1. Principle component transform

Input: $\mathbf{X}_{M\times N}$: a data set of M data points in N-D space
Input: n: the number of reduced dimensions
Output: $\mathbf{X}_{M\times n}$: data set in reduced n-D space

begin
    Normalize $\mathbf{X}$ – mean adjusting colums;
    Calculate the covariant matrix $\mathbf{C}$ of $\mathbf{X}$;
    Calculate the eigenvalues of $\mathbf{C}$;
    Calculate the eigenvectors $\mathbf{V}_N = \{v_i\}$ of $\mathbf{C}$;
    Select those v_i that correspond to the n highest eigenvalues;
    Let $\mathbf{V}_n$ be the matrix with the selected v_i as column vectors;
    Calculate $\mathbf{X}_{M\times n} = \mathbf{V}_n^t\, \mathbf{X}_{M\times N}^t$;
    return $\mathbf{X}_{M\times n}$;
end

The covariance matrix in step 3 above is calculated as

$$cov = \frac{1}{m-1}\sum_{i=1}^{m}(X_i - \bar{X})(X_i - \bar{X})^t$$

where X_i is the row vector of the data, $\bar{X}$ is the mean vector, and m is the number of data items in the data set.

The principle components are the n eigenvectors (column vectors in the matrix) with the highest eigenvalues. These n eigenvectors represent the n directions where the data set has the highest spread, and hence retain the most of the discriminating power in the original data set for classification. With an appropriate n, we can achieve a similar result in classification as using the original data, while significantly reducing the dimensionality to improve the run-time performance.

4 Experimental Results

As mentioned before, the experiments we performed were for 402 cases, each of which contained an image and some measurements from patients records. An image processing engine was used to generate 520 low-level image features. Of the 520 features, 218 were all zero (for the 402 images we had) and were removed. Hence, the feature space was immediately reduced to 302-D.

Several parameters were used in our experiments to find the "optimal" classification model, including:

- n — the number of dimensions to reduce to.
- k — the number of cases in the training set closest to the testing case. That is, the k in the k-NN algorithm.
- m — the size of the testing set; and the size of the training set is 402–m.

We tested various values for m. In this paper, we just show the results for $m = 35$. That is, 35 cases were randomly selected as the testing set and the remaining 367 cases as the training set. The test was repeated for 8 different 35-case sets, listed in Table 1. In the table, the numbers are the case id.

The software tool *MatLab* was used to perform the PCA process to reduce the dimensionality of the feature space. After several experiments, we found that $n =$ 60 gave a close-to-optimal results – significant reduction of the feature space and reasonable accuracy in the classification results.

We did not find a particular value of k yielding a better result (classification accuracy) than other values of k. A value of k might generate better result for a particular random set, but worse for another random set. However, it appeared that a smaller k tends to give a little bit more accurate results. The reason might be that a large k would include many of images from different groups, that makes it harder to determine the correct group for the testing image.

Here we show the results using k = 5, 10, 15, 20, 25, 30, 50. For each random set, the classification accuracy is calculated as the average accuracy of each case in the set. We ran the tests on the original 302-D data set and the transformed data of 60-D, and the classification accuracy results for the various k values are shown in Figures 1–7.

Table 1. Eight randomly selected sets of cases

Set							
1	2	3	4	5	6	7	8
355	332	6	322	43	165	350	387
246	322	159	111	312	364	18	255
326	245	82	37	299	235	143	134
233	173	304	303	332	246	108	381
367	316	312	127	30	267	287	165
278	97	324	265	269	392	116	401
102	48	315	73	257	73	198	352
307	250	267	402	109	270	88	199
111	70	93	75	329	200	296	300
69	51	343	283	162	383	313	132
103	105	370	260	369	229	79	284
346	184	295	232	8	15	219	48
337	386	402	324	217	106	113	129
163	6	107	8	282	338	211	178
281	114	145	314	45	98	360	340
197	140	8	312	89	277	367	232
368	355	398	161	339	25	247	35
300	44	1	198	117	88	134	62
251	348	303	362	101	259	141	334
132	96	184	61	97	344	173	133
328	216	126	14	80	156	239	250
213	85	161	239	355	230	53	160
268	273	71	233	333	21	157	217
294	27	337	118	258	176	368	257
266	56	122	282	52	36	5	22
19	108	368	51	156	205	174	39
41	13	92	311	295	92	221	118
95	95	224	292	36	128	233	113
254	110	18	279	271	29	394	174
65	73	209	18	395	321	44	17
369	306	356	147	283	297	222	36
382	307	153	357	274	248	208	260
106	212	274	179	387	373	75	53
211	291	194	351	316	339	26	135
354	234	60	2	361	285	371	211

From these results, we can see that the the average accuracy on the data of reduced feature space were a little bit lower than on the original data, simply because some information in the original data was lost due to the dimensionality reduction. However, the accuracy was still comparable to the original data and the run-time performance was significantly improved. We do not include the performance results in this paper because it is not hard to be convinced that the 80% dimensionality reduction (from 302 to 60) would make the process run much faster.

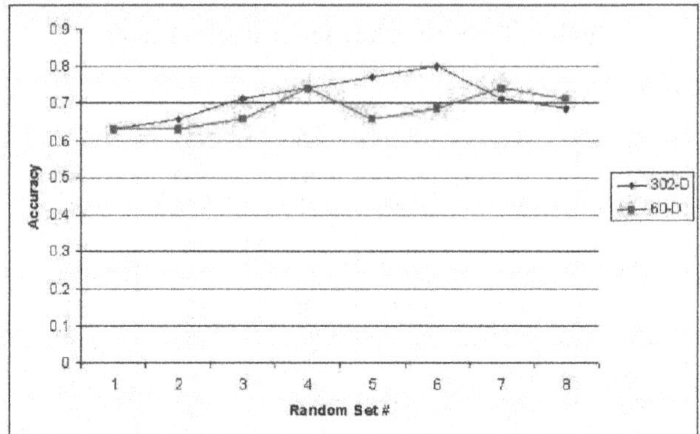

Fig. 1. Accuracy measures for original vs. reduced feature spaces, k = 5

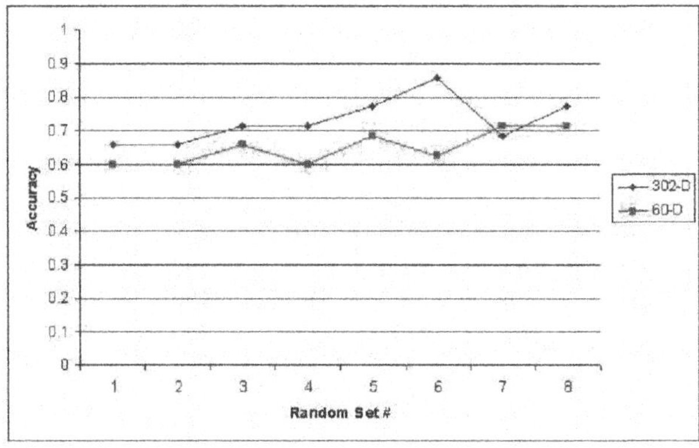

Fig. 2. Accuracy measures for original vs. reduced feature spaces, k = 10

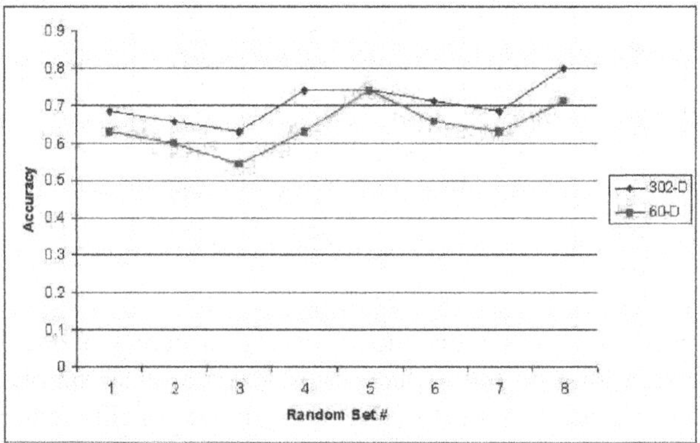

Fig. 3. Accuracy measures for original vs. reduced feature spaces, k = 15

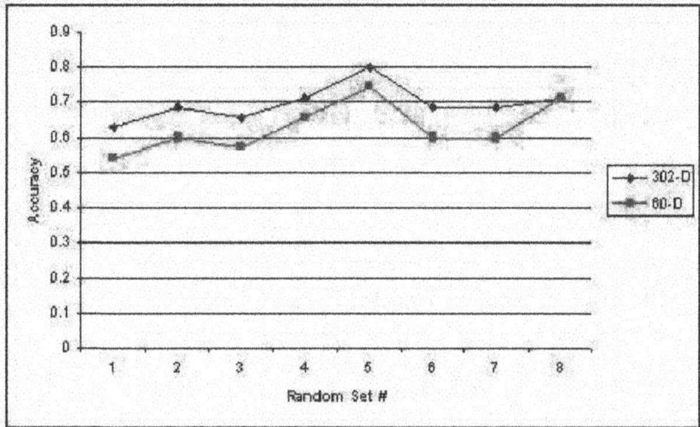

Fig. 4. Accuracy measures for original vs. reduced feature spaces, k = 20

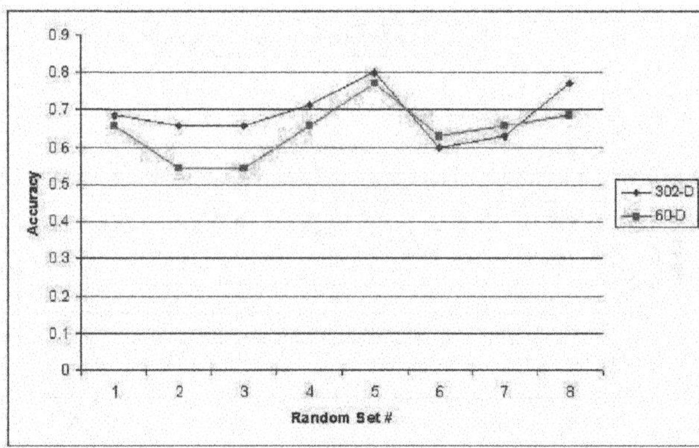

Fig. 5. Accuracy measures for original vs. reduced feature spaces, k = 25

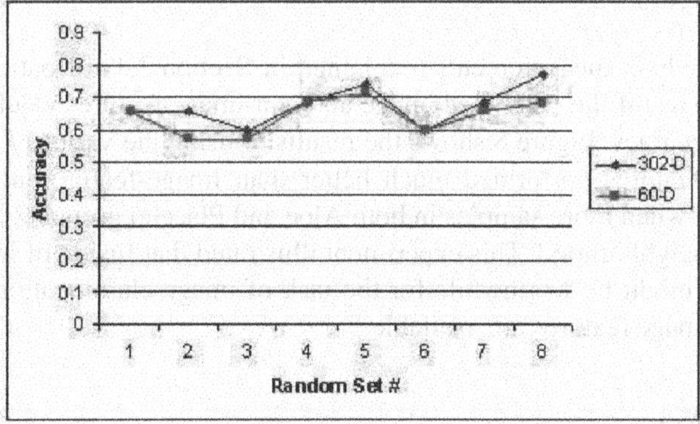

Fig. 6. Accuracy measures for original vs. reduced feature spaces, k = 30

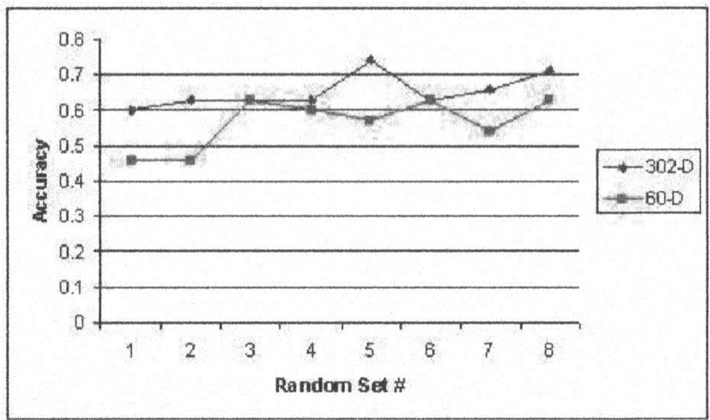

Fig. 7. Accuracy measures for original vs. reduced feature spaces, k = 50

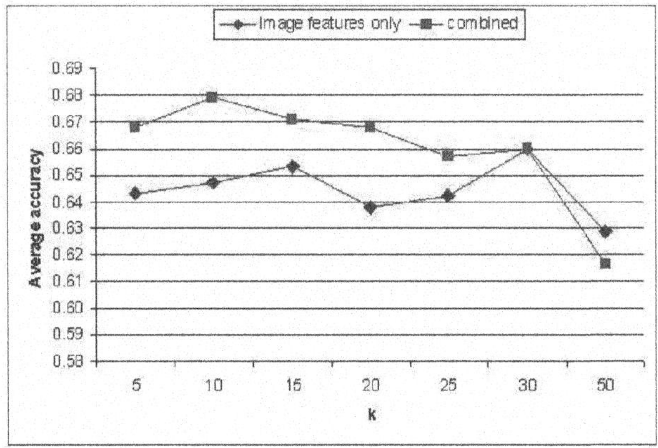

Fig. 8. Comparison of image-feature-only vs. combining image features and patient data

We also added six measurements (explained in Section 3.1) of patient data to the feature space to test if the combined image and non-image features would improve the classification accuracy. Figure 8 shows the results of using the various k values. Using the combined features performed much better than image-feature-only in all cases except for $k = 5$ when more samples in both Aloe and Placebo groups were included in the 50 nearest neighborhood. This experiment illustrated that fusion of image and non-image features might be worthwhile for the task of image classification and retrieval when the non-image features are available.

5 Conclusion

Content-based image retrieval is a technology to retrieve images from an image repository that are similar to a given test image based on the features extracted from the images. An important application area of CBIR is the medical field. To help improve the previously existing medical image system at the Mid-Michigan Medical Center, we developed adds-on modules to build an model for image classification of the Aloe and Placebo groups. The main improvements were the feature space reduction using the principle component analysis and the combination of the low-level image features with some measurements from the patient records.

Our experimental results show that the reduction of the feature space from 302-dimension to 60-dimension achieved significant run-time speed-up while keeping the classification effectiveness at the level close to the ones for the original data. And, with a few non-image patient data added to the feature space, the accuracy improved for all random sets and all values of k except for a large k, indicating that the approach of combining image and non-image features appears promising.

As the future work, we plan to run more tests to find (or attest) the optimal parameters n, m, k for the model, and to extend the system to include more categories (not just Aloe and Placebo) that would use some variation of the k-NN method such as (k,l)-NN. And, we also plan to adjust the non-image features to find the relationships between the non-image features and the image features that affect the classification accuracy. We also plan to develop a measurement on the degree of dimensionality reduction on the classification accuracy.

References

1. La Cascia, M., Sethi, S., Sclaroff, S.: Combining textual and visual cues for content-based image retrieval on the www. In: Proc. IEEE Workshop Content-Based Access of Image and Video Libraries, pp. 24–28. IEEE, Los Alamitos (1998)
2. Chang, N.-S., Fu, K.-S.: Chapter A relational databse system for images. In: Pictorial Information Systems. LNCS, vol. 80, pp. 288–321. Springer, Heidelberg (1980)
3. Chang, S.-K., Yan, C.W., Dimitroff, D.C., Arndt, T.: An intelligent image database system. IEEE Transactions On Software Engineering 14, 681–688 (1988)
4. Datta, R., Joshi, D., Li, J., Wang, J.Z.: Image retrieval: Ideas, influences, and trends of the new age. Technical Report CSE 06-009, Penn State University (2006)
5. Flickner, M., Sawhney, H., Niblack, W., Ashley, J., Huang, Q., Dom, B., Gorkani, M., Hafner, J., Lee, D., Petkovic, D., Steele, D., Yanker, P.: Query by image and video content: The QBIC system. Computer 28(9), 23–32 (1995)
6. Jolliffe, I.T.: Principal Component Analysis, 2nd edn. Springer, Heidelberg (2002)
7. Lehmann, T.M., Wein, B., Dahmen, J., Bredno, J., Vogelsang, F., Kohnen, M.: Content-based image retrieval in medical applications:a novel multistep approach. In: Proceedings of SPIE 2000, pp. 312–320 (2000)
8. Long, F., Zhang, H., Feng, D.D.: Chapter Fundamentals of Content-Based Image Retrieval. In: Multimedia Information Retrieval and Management: Technological Fundamentals, Springer, Heidelberg (2003)
9. Mojsilovic, A., Gomes, J.: Semantic based categorization, browsing and retrieval in medical image databases. In: Proc. Int. Conf. Image Processing, ICIP 2002, pp. 145–148. IEEE, Los Alamitos (2002)
10. Smeulders, A.W.M., Worring, M., Santini, S., Gupta, A., Jain, R.: Content-based image retrieval at the end of the early years. IEEE Transactions on Pattern Analysis and Machine Intelligence 22(12), 1349–1380 (2000)

11. Vasconcelos, N.: From pixels to semantic spaces: Advances in content-based image retrieval. In: Computer, pp. 20–26 (2007)
12. Wang, D., Lee, C., Kotecha, R.P., Mehta, R.H., Kaiser, J., Bott, M.: An image retrieval system based on patient data and imgage content. In: Proc. of the 5th ACIS International Conference on Software Engineering, Artificial Intelligence, Networking, Parallel/Distributed Computing, pp. 241–246. ACIS (2004)
13. Zhou, X.S., Huang, T.S.: Edge-based structural features for content-based image retrieval. Pattern Recognition Letters 22(5), 457–468 (2001)

Enterprise Process Model for Extreme Programming with CMMI Framework

Sung Wook Lee[1], Haeng Kon Kim[2], and Roger Y. Lee[3]

[1] Department of Computer information & Communication Engineering, Catholic Univ. of Daegu, Korea
sojiro@cu.ac.kr

[2] Department of Computer information & Communication Engineering, Catholic Univ. of Daegu, Korea
hangkon@cu.ac.kr

[3] Software Engineering & Information Technology Institute, Central Michigan University, USA
lee1ry@cmich.edu

Summary. It is currently a critical issue that software development organization in small and medium sized enterprises tends to apply agile methodology and extreme programming (XP), the front-runner of agile methodologies, to release software through project management centered development, rather than process centered.

They also try to appraise into CMMI level acquisition that is mainly focus on process that is different from XP, which manages project-centered development. However, CMMI level acquisition is possible to achieve through the control principle of agile methodology. In this paper, we describe how to aggregate agile methodology into CMMI frameworks and suggest process-for-process appraisal. We also identify and define the process for CMMI and extreme programming through the many existing comparison data. We illustrate new Process Model which includes the process of CMMI level 2 as perform the extreme programming project. We develop the process modeling with UML.

Keywords: Software Process Improvement, CMMI, Agile Methodology, eXtreme Programming.

1 Introduction

Capability Maturity Model Integration (CMMI) is a process improvement approach that provides organizations with the essential elements of effective processes. It can be used to guide process improvement across a project, a division, or an entire organization. CMMI helps integrate traditionally separate organizational functions, set process improvement goals and priorities, provide guidance for quality processes, and provide a point of reference for appraising current processes[3]. On the other hand, Extreme Programming (or XP) is a software engineering methodology, possibly relative to agile software development, prescribing a set of daily stakeholder practices that embody and encourage particular XP values -called "extreme" levels-leads to a development process that is more responsive to customer needs ("agile") than traditional methods, while creating software of better quality[13]. Proponents of XP and agile methodologies in

R. Lee and H.-K. Kim (Eds.): Computer and Information Science, SCI 131, pp. 169–180, 2008.
springerlink.com

general regard ongoing changes to requirements as a natural, inescapable and desirable aspect of software development projects. The adaptability to changing requirements at any point during the project life of the agile and XP is a more realistic and better approach than attempting to define all requirements at the beginning of a project and then expending effort to control changes to the requirements.

It is may be time for agile practitioners to take a close look at the lesson learned from the CMMI as it relates to creating common processes across an organizations. CMMI organization should, theoretically, have no trouble in adapting the key recommendation of the agile and XP approaches. The agile and XP methodologies are focus at the project level, whereas the CMMI is focused the organizational level. It is not easy to apply CMMI to agile and XP methodology[11].

In this paper, we suggest the model for introducing the CMMI to agile and XP process. We present the enterprise process model for XP and agile project based on CMMI frameworks. We approve that the agile method can actually help the users the CMMI model as it was intended for real process improvement.

2 Related Work

2.1 Agile Methodology

The starting point was a document that summarizes the agile philosophy: the agile manifesto[11] that includes a set of principles and values that support the philosophy. The characteristics of agile methods are elaborately defined in the twelve principles behind the agile manifesto[1]:

- Our highest priority is to satisfy the customer through early and continuous delivery of valuable software.
- Welcome changing requirements, even late in development. Agile processes harness change for the customer's competitive advantage.
- Deliver working software frequently, from a couple of weeks to a couple of months, with a preference to the shorter timescale.
- Business people and developers must work together daily throughout the project.
- Build projects around motivated individuals. Give them the environment and support they need, and trust them to get the job done.
- The most efficient and effective method of conveying information to and within a development team is face-to-face conversation.
- Working software is the primary measure of progress.
- Agile processes promote sustainable development. The sponsors, developers, and users should be able to maintain a constant pace indefinitely.
- Continuous attention to technical excellence and good design enhances agility.
- Simplicity–the art of maximizing the amount of work not done–is essential.
- The best architectures, requirements, and designs emerge from self-organizing teams.
- At regular intervals, the team reflects on how to become more effective, then tunes and adjusts its behavior accordingly.

2.2 Extreme Programming

Extreme Programming is a lightweight software methodology (or process) that is usually attributed to Kent Beck, Ron Jeffries, and Ward Cunningham [2, 3, 5]. XP is targeted toward small to medium sized teams building software in the face of vague and/or rapidly changing requirements. XP teams are expected to be co-located, typically with less than ten members. The XP Map shows how they work together to form a development methodology in figure 1.

- ***User Stories*** - User stories serve the same purpose as use cases but are not the same. They are used to create time estimates for the release planning meeting. They are also used instead of a large requirements document. User Stories are written by the customers as things that the system needs to do for them. They are similar to usage scenarios, except that they are not limited to describing a user interface. They are in the format of about three sentences of text written by the customer in the customer terminology without techno-syntax.

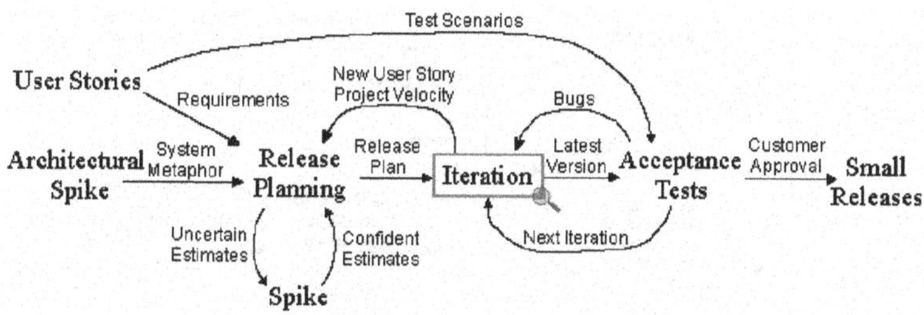

Fig. 1. Extreme programming project

- ***Create a Spike Solution*** - Create spike solutions to figure out answers to tough technical or design problems. A spike solution is a very simple program to explore potential solutions. Build a system which only addresses the problem under examination and ignore all other concerns. Most spikes are not good enough to keep, so expect to throw it away.
- ***Release planning*** - A release planning meeting is used to create a release plan, which lays out the overall project. The release plan is then used to create iteration plans for each individual iterations. Individual iteration is planned in detail just before each iteration begins and not in advance.
- ***Iterative Development*** - Iterative Development adds agility to the development process. Divide your development schedule into about a dozen iterations of 1 to 3 weeks in length. Keep the iteration length constant through out the project. This is the heart beat of your project. It is this constant that makes measuring progress and planning simple and reliable in XP.
- ***Acceptance tests*** - Acceptance tests are created from user stories. During iteration the user stories selected during the iteration planning meeting will be translated into acceptance tests. The customer specifies scenarios to test when a user story has been

correctly implemented. A story can have one or many acceptance tests, what ever it takes to ensure the functionality works.

2.3 CMMI Framework

CMMI is a process improvement maturity model for the development of products and services[2]. This paper refers to focus on relationship among process area. Process areas can be grouped into four categories.

- Process Management
- Project Management
- Engineering
- Support

We suggest observes project management and support which related extreme programming in this paper.

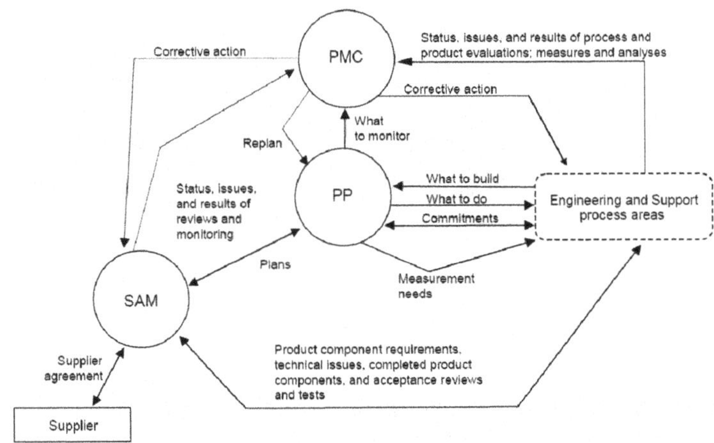

Fig. 2. Basic project management process areas

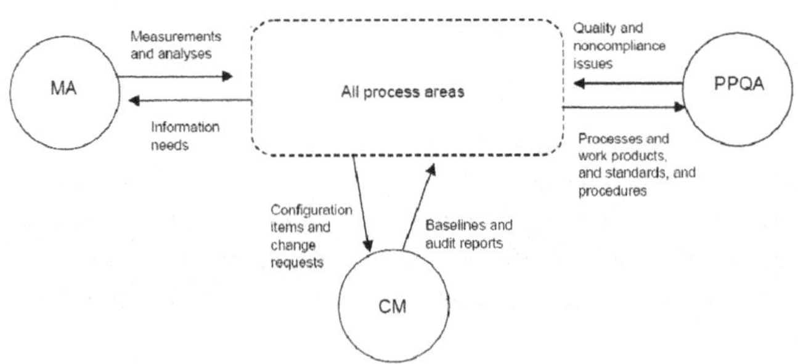

Fig. 3. Basic support process areas

Figure 2 addressed that related among each project management process area. The Project Planning process area includes developing the project plan, involving stakeholders appropriately, obtaining commitment to the plan, and maintaining the plan.

Figure 3 shows the related among each project management process area. The support process areas cover the activities that support product development and maintenance. The support process areas address processes that are used in the context of performing other processes. In general, the support process areas address processes that are targeted toward the project and may address processes that apply more generally to the organization.

3 Enterprise Process Model on CMMI Frameworks

Modeling is a concerned with the graphical representation of multiple views of structured information offering the ability to check coherence across the whole enterprise systems. Figure 4 shows a typical three layered approach to enterprise process model. We propose that the Enterprise process model is located layer between CMMI framework and extreme programming project.

Enterprise Process Model uses CMMI framework on process-centered guideline to improvement the software process management capability for extreme programming. We also use our model provide CMMI level 2 or 3 with agile methodology in small and medium development teams.

CMMI organization should, theoretically, have no trouble in adapting the key recommendation of the agile and XP approaches. The agile and XP methodologies are focus at the project level, whereas the CMMI is focused the organizational level. It is not easy

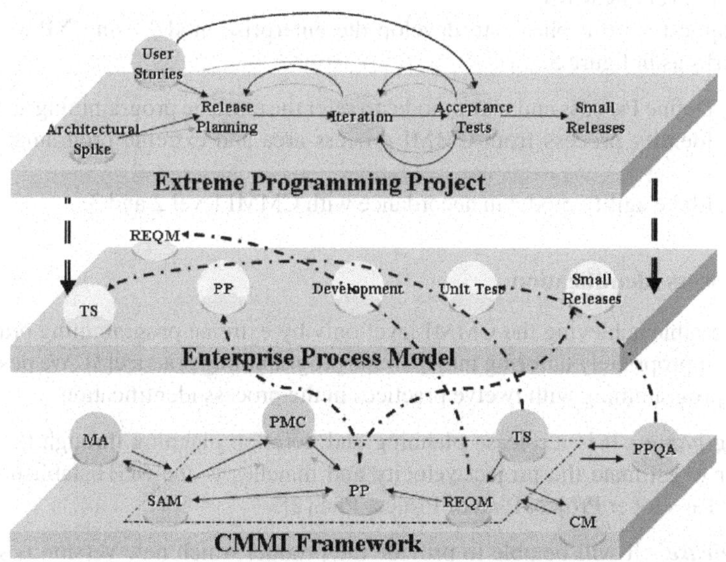

Fig. 4. Position of enterprise process model

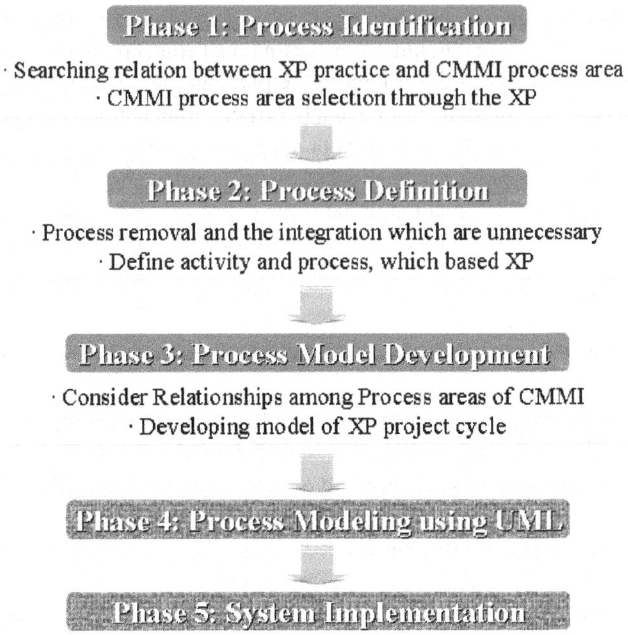

Phase 1: Process Identification

· Searching relation between XP practice and CMMI process area
· CMMI process area selection through the XP

Phase 2: Process Definition

· Process removal and the integration which are unnecessary
· Define activity and process, which based XP

Phase 3: Process Model Development

· Consider Relationships among Process areas of CMMI
· Developing model of XP project cycle

Phase 4: Process Modeling using UML

Phase 5: System Implementation

Fig. 5. Development procedure of enterprise process model

to apply CMMI to agile and XP methodology[11]. We suggest that it can be resolved the limitation using agile methodology in small and medium development team to achieve the certain level of CMMI.

We suggest several phases to develop the enterprise model using XP and CMMI frameworks as in figure 5.

Phase 1. Define Process and make model to refer the extreme programming and CMMI.
Phase 2. Identify process from CMMI process area and extreme programming practices.
Phase 3. Make agility model in accordance with CMMI level 2 and 3.

3.1 Process Identification

It is impossible achieving the CMMI level only by extreme programming practices. It needs do appropriately tailoring into extreme programming practice[3]. We describe the extreme programming with twelve practices in the process identification.

Planning game - It is a release planning and iteration planning through the story of customer to estimate the project velocity and matches to the SG1Establish Estimate and SG2 Developer Project Plan of Project Plan[2].

Small release - It will be able to provide the product which new version is simple to the customer. During short period, it matches to SP1.3 in Define Project Lifecycle of Project Plan[2].

Metaphor, Simple design - Metaphor is prototype to describe how a whole system is operating on architectural spike and spike in extreme programming project. Simple design means a simplicity designs to make a possible satisfying story. There is no exactly matching to CMMI process to complete the stories. SG2, Develop the Design of Technical Solution, refers to match about the design metaphor.

Testing - There are acceptance test and unit test in extreme programming project. Acceptance test is black box system test, which created by customer story. It assures the quality testing by customers after release with iteration. Quality assurance is an essential part of the XP process. It is related to the SG1 as Objectively Evaluate Process and Work Products of Process Product Quality Assurance. Unit test is perfectly executed by automation. SG3 implements the Product Design of Technical Solution explicitly by describing the unit test contents. Collective ownership and refactoring are second issues to be solved in the unit test.

Refactoring - It is started with unit test to improve design. This practice is related to SP2.4 Perform Make, Buy, or Reuse Analyses of Technical Solution cause by unit test for reuse.

Pair programming and Collective ownership - Pair programming is a software development technique in which two programmers work together at one keyboard. One types in code while the other reviews each line of code as it's typed in. The person typing is called the driver. The person reviewing the code is called the observer or navigator. The two programmers switch roles frequently. While reviewing, the observer also considers the strategic direction of the work, coming up with ideas for improvements and likely future problems to address. This frees the driver to focus all of his or her attention on the "tactical" aspects of completing the current task, using the observer as a safety net and guide. Collective Code Ownership encourages everyone to contribute new ideas to all segments of the project. Any developer can change any line of code to add functionality, fix bugs, or refactoring. Collective ownership is encouraged value of extreme programming[6]. It is appeared by SG2 Perform Peer Reviews of Verification and is

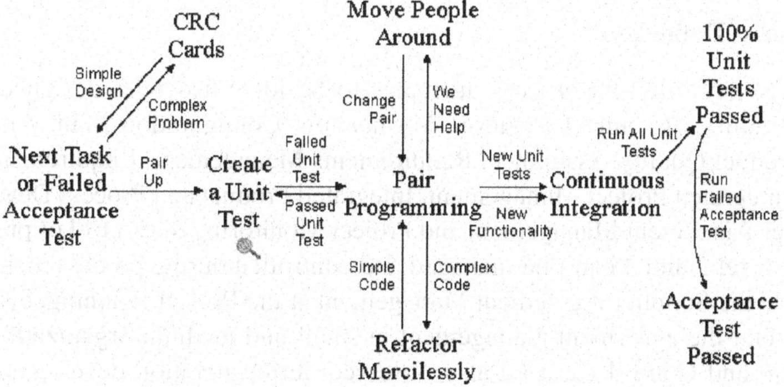

Fig. 6. Collective Code Ownership

described by team and operation on Integrated Project Management, Integrated Product and Process Development.

Continuous integration - Continuous Integration is a software development practice where members of a team integrate their work frequently, usually each person integrates at least daily - leading to multiple integrations per day. Each integration is verified by an automated build (including test) to detect integration errors as quickly as possible. Asynchronous method builds day by day and synchronous method integrates when finished episode by pair programming. This practice is supported by tools through SP1.2 Establish a Configuration Management System of Configuration Management.

Sustainable pace - Extreme Programming teams are in it for the long term. It's pretty well understood these days that death march projects are neither productive nor produce quality software. CMMI process does not include about working time. However, this practice is supported by SG1 Align Measurement and Analysis Activities of Measurement and Analysis. It's practices measure and analysis the defined process with this phase.

On-site customer - As part of presenting each desired feature, the XP Customer defines one or more automated acceptance tests to show that the feature is working. The team builds these tests and uses them to prove to themselves, and to the customer, that the feature is implemented correctly. This practice is described the Process management area in CMMI framework explicitly.

Coding standards - XP teams follow a common coding standard, so that all the code in the system looks as if it was written by a single - very competent - individual. The specifics of the standard are not important: what is important is that all the code looks familiar, in support of collective ownership. Code must be formatted to agree coding standards. This practice is added by Technical Solution because agreement CMMI process. Except these twelve practices, Requirement of Customer is discovered by User Stories in extreme programming project. This practice is identified by Requirements Management and Requirements Development of CMMI process.

3.2 Process Definition

There are many different process attributes to be identified over designed, such as Project Planning, Technical Solution, Verification, Configuration Management, Process & Product Quality Assurance, Requirements Management, Requirements Development, Integrated Project Management, Integrated Product and Process Development.

Supplier Agreement Management and Project Monitoring & Control of process area in CMMI level 2 and 3 can't be managed and controlled during process identification. We include the Supplier Agreement Management at the Project Planning, because it is not important the agreement management in small and medium organization. Project Monitoring and Control can't be also observed during iteration development. But it defines the management process with created preplan at unit test. This practice is described at SG1 Monitor Project against Plan in CMMI Project Monitoring & Control explicitly. Table 1 shows the activity and practices of enterprise process during development life cycle.

Table 1. Activity or Practice of process in our model

Process Area	Activity or Practice
Requirements Management	1. User Stories written
Project Planning	1.Planning Game
	2.Small Releases
	3.Supplier Agreement Management
Project Monitoring and Control	1. Iteration development monitoring and replan
Technical Solution	1. Simple Design
	2. Refactoring
	3. Coding Standards
	4. Verification
Configuration Management	1.Pair Programming
	2.Collective Ownership
	3.Continuous Integration
Configuration Management	1. Working hour per week
Measurement and Analysis	1. Stand Up Meeting
Unit Test	1. Process & Product Quality Assurance
	2. Acceptance Test
Organization Management	1. On-Site Customer

3.3 Proposal Process Model

We illustrate the enterprise process model through definition process in figure 7. This model includes the process of CMMI level 2 to perform the extreme programming project.

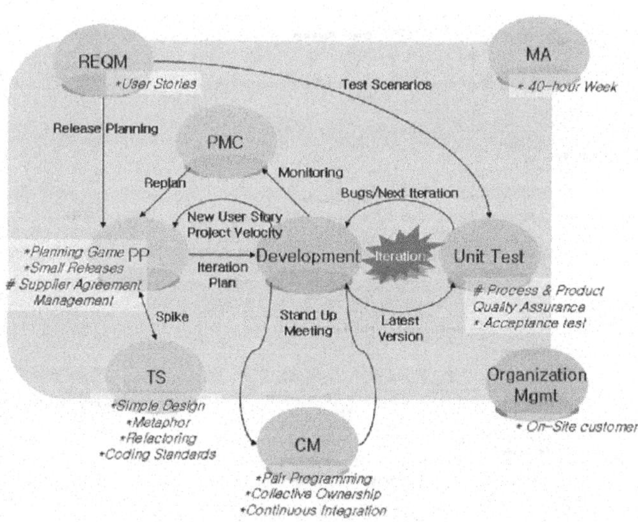

Fig. 7. Enterprise process model

The process starts from user stories written in requirements management process. Project planning is started by requirement development process or directly after gathering story card. Developer implements the related source code, and unit test is following for collected and created code through planning process. The project monitoring and control process observe these processes. At this time, configuration management process support to build a related tool. Finally, measurement and analysis process perform a measurement analysis about all process activity. Management about whole team or organization is performed by organization management process.

4 Enterprise Process Modeling

To verify our model, we design the use case diagram based the model in small and medium development organization.

Stakeholders are developer and customer. It consists of three different subsystems as requirements tool, development tool and organization management tool. We mainly focus on modeling the development tool and requirement tool not the organization management tool which focus on CMMI level 3.

Figure 8 illustrates requirement tool and development tool using UML.

Requirement management tool consists of project planning process, technical solution process and unit test process.

Story card is the message unit to communicate between developer and customer in requirement management tool. It also uses at the planning process and unit test process.

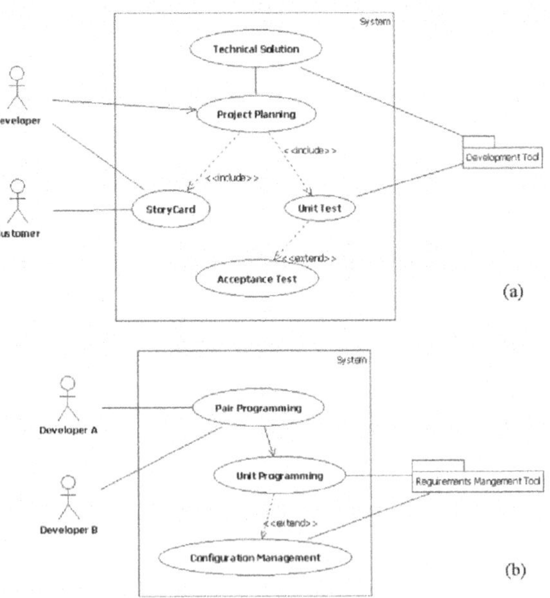

Fig. 8. Use case diagram

Unit test process is related with developer tool, which created class unit. Acceptance test is performed by story card and class unit.

Development tool supports pair and unit programming on criterion story card, which defined requirements management process. All source code is integrated by configuration management process every day. All documents in requirement management tool is created and very by managed with it.

We are studying for implementing system to the proposal process model. It is under implementing the system based on the model for verification.

5 Conclusion and Future Works

We introduce the extreme programming with CMMI level in small and medium enterprise. The CMMI process and guideline are too heavy too and also expensive to apply to extreme programming. Extreme programming is widely used by small organization. It can apply to the shorted project duration in developing software.

We describe the CMMI level through the extreme programming with comparison CMMI data. Consequently, we identify and define the process of for CMMI and extreme programming through the existing many comparison data. We illustrate new Process Model with process of CMMI level 2 to perform the extreme programming project. We also develop the process modeling using UML as a case study.

In the future, we will implement the system to apply practical case to the model. We also will study the verification strategy for this model.

References

1. Beck, K., Beedle, M., van Bennekum, A., et al.: Manifesto for Agile Software Development (2006), `http://www.agilemanifesto.org/`
2. Carnegie Mellon SEI. CMMI for Development, Version 1.2. CMU/SEI-2006-TR-008 (2006)
3. Jarvis, B., Gristock, S.P.: Extreme Programming (XP) Six Sigma CMMI (2005), `http://www.sei.cmu.edu/cmmi/presentations/sepg05.presentations/jarvis-gristock.pdf/`
4. Kent, B.: Extreme Programming Explained: Embrace change 2/E. Pearson Education, London (2005)
5. Siddiqi, J.: eXtreme Programming Pros and Cons: What Questions Remain? IEEE Computer Society Dynabook (2000), `http://computer.org/seweb/dynabook/Index.htm/`
6. Wells, D.: Extreme Programming: A gentle introduction (2006), `http://www.extremeprogramming.org/`
7. Anteon Corporation, integrate IT architects LLC, Realizing CMMI using Enterprise Architect and UML for Process Improvement (2006), `http://www.sparxsystems.com.au/downloads/whitepapers/RealizingCMMIusingEnterpriseArchitect.pdf`
8. Misic, V.B., Gevaert, H., Rennie, M.: Extreme Dynamics: Modeling the Extreme Programming Software Development Process. In: Proceedings of ProSIm 2004 workshop on Software Process Simulation and Modeling, pp. 237–242 (2004)
9. Acuna, S.T., Ferre, X.: Software Process Modeling. ISAS-SCI, pp. 237–242 (2001)

10. Robillard, P.N., Kruchten, P.: A Process Engineering Metamodel. Software Processes with the Unified Process for Education (UP/EDU), p. 350. Addison Wesley Longman, Amsterdam (2001)
11. Paulk, M.C.: Extreme Programming from a CMM Perspective. IEEE Software 18(6), 19–26 (2001)
12. Beck, K.: Embracing Change with Extreme Programming. IEEE Computer 32(10), 70–77 (1999)
13. Fritzsche, M., Keil, P.: Agile Methods and CMMI: Compatibility or Conflict? e-Informatica Software Engineering Journal 1 (2007)
14. Ambler, S.: Agile Modeling: Effective Practices for eXtreme Programming and the Unified Process. Wiley Computer Publishing, Chichester (2002)
15. Boehm, B., Turner, R.: Balancing Agility and Discipline: A Guide for the Perplexed. Addison-Wesley, Reading (2003)

Lecture Sequence Management System Using Mobile Phone Messages

Toshiyuki Maeda[1], Tadayuki Okamoto[2], Yae Fukushige[3], and Takayuki Asada[4]

[1] Faculty of Management Information, Hannan University, 5-4-33, Amamihigashi, Matsubara, Osaka 580-8502 Japan
maechan@hannan-u.ac.jp

[2] Faculty of Law and Letters, Ehime University, 3 Bunkyocho, Matsuyama, Ehime 790-8577 Japan
tadayuki@ll.ehime-u.ac.jp

[3] Graduate School of Economics, Osaka University, 1-7, Machikaneyama, Toyonaka, Osaka 560-0043, Japan
ppr-mint@r7.dion.ne.jp

[4] Graduate School of Economics, Osaka University, 1-7, Machikaneyama, Toyonaka, Osaka 560-0043, Japan
asada@econ.osaka-u.ac.jp

Summary. We present an e-mail-based lecture support system; This system consists of; Attendance management subsystem, Attendance history management subsystem, Short examination management subsystem, Questionnaire subsystem, and Assignment delivery subsystem. Both students and teachers mainly use only e-mail functions and it can access the serve. This system can be used regardless of terminal models only if mails that can be sent and received through the Internet. As to managing lecture sequences, we introduce session-based e-mail communication architecture. In this paper, the outline of this system is described, and the functions and effects are discussed.

1 Introduction

A major problem of universities in Japan is reduction of 18 year-old population. The rate of aged people is getting higher and higher, and accordance with the changes of this population composition, 18 year-old population is decreasing as well. The population of people of 18-year-old was 1,750,000 in 1995, and it is predicted that the population of 18 year-old will be 1,210,000 in 2008. Thus the decreasing rate of population of 18 year-old people will be around 30 percent. The population decrement of 18 year-old people significantly influences education of universities in Japan. The most significant effect is that average academic abilities of students are falling down. Therefore teaching staffs should work variously because of various levels of students. Teachers have to arrange contents of lectures corresponding to the various levels. Lectures should be cared for various abilities while keeping general academic level. Teachers thus should do no longer simple lectures. Especially, keeping up motivation of students is very important in education at universities. Network-based learning may be a solution to keep the motivation of the students with various abilities. In addition, students should

R. Lee and H.-K. Kim (Eds.): Computer and Information Science, SCI 131, pp. 181–187, 2008.
springerlink.com

be provided various education opportunities not only in lecture time but also in any time.

We have therefore studied several systems (for instance [5, 4]), and here present an advanced e-mail-based education system named as A-mobi (version 2). This system consists of Attendance management subsystem, Attendance history management subsystem, Short examination (quiz) management subsystem, Questionnaire subsystem, and Assignment delivery subsystem. Both students and teachers use whole system functions by e-mails, or mobile phone messages, including the server configuration. Mobile phone messages can be used regardless of carriers if those can be sent and received through the Internet. In the following, the outline of this system is described, and the functions and effects are discussed.

2 Related Works

There are many useful various education systems supported by computers. A virtual collaboration space; EVE (Educational Virtual Environment) has been developed including synchronous and asynchronous e-learning services[1]. A laboratory has been built around a web-based digital model railroad platform controlled by a client-server system for education of computer science[6]. Also, a web-based system has been developed for control engineering education[7]. To put communication skills into engineering curriculum, a web-based system to integrate workplace has been developed. The purpose of the web-based systems is to establish efficient education, and to communicate sufficiently among teachers and students. In various education areas, various education problems are solved using web-based systems[2]. The most critical problem is, however, that web-based systems cannot use lecture rooms where computers are not settled for all students, and is essential for many cases. We have thus developed an e-mail-based system using mobile phones, which almost all students have in Japan. There are only few e-mail-based systems for similar purpose, such as[3].

3 E-Mail-Based System Architecture

3.1 Background

There is a positive correlation between attendance frequency and marking result of students, and a lot of teachers want to check attendance for each class as usual as possible. Not a few teachers require to check attendance every time because they think it may improve the study achievement level. With the view of higher educational service, teachers confirm attendance and the execution of the questionnaire as one of the educational policies is in increasing tendency as well. It is expected that the number of universities that make the classification according to the proficiency, and that increases basic subjects (e.g. science course).

Interactivity in a class improves students' motivation and it is difficult to improve interactivity especially at a large classroom. Equipments (response analyzer, etc.) improve interactivity in the class though it is economically difficult to introduce those equipments, though most students have mobile phones.

To check attendance, usage of attendance cards spends much time and the cost for not only the reduction of the equivalent extent at valuable class time but also the data collection and arrangement, and that causes the mistake easily in a lot of classes of the participants. Furthermore, the teacher cannot get the current attendance information in real time in any way. Spent time and cost of questionnaires are similarly required by the collection and the analysis, and real-time information is not obtained. It is difficult for both teachers and students to understand the study achievement level, the motivation, and the problem in real time.

As a prototype, the concept of our system is that a mobile phone is used as an input device. It includes several subsystem and utilizes at a whole university, and solves some sort of problems previously shown.

The system must be cost-effective and be fair in any places in a classroom because neither the personal computer nor the equipment of information processing on networks are needed. A wide application must be possible to the field of a basic education by the repetition study and improving the proficiency etc. Integration with various subsystems including the educational evaluation system must be easy. It is possible to fit needs of students of seat reservation at the library etc.

It cannot be expected to make students use information instruments in lectures in a large classroom though the introduction of educational support system. It is easy to some degree, and already has various systems in a classroom where personal computers are maintained. It may be possible that it takes time for the data collection and the arrangement when cards are used for the attendance check and short examinations, etc. in case of such lectures and a mistake may be caused. Moreover, the possibility of illegal registration comes out for independent short examinations as well.

3.2 System Concept

Figure 1 shows the basic flow of our system. The controlling function of a short examinations setting questions, and the history that synchronized with attendance, and this integration as information switching system is realized by using only e-mail functions of mobile phones. As almost all of university students have mobile phones recently, each student's extra cost (payment) is not necessary at all. As the communication for this system, it requires at most several tens characters, and so the cost is not so much high. The annual cost is a few hundreds of yen, and in the future the packet transmission of the fixed charge system will realised and then the cost will be negligible. Because e-mail can exchange information interactively, it processes by the server by the students' attendance, and using the mobile phone mail function for short examinations and the questionnaire respectively, and the system immediately replies the result to each student. A teachers also sends the instruction to the server by using the mail function of mobile phones, and receives the result of the total result with each system by the mobile phone. The personal computer is not required and so this system is applicable for all classrooms. It is only required a mobile phone as a terminal that can send and receive the e-mail, though you may use the personal computer.

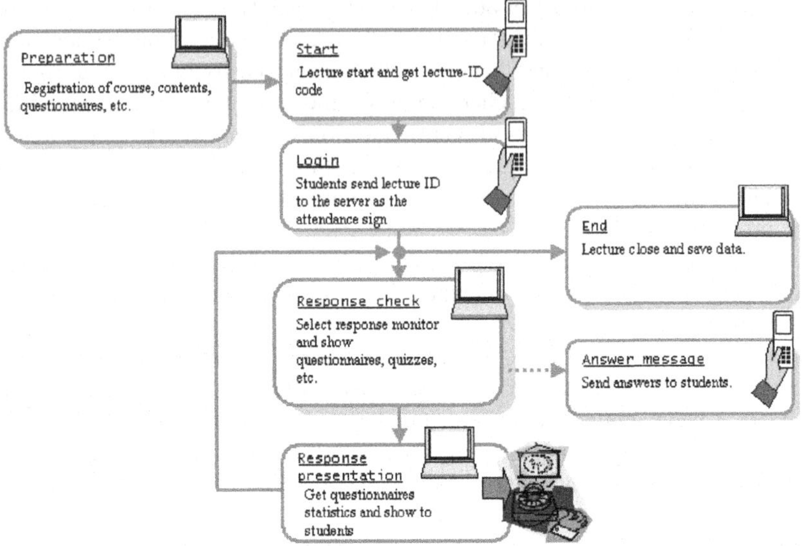

Fig. 1. System flow

3.3 Attendance Management Subsystem

Because sending mails by students themselves automatically does attendance check, teachers can understand the situation in real time. It is recorded in the attendance database that the teacher receives, not only attendance and absence but also students' registration, and because it can be received as the registration order list of names, it can be used for the nomination etc. in the class. Moreover, latecomer(s) can be easily recognized by a reverse-order list (order with late registration) of student names. Attendance can be confirmed in the registration result because it individually replies to an individual student.

Moreover, if the mistake is found in sending mails, the student gets replies of error mail, and the registration mistake is not generated because the attendance registration can be done again. Using it together with Short examination and Questionnaire subsystem can prevent injustice.

3.4 Short Examination Subsystem

Teachers make short examinations by using Short examination subsystem, as delivering as e-mails or printing the necessary number of copies. As for printed matters, when short examinations are executed, printed matters are turned upside down and distributed. The answer begins simultaneously for all students. Students fill in the student ID number, the short examination ID number, and the answer in the text of e-mail by one line and send the answer result to the server. The server automatically makes the grade and replies the result.

Because students can challenge questions repeatedly until full marks. They can put the order of examinations in the short examinations management subsystem, working on examinations for the student in a game-like sense , and as a result the improvement of the understanding level can be expected. It is possible to study short examinations at home after the lecture time even when missing it, and then continuousness of study can be maintained.

3.5 Session Management Feature

In order to manage lecture sequences, we introduce session-based e-mail communication architecture. In a general sense, several examinations and/or questionnaires are done by one lecture. In that case, the system must acknowledge the sequence of the lecture, as the examinations are related to each other. We install the feature using e-mail addresses as pseudo "cookie" data. The system confirms a sequence by sending to and receiving from one address, which can be a verification of the user. Even though "From" address may be faked, this system sends an e-mail to the indicated (=pre-registered) address and the faking is in vain.

4 Discussion

We have already used some variations of this system for 2 years and we have got some practice data from students. In this section we discuss those.

Table 1 shows some comments about students' mobile phone charges. In Japan, fixed-charge fare plan is spreading and so cost problem is getting negligible.

Table 1. Phone charges

	Jan-05	Jul-05	Jan-06	Jul-06
Fixed charge	13%	16%	30%	44%
Up to inclusive basic charge	23%	18%	28%	22%
Limit to some arbitrary price	13%	6%	6%	2%
Don't care	51%	60%	36%	31%
(Blank)	1%	0%	0%	1%

Table 2 shows students' opinions about attendance registration. Students make it by registering from the beginning as long as the attendance record is seen to fixation. Illegal attendance or late attendance of students are always monitored.

Table 3 shows opinions about short examination.

These results correspond to;

- This system deals with some sensitive questions, which is hard for students to make their own opinions public in classes,
- Students can recognize gaps between other students' understandings,
- Teachers can pick up unveiled dissatisfaction of students.

The correlation with the understanding level by the examination result (final examination and short examinations) must be investigated.

Table 2. About attendance registration (multiple answers allowed)

	Jan-05	Jul-05	Jan-06	Jul-06
Good to register by oneself	34%	56%	54%	67%
Good to check the attendance statistics by oneself	47%	46%	54%	64%
Good to be surplus attendance point	28%	16%	24%	38%
Interesting as a IT tool	23%	52%	19%	35%
Seems more interesting if improved	20%	24%	28%	40%
Need to proceed strictly about unfair registration	30%	12%	20%	21%
Dislike because of troublesome registration	20%	8%	9%	2%
Seems unnecessary on university lectures	16%	4%	6%	1%
Get harder to escape from lectures	6%	6%	4%	4%
Good to use this system at other lectures	15%	24%	24%	30%
(Others)	9%	4%	0%	0%

Table 3. About short examination

	Jan-05	Jul-05	Jan-06	Jul-06
Good to check understanding level	15%	34%	46%	51%
Good to know other students' levels and thoughts	37%	46%	37%	25%
Wants more frequently	6%	12%	6%	23%
Seems more interesting if arranged use	23%	28%	22%	38%
Interesting as a IT tool	18%	28%	11%	20%
Seems more interesting if improved	14%	24%	20%	36%
Dislike because of troublesome use	23%	10%	17%	16%
Seems unnecessary on university lectures	8%	4%	6%	5% 2
Interesting rather than talk-only lectures	17%	18%	15%	18%
Good to use this system at other lectures	3%	12%	11%	16%
(Others)	3%	6%	2%	0%

This system stimulates the participation motivation for attendance. If it is monitored, the deterrent effect is high even though some students try illegal attendance temporarily. It becomes a chance to change lecture style by touching off students because of interactive class performance. Even if teachers prepare many questionnaires, that is not a heavy task. The problem is that of the intention of teachers who try to use.

Furthermore, especially in a large classroom, it is very hard to check understanding levels of students without this system, as it is very difficult to add up answers of short examinations. This system also makes stimuli of answering for students, because the system lets students avoid pending to answer and motivate to answer quickly and correctly.

5 Conclusions

We present an e-mail-based education system and discuss some field tests, or feasibility experiments. We certify the effectiveness of our system, which consists of several subsystems.

As the future plan, we need more field tests to refine our system, and concurrently improve each subsystems, for instance, short examination functions, and so on.

Acknowledgement

This research was partially supported by the Ministry of Education, Science, Sports and Culture, Japan; Grant-in-Aid for Scientific Research (A), 19201032, 2007. This study is also supported and collaborated by A-Live Co., Ltd. The authors greatly appreciate those.

References

1. Bouras, C., Giannaka, E., Tsiatsos, T.: Virtual collaboration spaces: the eve community. In: Proceedings of the 2003 International Symposium on Applications and the Internet, Orlando, FL (USA), pp. 48–55 (2003)
2. Hanakawa, N., Goto, K., Maeda, T., Akazawa, Y.: Discovery Learning for Software Engineering –A Web based Total Education System: HInT. In: Proceedings of The International Conference on Computers in Education (ICCE 2004), Melbourne (Australia), pp. 1929–1939 (2004)
3. Johansen, D., van Renesse, R., Schneider, F.B.: Supporting Broad Internet Access to TACOMA. In: Proceedings of the 7th SIGOPS European Workshop, Connemara, Ireland, pp. 55–58 (1996)
4. Maeda, T., Okamoto, T., Miura, T., Fukushige, Y., Asada, T.: Interactive Lecture Support Using Mobile-Phone Messages. In: Proceedings of World Conference on Educational Multimedia, Hypermedia & Telecommunications (ED-MEDIA 2007), Vancouver (Canada), pp. 3659–3665 (2007)
5. Maeda, T., Tomo, M., Asada, T.: Integrated lecture-support system using e-mail. In: Proceedings of National University Symposium on Information Edication Methods (in Japanese), Tokyo (Japan), vol. 7, pp. 26–27 (2004)
6. Sanchez, P., Alvarez, B., Iborra, A., Fernandez-Merono, J.M., Pastor, J.A.: Web-based activities around a digital model railroad platform. Journal of IEEE Transaction on Education 46(2), 302–306 (2003)
7. Schmid, C., Ali, A.: A web-based system for control engineering education. In: Proceedings of the 2002 American Control Conference, Chicago, IL (USA), pp. 3463–3467 (2002)

A Robust Authentication System Using Multiple Biometrics

Md. Maruf Monwar and Marina Gavrilova

Computer Science University of Calgary Canada

Summary. In this work, a multibiometric system has been developed to overcome the drawbacks associated with monomodal biometric systems, such as noise, intra-class variability, distinctiveness, non-universality and spoof attacks. Information from three different Fisher's Linear Discriminant driven monomodal experts based on face, ear and signature biometric traits are combined through decision level fusion method. AND/OR, majority voting, weighted majority voting and behavioural knowledge space approaches of decision level fusion method are examined to achieve a higher recognition accuracy. Experimental results indicate that fusing information from multiple biometric traits can results in higher recognition rates. The system can be a contribution to homeland security or other intelligence departments.

1 Introduction

The development of a multimodal biometric system is one of the newest areas of research, where more than one biometric traits are used for identification or verification purposes. These systems overcome the drawbacks associated with unimodal biometric systems, such as noise, intra-class variability, distinctiveness, non-universality, spoof attacks, unacceptable error rates etc. [1]. As a rule, a multimodal biometric system performs significantly better than a single biometric system in most cases. But on the other side, these systems need more storage and faster processors than unimodal systems.

Information fusion is the key part in a multibiometric system. Information can be fused at all main levels of a biometric system sensor level, feature level, match score level and decision level. Fusion can also be done in rank level, after getting ranks from different matchers. Among these fusion approaches, feature level and match score level fusion methods have been deployed in many multibiometrics system by a number of researchers. However, accessing features is not allowable every time and combining different match scores from different experts is not always an easy task. In this work, we choose decision level fusion method to combine information from three different monomodal experts. This fusion method is relatively easy to implement as well as gives good recognition performance. 'AND'/'OR' method, majority voting method, weighted majority voting method, behavioral knowledge space method of the decision level fusion approach are used in this system. Then performances of these systems are compared to the performance of monomodal systems.

The main goal of our research is to improve the recognition performance of monomodal biometric system by incorporating multiple biometric traits. Usually,

R. Lee and H.-K. Kim (Eds.): Computer and Information Science, SCI 131, pp. 189–201, 2008.
springerlink.com

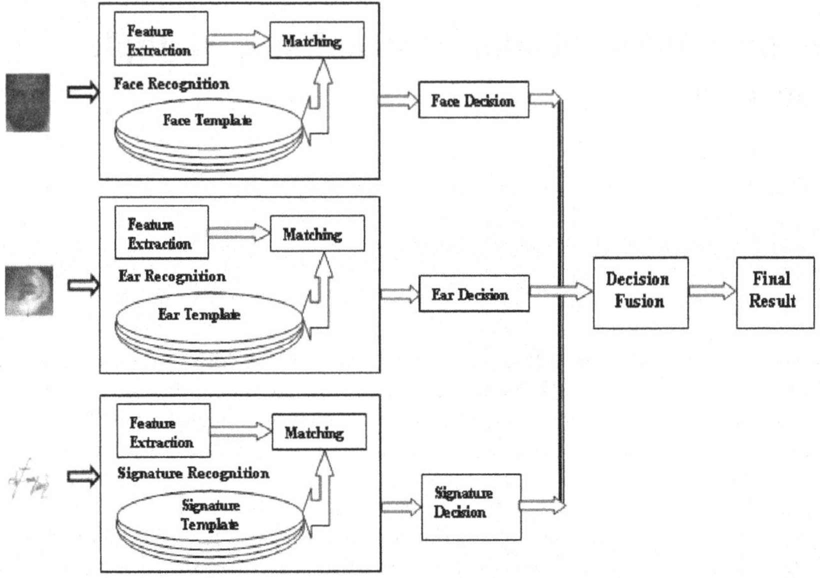

Fig. 1. Block diagram of the proposed multibiometric system

performance of a biometric system is expressed by some parameters, such as false acceptance rate (FAR), which is defined as the probability of an impostor being accepted as a genuine individual, false rejection rate (FRR), which is defined as the probability of a genuine individual being rejected as an impostor, genuine acceptance rate (GAR) etc. [2]. In our system, we documented our results through FAR and FRR for monomodal and multibiometric systems.

The block diagram of the proposed multibiometric system is given in figure 1. The system has three recognition units for face, ear and signature. After enrolling the biometric images, features are extracted and then matching is performed. Three predefined templates (governed by the Fisher Linear Discriminant Analysis method) are used to compare the test images. After individual matching process, three decisions come out from three matching units. At the end, these three decisions are consolidated through various decision level fusion approaches for the consensus recognition decision.

2 Related Research

Multibiometric systems have a number of benefits, such as, lower false acceptance and false rejection error rates, robustness against individual sensor or subsystem failures, handling one or more noisy traits etc. Among all of these benefits, the increase in accuracy of recognition, measured by the false acceptance and false rejection error rates, is the main focus of the majority of research done in the field.

Several approaches have already been proposed and developed for multibiometric authentication system. In 1998, Hong, L. and Jain, A. developed a PCA-based face and a minutiae-based fingerprint identification system with a fusion method at the decision level [3]. In 1999, Jain, A., Hong, L. and Kulkarni, Y. proposed a multimodal approach using fingerprint, face and speech, with fusion at the decision level [4]. In 2000, Frischholz, R. and Dieckmann, U. developed a system in which face, voice and lip movement are used. They used match score and decision level fusion with weighted sum rule and majority voting algorithm respectively [5]. Fierrez-Aguilar, J. et al. proposed a face verification system in 2003 based on a global appearance representation scheme, a minutiae-based fingerprint verification system and an on-line signature verification system based on HMM modeling of temporal functions, with fusion methods, sum-rule and support vector machine (SVM) user-independent and user-dependent, at the score level [6]. Also in that year, Kumar, A. et al. proposed a multimodal approach for palmprint and hand geometry with fusion methods at the features level by combining the feature vectors by concatenation, and the matching score level by using max-rule [7]. Ross, A. and Jain, A. again in 2003 proposed a multimodal system for face, fingerprint and hand geometry, with three fusion approaches at the matching score level [8]. There were also some PCA based multimodal biometric systems proposed in 2003. Among them, the system of Wang, Y. et al. [9], which used face and iris, and the system of K. I. Chang et. al. [27], which used 2D and 3D face images are noticeable. In 2004, Toh, K. A. et al. developed a system using hand geometry, fingerprint and voice and with weighted sum rule based match score level fusion [10]. In 2005, Snelick, R. et al. [11] developed a multimodal approach for face and fingerprint, with fusion methods at the match score level. In the same year Jain, A. and others showed some techniques for score normalization with the example of face, fingerprint and hand geometry [12].

From the above discussion, it can be concluded that many multibiometric systems with various methods and strategies have been proposed over the last decade to achieve higher accuracy performance from the multibiometric systems. Also, according to our literature review on this topic, practically no research has concentrated on investigating FLD methods performance in multibiometric systems. Thus, aiming at the same issue, i.e., to reduce false acceptance and false rejection error rates, we fill the niche and develop a multibiometric system incorporating three unimodal experts for face, ear and signature. The system is based on FLD methods and decision level fusion approach to consolidate the outputs produced by three unimodal experts. The results obtained after applying decision level fusion method are then compared to the results obtained from monomodal experts.

3 Decision Level Fusion

The key to a successful multibiometric system is in an effective fusion scheme, necessary to combine the information presented by multiple domain experts. The goal of fusion is to determine the best set of experts in a given problem domain and devise an appropriate function that can optimally combine the decisions rendered by the individual experts. Fusion can be employed in different levels of a multimodal biometric system. In this paper, we use fusion at the decision level.

In decision level fusion, the outcomes of various matchers are consolidated in the decision level to find out the final decision or authentication result. Specifically, when

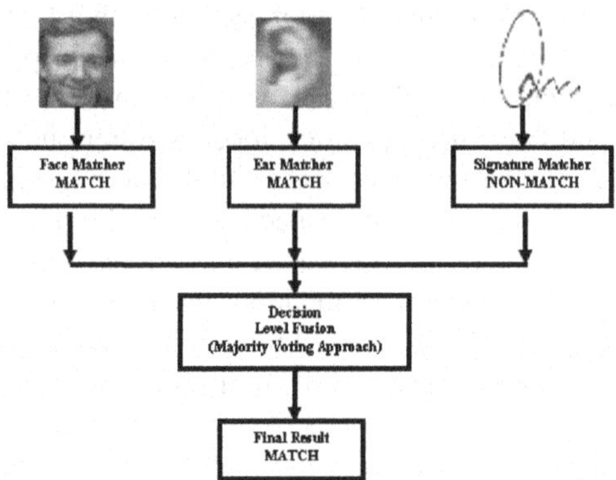

Fig. 2. Example of the majority voting approach of decision level fusion method

only the final outcomes of individual matchers are available, decision level fusion is the only feasible fusion method at that time. Many researchers proposed various methods for decision level fusion. In this system, we have used 'AND'/'OR' rules [13], majority voting rules [14], weighted majority voting rules [15] and behavior knowledge space rules [16].

'AND' and 'OR': These are the simplest method used in combining decisions of various matchers in the final decision level fusion module. In 'AND' approach, when all the matchers output 'match' for a certain sample, the final output will be 'match'. In case of 'OR' approach, when one of the matching modules output 'match', the final output is a 'match' for that certain sample. We will have very high 'False Reject Rate' (FRR) and relatively low 'False Acceptance Rate' if we use 'AND' approach for decision fusion. For 'OR' approach, FRR is significantly lower than FAR. Individual matchers with different recognition rate can be problematic for this system [2].

Majority voting: This is the most common and intuitive method in which the input sample is assigned to that identity on which majority of the matchers agree. Figure 2 is a simple illustration of this approach, where, two of the three matchers agree that the input sample is matched with the template. So, the output will be 'genuine' in this case. For odd number of matchers, the final result will be the result of the maximum matchers, i.e. results of at least half of (the total no. of matchers + 1) matchers. For even matchers, the final output will be the result of at least half of the total no. of matchers plus one. When the match and mismatch output comes out from same no. of matchers, we can then choose any one from match or mismatch [2].

Weighted majority voting: If the recognition accuracy of different matchers varies significantly, weighted majority voting approach is used. In this approach different weights are assigned to the decision of different matchers. Higher weights are assigned to the decisions made by the more accurate matchers. The recognition procedure is almost similar to the majority voting approach procedure, except that the weights of the decision of individual matchers are also considered in this case [2].

Behavior Knowledge Space: In this method, a look-up table that maps the decisions of the multiple matchers to a single decision is developed using the training data. The advantage of this method is that it takes into account the relative performance of the matchers and the correlation between the matchers. Large number of training samples required is the limitation of this approach. The BKS method is widely used when we have a large number of training samples [2].

4 Multibiometric System Development

This section deals with the development procedures of the proposed multimodal biometric system. Fisherimage technique is used in this system for enrollment and recognition of biometric traits.

We have considered all biometric traits as images. For images, there are basically two types of recognition approaches – appearance-based and model-based. Among these, appearance-based systems are more popular. Principal Component Analysis (PCA) [17,18], Independent Component Analysis (ICA) [19], Linear Discriminant Analysis (LDA) [20,21] are the examples of appearance- based recognition approaches.

Among all of the above approaches, PCA, which is a statistical method involving analysis of n-dimensional data, has been studied extensively. Although, one popular technique of PCA, 'eigenimage' is used by many researchers in the last two decades for biometric system development, but it has some limitations too. This technique is very sensitive to image conditions such as background noise, image shift, occlusion of objects, scaling of the image, and illumination change. When substantial changes in illumination and expression are present in face image, much of the variation in the data is due to these changes [22]. The eigenimage technique, in this case, cannot give the highly reliable results.

For the above reasons, we propose to utilize fisherimage approach introduced by Belhumeur et al. [20] in order to achieve higher recognition rate. Due to certain illumination changes in the images of the databases used in this work, a fisherimage based recognition method is developed for three monomodal experts. The fisherface method uses both PCA and LDA to produce a subspace projection matrix, similar to that used in the eigenface method. However, the fisherface method is able to take advantage of within-class information, minimizing variation within each class, yet still maximizing class separation [23].

Recognition using fisherimage

Fisherimage technique, which is a very powerful tool for recognition, used in our system for training and recognition of biometric images and accomplished by both PCA and FLDA (Fisher's Linear Discriminant Analysis) approaches. The reason for considering fisherimage technique is to be able to receive the best recognition performance based on intra-class variations. This subsection illustrates the fisherimage technique as monomodal recognizer.

For this method, we expand our training set of biometric images (face or ear or signature) to contain multiple images of each person, providing examples of how a person's image may change from one image to another due to variations in lighting conditions, and even small changes in orientation. We define the training set as

$$\text{Training set} = \left\{ \underbrace{\Gamma_1\Gamma_2\Gamma_3\Gamma_4\Gamma_5}_{X_1}\underbrace{\Gamma_6\Gamma_7\Gamma_8\Gamma_9\Gamma_{10}}_{X_2}\underbrace{\Gamma_{11}\Gamma_{12}\Gamma_{13}\Gamma_{14}\Gamma_{15}}_{X_3}\underbrace{\Gamma_{16}\Gamma_{17}\cdots}_{X_4}\cdots\underbrace{\Gamma_M}_{X_c} \right\}$$

Where Γ_i an biometric image and the training set is partitioned into c classes, such that, all the images in each class X_i are of the same person and no single person is present in more than one class (according to Belhumeur [20]).

Then we compute two scatter matrices, representing the within-class (S_w), between-class (S_b) and total (S_t) distribution of the training set through image space.

$$S_W = \sum_{i=1}^{c} \sum_{\Gamma_i \in X_i} (\Gamma_k - \Psi_i)(\Gamma_k - \Psi_i)^T$$

$$S_B = \sum_{n=1}^{M} |X_i| (\Gamma_i - \Psi)(\Gamma_i - \Psi)^T$$

$$S_T = \sum_{n=1}^{M} (\Gamma_n - \Psi)(\Gamma_n - \Psi)^T$$

Where $\Psi = \frac{1}{M} \sum_{n=1}^{M} \Gamma_n$ is the average image vector of the entire training set,

and $\Psi_i = \frac{1}{|X_i|} \sum_{\Gamma_i \in X_i} \Gamma_i$ is the average of each individual class X_i (person).

By performing PCA on the total scatter matrix S_t, and taking the top $M - c$ principal components, we produce a projection matrix U_{pca}, which is used to reduce the dimensionality of the within-class scatter matrix before computing the top c-1 eigenvectors [18] of the reduced scatter matrices, U_{fld} as shown below:

$$U_{fld} = \max \left\{ \frac{\left| U^T U_{pca}^T S_B U_{pca} U \right|}{\left| U^T U_{pca}^T S_W U_{pca} U \right|} \right\}$$

Finally, the matrix U_{ff} is calculated to project a biometric image into a reduced space of $c - 1$ dimensions, in which the between class scatter is maximized for all c classes, while the within-class scatter is minimized for each class X_i:

$$U_{ff} = U_{fld} U_{pca}.$$

Once the U_{ff} matrix has been constructed, it is used as the projection matrix (fisherspace). The components of the projection matrix can be viewed as images, referred to as fisherimages. In our system, three projection matrices have been created for face, ear and signature. Figure 3 shows the components of the projection matrix for face images, which is defined as fisherfaces.

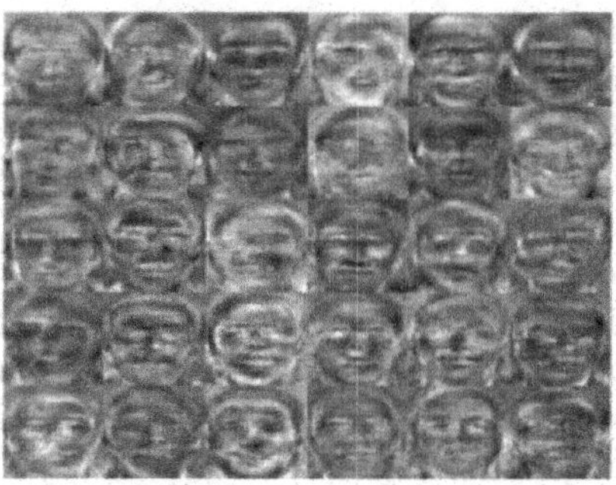

Fig. 3. Sample fisherfaces

For recognition, we project the test biometric image into the fisherspace. Then we measure the distance between the unknown image's position in fisherspace and all the known images' positions in fisherspace. At last, we select the image closest to the unknown image in the fisherspace (the distance is the lowest and less than a predefined threshold value) as the match. If no image is found whose distance is less than the threshold, a mismatch occurs.

Fusing decision outputs

After obtaining the results from three monomodal experts, various decision level fusion approaches are used to consolidate those results. In this system, 'AND'/'OR' method, majority voting method, weighted majority voting method, behavioral knowledge space method of the decision level fusion approach are used. For weighted majority voting approach, different weights are assigned to the result of different matchers. Higher weights are assigned to the decisions made by the more accurate matchers. We use 0.9 as weights of the decisions from signature matchers, 0.8 for decisions from face matchers and 0.5 for weights for decisions from ear matchers. These weights are assigned after examining the outputs of the monomodal experts and can b changed by the user in run time. For behavioural knowledge space methods, the number of training samples is also considered for fusion.

5 Database Used

Training dataset plays a very important role in achieving better recognition performance from a biometric system. In a multibiometric system, it is quite often that the dataset used is not the true dataset ((because the cost and effort associated with it), i.e., different biometric traits are collected from the same person, rather they are virtual dataset which contains records which are created by consistently pairing a user from one unimodal database (e.g., face) with a user from another database (e.g., iris) [2]. The creation of virtual users is based on the assumption that different biometric traits of the same person are independent. In this work, we use a virtual dataset which contains data from three unimodal datasets fro face, ear and signature respectively.

Among these three separate databases, the databases which are used for face and ear are from public domain and available from the web. For face, we have used the Olivetti Research Lab (ORL) Database [24], which contains 400 images, 10 of each 40 different subjects. The subjects are either Olivetti employees or Cambridge students (both male and female) and have age ranges from 18 to 81 (but majority is from age 20-35). This database was collected between 1992 and 1994 with no restrictions imposed on expression (only limited side movement and limited tilt were accepted). Most of the subjects were photographed at different times and with different lighting conditions, but always with a dark background. Some subject image captured with or without glasses. The images were 8-bit grey scale image and have 92 x 112 pixels resolution.

For ear, another public domain database [25] is used. The database contains 102 grayscale images (6 images for 17 subjects) in PGM format. The images were captured in May 1995 with a grey scale CCD camera Kappa CF 4 (focal 16 mm, objective 25.5 mm, f-number 1.4-16) and a personal computer (with Intel 486 processor) using the program Vitec Multimedia Imager for VIDEO NT v1.52 Jan 15th 1995 for Windows. Each raw image had a resolution of 384 x 288 pixels and 256-bit grey scales. The camera was at around 1.5 m from the subject. Six views of the left profile from each subject were taken under uniform, diffuse lighting. Slight changes in the head position were encouraged from image to image.

There were 17 different subjects, all students or professors at the Faculty of Informatics of the Technical University of Madrid. The raw images were then cropped and rotated for uniformity (to a ratio height:width of 1.6), and slightly brightened (gamma = 1.5 approx.), using the xv program in a Linux system.

For signature, we have used University of Rajshahi signature database. The database consists of 300 signatures with 10 signatures of 30 individuals. Then those signatures were scanned with an Epson scanner. In a scanned image, it is common to have some pixels of lower concentration in the position of white portion of the image due to various reasons such as fault of scanner. For this, high pass filtering is used to remove this noise from the input image [26]. Then only signature portion is cropped automatically by scanning each pixels from left, right, up, bottom of the captured signature images and saved into image file (.bmp and .jpg) of equal size (100x100 pixels).

To build our virtual multimodal dataset, we have randomly chosen 102 face images from 17 randomly chosen persons (6 from each person). For each person, 4 face images are randomly sampled as training sample and the remaining 2 are for test sample. The technique is also applied for ear and signature database to collect 68 training samples and 34 test samples each for ear and signature. Then each sample of the face dataset

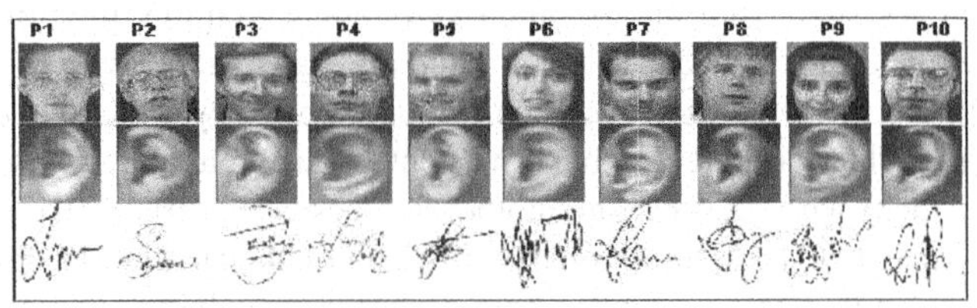

Fig. 4. Sample of our virtual multimodal database

(previously chosen) is randomly combined with one sample of ear dataset and one sample of signature dataset. Thus we obtain a virtual multimodal dataset which then is used for training and recognition purposes. Figure 4 shows sample of our virtual multimodal database based on face, ear and signature images.

6 Experiments and Results

We have implemented our multibiometric system in MATLAB 7.0 on a PENTIUM-IV windows XP workstation. The system is Graphical User Interface (GUI)-based and menu driven. The necessary image preprocessing can be easily done by just selecting the image directory. Also the threshold for recognition of face, ear and signature can be changed in run time by selecting proper menu. For convenient use of the system, the proper database, consisting of different subdirectories of training faces, ears and signatures, will be automatically connected to the system after execution. The multiple biometrics of a single identity for final result can also be chosen by only selecting the directory containing face, ear and signature images of that identity. To make the system robust, thresholds are chosen in such a way that the system can differentiate between a face and a non face image. For efficient use in later time, the system has also an action button driven option to free the used memory and clear all the selected images. Figure 5 gives the FAR chart for our multibiometric system.

From the graph, it is clear that, fusing information in a multibiometric system has significant influences on recognition performances of the system. The 'AND' method of combining various decisions outputted by various matchers have the lowest false acceptance rate (5%) among the decision level fusion approaches. Also majority voting method (9%) and weighted majority voting method (7%) have noticeable false acceptance rate and 'OR' method (22%) gives us the worst result among these techniques.

Figure 6 gives the FRR chart for our multimodal biometric system. Among the decision level fusion approaches, this time 'OR' method gives us the best performance (3%).

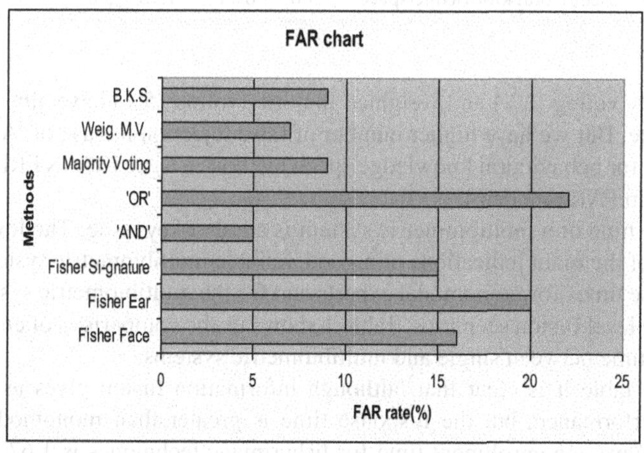

Fig. 5. Comparison between various methods in terms of FAR

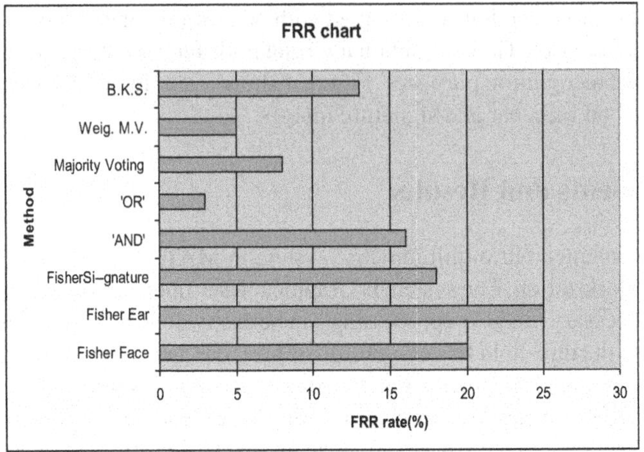

Fig. 6. Comparison between various methods in terms of FRR

Table 1. Response time comparison

Approaches	Enrollment Time (min)	Recognition Time (min)
Fisherimage	1.67 ± 0.36	0.48 ± 0.22
'AND' method	3.67 ± 0.11	0.82 ± 0.16
'OR' method	3.67 ± 0.11	0.65 ± 0.06
Majority Voting	3.67 ± 0.11	0.72 ± 0.1
Weighted majority voting	3.67 ± 0.11	0.80 ± 0.12
Behavioral knowledge space	3.67 ± 0.11	1.23 ± 0.63

Also majority voting (8%) and weighted majority voting (5%) have almost same performance rate. But we have higher number of false rejection in case of 'AND' method (16%). Also for behavioural knowledge space method, we get 13% as FRR rate, which is greater than FAR rate (9%).

Response time of a multibiometric system is another key issue. The lower response time is one of the main indications of a good, reliable multibiometric system. We compare response times for monomodal experts and for the multibiometric system for various decision level fusion scenarios. Table 1 shows us the comparison of enrollment and recognition time between single and multibiometric systems.

From the table it is clear that, although information fusion gives us better identification performance, but the response time is greater than monomodal biometric system. The average enrollment time for fisherimage techniques is 1.67 ± 0.36 minutes, whereas the enrollment time for the multimodal biometric system is 3.67 ± 0.11

minutes. Among the decision level fusion approaches, 'AND' method and weighted majority method take more time for recognition.

7 Conclusion

Recently, more investigations have been carried out in the domain of multibiometrics. Investigation of good combination of multiple biometric traits and various fusion methods to get the optimal identification results are at the focus of current research. In this paper, comparison between various FLD based multimodal biometric systems and differences between the results obtained before and after using the decision level fusion is presented. In addition, various decision level fusion methods are compared against each other.

Among all the decision level fusion approaches tested, weighted majority voting method gives us the best performance in terms of the average of false acceptance rate and false rejection rate. 'AND' method is good in case of only false acceptance rate and 'OR' method is good for only false rejection rate. But all the approaches take longer response time than unimodal biometric system.

It is worth noting that the face, ear and signature of an identity considered for training and recognition are taken from three different databases. Also, the background and the illumination conditions are varied across the databases used. As these considerations have significant influence on the effectiveness of various recognition approaches, using the unique database for all biometric traits in real time environment and incorporating dual or tri-level fusion approaches are promising future directions of research in this domain.

Acknowledgements

The research was conducted in the Biometric technologies Laboratory, University of Calgary, with support from NSERC and GEOIDE funding agencies.

References

[1] Bubeck, U.M., Sanchez, D.: Biometric authentication: Technology and evaluation, Technical report, San Diego State University, USA (2003)

[2] Ross, A., Nandakumar, K., Jain, A.K.: Handbook of Multibiometrics. Springer, New York (2006)

[3] Hong, L., Jain, A.K.: Integrating faces and fingerprints for personal identification. IEEE Transaction on Pattern Analysis and Machine Intelligence 20(12), 1295–1307 (1998)

[4] Jain, A.K., Hong, L., Kulkarni, Y.: A multimodal biometric system using fingerprint, face and speech. In: Proc. of Second Int. Conf. on Audio- and Video-based Biometric Person Authentication, Washington D.C., USA, pp. 182–187 (1999)

[5] Frischholz, R., Dieckmann, U.: BioID: A multimodal biometric identification system. IEEE Computer 33(2), 64–68 (2000)

[6] Fierrez-Aguilar, J., et al.: A comparative evaluation of fusion strategies for multimodal biometric verification. In: Kittler, J., Nixon, M.S. (eds.) AVBPA 2003. LNCS, vol. 2688, pp. 830–837. Springer, Heidelberg (2003)

[7] Kumar, A., et al.: Personal verification using palmprint and hand geometry biometric. In: Kittler, J., Nixon, M.S. (eds.) AVBPA 2003. LNCS, vol. 2688, pp. 668–678. Springer, Heidelberg (2003)

[8] Ross, A., Jain, A.K.: Information fusion in biometrics. Pattern Recognition Letters 24, 2125–2215 (2003)

[9] Wang, Y., Tan, T., Jain, A.K.: Combining face and iris biometrics for identity verification. In: Kittler, J., Nixon, M.S. (eds.) AVBPA 2003. LNCS, vol. 2688, pp. 805–813. Springer, Heidelberg (2003)

[10] Toh, K.A., Jiang, X.D., Yau, W.Y.: Exploiting global and local decisions for multi-modal biometrics verification. IEEE Transacton on Signal Processing (Supplement on Secure Media) 52(10), 3059–3072 (2004)

[11] Snelick, R., et al.: Large scale evaluation of multimodal biometric authentication using state-of the-art systems. IEEE Transaction on Pattern Analysis and Machine Intelligence 27(3), 450–455 (2005)

[12] Jain, A.K., Nandakumar, K., Ross, A.: Score normalization in multimodal biometric systems. Pattern Recognition 38, 2270–2285 (2005)

[13] Daugman, J.: Combining multiple biometrics (Retrieved on Mar 6, 2007) (2000), `http://www.cl.cam.ac.uk/users/jgd1000/combine/combine.html`

[14] Lam, L., Suen, C.Y.: Application of majority voting in pattern recognition: An analysis of its behavior and performance. IEEE Transaction on System, Man and Cybernetics - Part B: Cybernetics 34(1), 621–628 (1997)

[15] Kuncheva, L.I.: Combining pattern classifiers: Methods and algorithms. Wiley, Chichester (2004)

[16] Huang, Y.S., Suen, C.Y.: Method of combining multiple experts for the recognition of unconstrained handwritten numerals. IEEE Transaction on Pattern Analysis and Machine Intelligence 17(1), 90–94 (1995)

[17] Kirby, M., Sirovich, L.: Application of the Karhunen-Loeve procedure for the characterization of human faces. IEEE Transaction on Patten Analysis and Machine Intelligence 12(1), 103–108 (1990)

[18] Turk, M., Pentland, A.: Eigenfaces for recognition. Journal of Cognitive Science, 71–86 (1991)

[19] Bartlett, M.S., Lades, H.M., Sejnowski, T.J.: Independent component representations for face recognition. In: Proc. of Conf. on Human Vision and Electronic Imaging III, San Jose, California (1998)

[20] Belhumeur, P.N., Hespanha, J.P., Kriegman, D.J.: Eigenfaces vs. Fisherfaces: Recognition using class specific linear projection. IEEE Transaction on Pattern Analysis and Machine Intelligence 19(7), 711–720 (1997)

[21] Etemad, K., Chellappa, R.: Face recognition using discriminant Eigenvectors. In: Proc. of the IEEE Int. Conf. on Acoustics, Speech and Signal Processing, Atlanta, USA, pp. 2148–2151 (1996)

[22] Rahman, M.M., Ishikawa, S.: A robust recognition method for partially occluded/destroyed objects. In: Proc. of the 6th Asian Conf. on Computer Vision, Jeju, Korea, pp. 984–988 (2004)

[23] Heseltine1, T., et al.: Face recognition: A comparison of appearance-based approaches. In: Sun, C., et al. (eds.) Proc. of the 7th Digital Image Computing: Techniques and Applications, Sydney, Australia (2003)

[24] Samaria, F., Harter, A.: Parameterization of a stochastic model for human face identification. In: Proc. of the 2nd IEEE Workshop on Applications of Computer Vision, Sarasota, Florida (1994)

[25] Perpinan, C.: Compression neural networks for feature extraction: Application to human recognition from ear images, MSc thesis, Faculty of Informatics, Technical University of Madrid, Spain (1995)
[26] Gonzalez, R.C., Wintz, P.: Digital image processing, 2nd edn. Pearson Education Pvt. Ltd., London (2002)
[27] Chang, K.I., Bowyer, K.W., Flynn, P.J.: Face recognition using 2D and 3D facial data. In: Workshop in Multimodal User Authentication, Santa Barbara, California, pp. 25–32 (2003)

Approximate Reasoning in a Symbolic Multi-valued Framework

Amel Borgi[1,3], Saoussen Bel Hadj Kacem[2,3], and Khaled Ghedira[2,3]

[1] National Institute of Applied Science and Technology, INSAT Centre Urbain Nord de Tunis, BP 676, 1080 Tunis, Tunisia
Amel.Borgi@insat.rnu.tn

[2] National School of Computer Science, University campus Manouba, Manouba 2010, Tunisia
Saoussen.BHK@fst.rnu.tn, Khaled.Ghedira@isg.rnu.tn

[3] Research Unit SOIE-MIAD, University campus Manouba, Manouba 2010, Tunisia

Summary. We focus in this paper on approximate reasoning in a symbolic framework, and more precisely in multi-valued logic. Approximate reasoning is based on a generalization of Modus Ponens, known as Generalized Modus Ponens (GMP). Its principle is that from an observation different but approximately equal to the rule premise, we can deduce a fact approximately equal to the rule conclusion. We propose a generalization of the approximate reasoning axiomatic introduced by Fukami, and we show the weakness of GMP approaches in the multi-valued context towards this axiomatic. Moreover, we propose two rules of symbolic GMP that check the axiomatic. One is based on the implication operator and the second on linguistic modifiers.

1 Introduction

Representation and treatment of imperfect knowledge hold a major place in Artificial Intelligence research. The main reason is that most information pro- vided by the human experts are often uncertain, vague or imprecise. In the literature, several approaches have been proposed for the representation of these types of knowledge. Two of them dominate. Firstly, fuzzy logic [17] is recognized as a good tool to deal with imperfect information in knowledge based systems. In this logic, imperfect information is represented in a numerical way. Indeed, each fuzzy set A is defined on a set of reference U which must be numerical. However, we often manipulate abstract and qualitative terms that must be represented in a symbolic way. For example, the linguistic term *beautiful*, can't be represented by a fuzzy set. For this reason, another formalism is used in this case, which is multi-valued logic [1, 15]. This logic provides a symbolic representation of terms.

To perform inferences with imperfect data, Zadeh [18, 19] introduced the concept of approximate reasoning within fuzzy logic. It allows having a flexible reasoning, similar to human reasoning. Indeed, human mind presents a strong reasoning model, since he is able to reason with perfect and imperfect knowledge. Thus, human inferences do not always require a perfect correspondence between facts or causes to conclude.

Approximate reasoning is based on a generalization of Modus Ponens (MP) known as Generalized Modus Ponens (GMP). Such a reasoning allows firing a rule with a fact

R. Lee and H.-K. Kim (Eds.): Computer and Information Science, SCI 131, pp. 203–217, 2008.
springerlink.com

not exactly equal to the rule premise: the inference is allowed even with facts similar or approximately equivalent to the premise. Nevertheless, the conclusion inference is not constructed by an arbitrary way. Indeed, a set of axioms is considered to have a coherent and logic result [4, 10, 12, 14].

In this paper, we propose a generalization of the axiomatic of approximate reasoning found in literature [14]. After that, we focus our work on approximate reasoning in the multi-valued framework. Thus, we discuss approaches of GMP in the multi-valued context [1, 15] and we demonstrate that these approaches don't check the axiomatic of approximate reasoning. Our aim is then to introduce new rules of GMP which answer to this axiomatic. Two inference rules for approximate reasoning are proposed in this paper. The first one is based on linguistic modifiers [3]. The second one is an amelioration of Khoukhi model [16].

This paper is organized as follows. Section 2 is devoted to approximate reasoning. We present first the approximate reasoning axiomatic, then, and after a brief introduction of the multi-valued logic basic concepts, we illustrate the main GMP rules within this logic and we show their weakness towards the approximate reasoning axiomatic. Section 3 covers the use of linguistic modifiers in approximate reasoning and presents the Generalized Symbolic Modifiers defined in multi-valued logic, the context of our work. In section 4, we propose a GMP in multi-valued logic framework based on linguistic modifiers. In section 5, we describe the amelioration of Khoukhi model leading to a GMP checking the approximate reasoning axiomatic. Finally, section 6 concludes the study.

2 Approximate Reasoning

Fuzzy logic allows describing vague concepts, imprecision, gradual properties or uncertain events. This logic introduced by Zadeh [17] constitutes the frame of the approximate reasoning in its first meaning of a word. Approximate reasoning is based on a generalization of Modus Ponens (MP) known as Generalized Modus Ponens (GMP), GMP was initially defined in a fuzzy context [18, 19]. This rule can be expressed in standard form as follows:

$$\frac{\begin{array}{l}\text{If } X \text{ is } \mathcal{A} \text{ then } Y \text{ is } \mathcal{B}\\ X \text{ is } \mathcal{A}'\end{array}}{\hfill Y \text{ is } \mathcal{B}'} \tag{1}$$

where X and Y are two variables and $\mathcal{A}$, $\mathcal{A}'$, $\mathcal{B}$ and $\mathcal{B}'$ are predicates whose interpretation depends on the used logic. In the case of fuzzy logic, these predicates are fuzzy sets, and in the case of multi-valued logic, they are multi-sets.

2.1 Axiomatic of Approximate Reasoning

The conclusion inference "Y is $\mathcal{B}'$" of the schema (1), is not constructed by an arbitrary way. Indeed, a set of axioms is considered to have a coherent and logic result. In the approximate reasoning literature, several propositions of axioms are found [4, 10, 12, 14]. In [14], Fukami, Mizumoto and Tanaka have proposed the following axioms:

$$\begin{array}{ll} \textbf{Criterion I} & A' = A \Rightarrow B' = B \\ \textbf{Criterion II-1} & A' = \text{very } A \Rightarrow B' = \text{very } B \\ \textbf{Criterion II-2} & A' = \text{very } A \Rightarrow B' = B \\ \textbf{Criterion III} & A' = \text{more or less } A \Rightarrow B' = \text{more or less } B \\ \textbf{Criterion IV-1} & A' = \overline{A} \Rightarrow B' \text{ is indefinite} \\ \textbf{Criterion IV-2} & A' = \overline{A} \Rightarrow B' = \overline{B} \end{array} \quad (2)$$

The terms *very* and *more or less* are Zadeh's linguistic modifiers. More precisely they are precision modifiers: $very\ A = A^2$, and $more\ or\ less\ A = A^{1/2}$. They respectively provoke a reinforcement or a weakening of the fuzzy set A. One can notice that, the same observed modifier is applied to the rule conclusion.

We propose the following generalization of criteria (2) appeared in [14]:

$$\begin{array}{ll} \textbf{Criterion I} & \mathcal{A}' = \mathcal{A} \Rightarrow \mathcal{B}' = \mathcal{B} \\ \textbf{Criterion II-1} & \mathcal{A}' \succ \mathcal{A} \Rightarrow \text{the more } \mathcal{A}' \succ \mathcal{A}\text{, the more } \mathcal{B}' \succ \mathcal{B} \\ \textbf{Criterion II-2} & \mathcal{A}' \succ \mathcal{A} \Rightarrow \mathcal{B}' = \mathcal{B} \\ \textbf{Criterion III} & \mathcal{A}' \prec \mathcal{A} \Rightarrow \text{the more } \mathcal{A}' \prec \mathcal{A} \text{ the more } \mathcal{B}' \prec \mathcal{B} \\ \textbf{Criterion IV-1} & \mathcal{A}' = \overline{\mathcal{A}} \Rightarrow \mathcal{B}' \text{ is indefinite} \\ \textbf{Criterion IV-2} & \mathcal{A}' = \overline{\mathcal{A}} \Rightarrow \mathcal{B}' = \overline{\mathcal{B}} \end{array} \quad (3)$$

with: $\mathcal{A}' \succ \mathcal{A}$ means that $\mathcal{A}'$ is a reinforcement of $\mathcal{A}$;
$\mathcal{A}' \prec \mathcal{A}$ means that $\mathcal{A}'$ is a weakening of $\mathcal{A}$.

These criteria are more general than those defined in [14] as they can be applied not only with fuzzy sets but also with other kinds of predicates, such as multi-valued predicates in the multi-set framework [1]. Moreover, they provide freedom in determining the intensity of changes: they do not restrict the used modifiers to Zadeh's linguistic modifiers *very* and *more or less*. Indeed, the relations $\succ$ and $\prec$ that we have defined generalize the reinforcement and weakening expressed by the modification terms (for instance *very*, *less*, etc). The modification terms can have any intensity, and this intensity is not necessarily the same for the premise and the conclusion.

Criterion I checks the case of traditional Modus Ponens. So, when $\mathcal{A}' = \mathcal{A}$ it is necessary that $\mathcal{B}' = \mathcal{B}$. In the case where $\mathcal{A}'$ is a reinforcement of $\mathcal{A}$, there are two exclusive cases: (1) $\mathcal{B}'$ is a reinforcement of $\mathcal{B}$, i.e criterion II-1, (2) $\mathcal{B}'$ is equal to $\mathcal{B}$, i.e criterion II-2. Criterion II-1 enables to have a gradual reasoning based on the following meta-knowledge "The more X is $\mathcal{A}$, the more Y is $\mathcal{B}$". For example, let us consider the rule "If the tomato is red then it is ripe", any human knows that the more the tomato is red the more it is ripe. However, we can consider this reasoning behavior only if we already know that there is a great causality between the premise and the conclusion of the rule [14]. If not, it is necessary to have a careful behavior with considering that $\mathcal{B}' = \mathcal{B}$, and thus to apply criterion II-2. When the observation A' is less precise than the rule premise A, B' must be less precise than the conclusion of the rule B. It represents criterion III and modelizes a natural human reasoning. Criteria IV-1 and IV-2 treat the case where the observation is equal to the premise negation. Thus, the inference conclusion can be either equal to the conclusion negation, or unspecified. In our work, we are not interested in criteria IV. We thus adopt criterion IV-1.

According to these criteria, two types of approximate reasoning can be released [14]:

Type 1: criteria I, II-1, III, IV-1
Type 2: criteria I, II-2, III, IV-1

The expert can choose his own strategy according to the manipulated knowledge. Our aim in this work is to provide a GMP for every type of approximate reasoning in the multi-valued logic context [1].

2.2 Discussion on GMP in Symbolic Multi-valued Logic

The multi-valued logic defined in [1] is based on multi-sets theory. As fuzzy sets, this theory is a generalization of classic sets theory. The notion of membership in a classic set is replaced here by that of the partial membership in a multi-set. In symbolic multi-valued logic, each linguistic term (such as *large*) is represented by a multi-set [1]. To express the imprecision of a predicate, a qualifier $v_\alpha A$ is associated to each multi-set (such as *rather*, *little*, etc). When a speaker uses a statement "X is $v_\alpha A$", v_α is the degree to which X satisfies the predicate A, denoted mathematically by "$X \in_\alpha A$": the object X belongs with a degree α to a multi-set A. Thus a truth-degree τ_α must correspond to each adverbial expression v_α so that:

$$\begin{aligned} X \text{ is } v_\alpha A &\Longleftrightarrow \text{“}X \text{ is } v_\alpha A\text{” is true} \\ &\Longleftrightarrow \text{“}X \text{ is } A\text{” is } \tau_\alpha\text{-true} \end{aligned}$$

For example, the statement "John is rather tall" means that John satisfies the predicate *tall* with the degree *rather*. This theory of multi-sets can be seen as an axiomatization of the theory of fuzzy-sets. However, in multi-valued logic, membership degrees are symbolic terms of the natural language, and not reals belonging to $[0, 1]$ as in fuzzy logic. The set of symbolic truth-degrees forms an ordered list $\mathcal{L}_M = \{\tau_0, ..., \tau_i, ..., \tau_{M-1}\}$[1] with the total order relation: $\tau_i \leq \tau_j \Leftrightarrow i \leq j$, its smallest element is τ_0 (false) and the greatest is τ_{M-1} (true) [1]. In practice, the number of truth-degrees is often close to 7. The expert can even propose his own list of truth-degrees; the only restrictive condition is that they must be ordered. For example, the list of truth-degrees used in [1] for $M = 7$ is $\mathcal{L}_7 = \{$not-at-all, very-little, little, moderately, enough, very, completely$\}$. Figure 1 shows a graphic representation of the scale $\mathcal{L}_7$.

In order to aggregate truth-degrees, T-norms[2], T-conorms[3] and implicators[4] are used as in fuzzy logic. In multi-valued logic, the aggregation functions of Lukasiewicz are often used. In this context and with M truth-degrees, they are defined by:

$$T_L(\tau_\alpha, \tau_\beta) = \tau_{max(0, \alpha+\beta-M+1)} \tag{4}$$

[1] With M a positive integer not null.

[2] In multi-valued logic, a T-norm is any symmetric, associative, increasing $\mathcal{L}_M \times \mathcal{L}_M \to \mathcal{L}_M$ mapping T satisfying $T(\tau_{M-1}, \tau_\alpha) = \tau_\alpha \; \forall \, \tau_\alpha \in \mathcal{L}_M$.

[3] In multi-valued logic, a T-conorm is any symmetric, associative, increasing $\mathcal{L}_M \times \mathcal{L}_M \to \mathcal{L}_M$ mapping S satisfying $S(\tau_0, \tau_\alpha) = \tau_\alpha \; \forall \, \tau_\alpha \in \mathcal{L}_M$.

[4] In multi-valued logic, an implicator is any $\mathcal{L}_M \times \mathcal{L}_M \to \mathcal{L}_M$ mapping $\mathcal{I}$, whose first and second partial mappings are decreasing, respectively increasing, satisfying $\mathcal{I}(\tau_0, \tau_0) = \mathcal{I}(\tau_0, \tau_{M-1}) = \mathcal{I}(\tau_{M-1}, \tau_{M-1}) = \tau_{M-1}$.

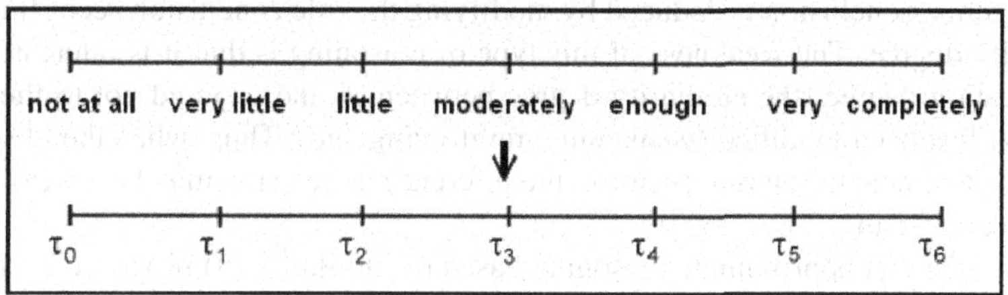

Fig. 1. A scale of symbolic truth-degrees

$$S_L(\tau_\alpha, \tau_\beta) = \tau_{min(M-1,\alpha+\beta)} \tag{5}$$

$$\mathcal{I}_L(\tau_\alpha, \tau_\beta) = \tau_{min(M-1,M-1-\alpha+\beta)} \tag{6}$$

In fuzzy logic, reinforcement and weakening of knowledge are modelled by fuzzy sets inclusion or translation. However in multi-valued logic, these notions are expressed by the modification of the truth-degree of the same multi-set. So the reinforcement of a multi-set A is represented by the increase of its truth-degree, and its weakening is represented by the reduction of its truth-degree.

The first generation of Generalized Modus Ponens in the multi-valued framework was proposed by Akdag and al. [1]:

$$\begin{array}{l} \text{if “}X\text{ is }A\text{” then “}Y\text{ is }B\text{” is }\tau_\alpha\text{-true} \\ \text{“}X\text{ is }A\text{” is }\tau_\beta\text{-true} \\ \hline \qquad\qquad \text{“}Y\text{ is }B\text{” is }\tau_\gamma\text{-true with }\tau_\gamma = T(\tau_\alpha, \tau_\beta) \end{array} \tag{7}$$

where A and B are multi-sets, τ_α, τ_β and $\tau_\gamma \in \mathcal{L}_M$ and T is a T-norm.

This rule of Generalized Modus Ponens allows inferring with an observation represented by the same multi-set as the premise, but with different truth-degrees. Let us notice that the premise of the rule in this case must be completely true (its truth-degree is τ_{M-1}). This type of rules is called free multi-valued rule. It takes the form: "X is A" then "Y is B" is τ_α-true. There are other types of rules such as strong rules whose premise and conclusion are accompanied by truth-degrees. A strong rule has the following form: If "X is $v_\alpha A$" then "Y is $v_\beta B$"[5]. We remark that strong rules whose premise is completely true can be regarded as free rules, so strong rules generalize free rules. For this reason, in our work on GMP we consider only strong rules. Khoukhi proposed a Generalization of Modus Ponens for this type of rules [16]:

$$\begin{array}{l} \text{If “}X\text{ is }v_\alpha A\text{” then “}Y\text{ is }v_\beta B\text{”} \\ \text{“}X\text{ is }v_\gamma A\text{”} \\ \hline \qquad\qquad \text{“}Y\text{ is }v_\lambda B\text{” with }\tau_\lambda = T(Sim(\tau_\alpha, \tau_\gamma), \tau_\beta) \end{array} \tag{8}$$

where Sim is a similarity relation of truth-degrees defined as follows:
$Sim(\tau_\alpha, \tau_\beta) = Min\{\mathcal{I}(\tau_\alpha, \tau_\beta), \mathcal{I}(\tau_\beta, \tau_\alpha)\}$, with $\mathcal{I}$ an implication operator.

The models (7) and (8) are based on similarity. Their principle is to calculate a degree which represents the similarity between the premise and the observation. Next,

[5] Denoted also: If "X is A" is τ_α-true then "Y is B" is τ_β-true.

the inference conclusion is deduced by modifying the rule conclusion according to this similarity degree. The weakness of this type of reasoning is that it is interested in the modification degree (the similarity degree between $\mathcal{A}$ and $\mathcal{A}'$) and not to the way of which A has been modified (weakening, reinforcing, etc). Thus, when the observation is a reinforcement of the rule premise, the inference conclusion may be a weakening of the rule conclusion.

We notice that approximate reasoning based on similarity (8) proposed by Khoukhi [16] does not check the axiomatic of approximate reasoning (3), more precisely criterion II is not verified. Indeed, when the observation is a reinforcement of the rule premise, the inference conclusion may be a weakening of the rule conclusion. More precisely, we find the following results:

Proposition 1. *The GMP with strong rules (8) proposed by Khoukhi [16] verifies criterion I of the approximate reasoning axiomatic (3).*

Proof. Criteria I concerns the case where A' is equal to A, i.e. $\tau_\gamma = \tau_\alpha$. Let us demonstrate that in that case, the degree of the inference conclusion B' is equal to B, i.e. $\tau_\lambda = \tau_\beta$. The degree τ_λ is obtained by $T(Sim(\tau_\alpha, \tau_\gamma), \tau_\beta)$. We know that every implication operator verifies the following property: if $\tau_\alpha \leq \tau_\beta$ then $\tau_\alpha \rightarrow \tau_\beta = \tau_{M-1}$.

As $\tau_\gamma = \tau_\alpha$, we have:
$Sim(\tau_\alpha, \tau_\gamma) = Min\{\mathcal{I}(\tau_\alpha, \tau_\gamma), \mathcal{I}(\tau_\gamma, \tau_\alpha)\} = Min\{\tau_{M-1}, \tau_{M-1}\} = \tau_{M-1}$
$\Rightarrow \tau_\lambda = T(Sim(\tau_\alpha, \tau_\gamma), \tau_\beta) = T(\tau_{M-1}, \tau_\beta)$.

The neutral element of every T-norm in the multi-valued logic is τ_{M-1}, so $T(\tau_{M-1}, \tau_\beta) = \tau_\beta$. We conclude that $\tau_\lambda = \tau_\beta$.

Proposition 2. *The GMP with strong rules (8) proposed by Khoukhi [16] does not verify criterion II-1 of the approximate reasoning axiomatic (3).*

Proof. Criteria II-1 concerns the case where A' is a reinforcement of A, i.e. $\tau_\gamma > \tau_\alpha$ according to (8). Let us show that the degree τ_λ of the inference conclusion B' is not higher than the degree τ_β, i.e. that B' is not a reinforcement of B.

The degree τ_λ is given by $T(Sim(\tau_\alpha, \tau_\gamma), \tau_\beta)$. Considering $\tau_\theta = Sim(\tau_\alpha, \tau_\gamma)$, we obtain $\tau_\lambda = T(\tau_\theta, \tau_\beta)$.

We know that for every T-norm T, $T(x, y) \leq min(x, y)$. Consequently: $T(\tau_\theta, \tau_\beta) \leq min(\tau_\theta, \tau_\beta) \leq \tau_\beta \Longrightarrow \tau_\lambda \leq \tau_\beta$. We deduce that $\tau_\lambda \leq \tau_\beta$ for every premise degree τ_γ, thus criterion II-1 can never be checked.

Proposition 3. *The GMP with strong rules (8) proposed by Khoukhi [16] does not verify criterion II-2 of the approximate reasoning axiomatic (3).*

Proof. Criteria II-2 concerns the case where A' is a reinforcement of A, i.e. $\tau_\gamma > \tau_\alpha$ according to (8). Let us show that the degree τ_λ of the inference conclusion B' is not equal to the degree τ_β, i.e. that B' is not equal to B. In other words, given $\tau_\gamma > \tau_\alpha$, can we conclude that $\tau_\lambda = \tau_\beta$?

The used relation Sim is a similarity relation that verifies the following condition:
$Sim(\tau_\alpha, \tau_\gamma) = \tau_{M-1}$ if $\tau_\alpha = \tau_\gamma$, else $Sim(\tau_\alpha, \tau_\gamma) < \tau_{M-1}$.

Our hypothesis is that $\tau_\gamma > \tau_\alpha$. Considering $\tau_\theta = Sim(\tau_\alpha, \tau_\gamma)$, we have: $\tau_\theta = Sim(\tau_\alpha, \tau_\gamma) < \tau_{M-1}$. We consider the particular case where $\tau_\beta = \tau_{M-1}$. We have demonstrate that $\tau_\theta < \tau_{M-1}$, so $\tau_\theta < \tau_\beta$. Also, we know that the Zadeh's T-norm min is the biggest T-norm, thus: $T(\tau_\theta, \tau_\beta) \leq min(\tau_\theta, \tau_\beta) = \tau_\theta$. In this particular case where the rule conclusion is precise, $\tau_\lambda = T(\tau_\theta, \tau_\beta) \leq \tau_\theta < \tau_\beta$. So $\tau_\lambda \neq \tau_\beta$. This particular case is enough to prove that criterion II-2 is not checked.

Proposition 4. *The GMP with strong rules (8) proposed by Khoukhi [16] verifies criterion III of the approximate reasoning axiomatic (3).*

Proof. Criteria III concerns the case where $\mathcal{A}'$ is a weakening of $\mathcal{A}$, i.e. $\tau_\gamma < \tau_\alpha$. This criterion ensures that the more A' is a weakening of A, the more B' is a weakening of B. Let us consider two different observations which weaken the premise and which have the degrees τ_{γ_1} and τ_{γ_2}, with $\tau_{\gamma_1} < \tau_{\gamma_2}$. Let us demonstrate that the degrees τ_{λ_1} and τ_{λ_2} of the inference conclusions $\mathcal{B}'_1$ and $\mathcal{B}'_2$ are so that $\tau_{\lambda_1} \leq \tau_{\lambda_2} \leq \tau_\beta$.

The degree τ_{λ_1} is obtained by $T(Sim(\tau_\alpha, \tau_{\gamma_1}), \tau_\beta)$ with $Sim(\tau_\alpha, \tau_{\gamma_1}) = Min\{\mathcal{I}(\tau_\alpha, \tau_{\gamma_1}), \mathcal{I}(\tau_{\gamma_1}, \tau_\alpha)\}$. Since $\tau_{\gamma_1} < \tau_\alpha$, so $\mathcal{I}(\tau_{\gamma_1}, \tau_\alpha) = \tau_{M-1}$. So we obtain $Sim(\tau_\alpha, \tau_{\gamma_1}) = \mathcal{I}(\tau_\alpha, \tau_{\gamma_1})$. With the same way for τ_{γ_2}, one can obtain $Sim(\tau_\alpha, \tau_{\gamma_2}) = \mathcal{I}(\tau_\alpha, \tau_{\gamma_2})$. We know that every implication operator and T-norm are monotonous[6] :

$$\begin{aligned}\tau_{\gamma_1} < \tau_{\gamma_2} &\Rightarrow \mathcal{I}(\tau_\alpha, \tau_{\gamma_1}) \leq \mathcal{I}(\tau_\alpha, \tau_{\gamma_2})\\ &\Rightarrow Sim(\tau_\alpha, \tau_{\gamma_1}) \leq Sim(\tau_\alpha, \tau_{\gamma_2})\\ &\Rightarrow T(Sim(\tau_\alpha, \tau_{\gamma_1}), \tau_\beta) \leq T(Sim(\tau_\alpha, \tau_{\gamma_2}), \tau_\beta)\\ &\Rightarrow \tau_{\lambda_1} \leq \tau_{\lambda_2}\end{aligned}$$

We know that for any T-norm T, $\forall x, y\ T(x,y) \leq min(x,y)$, so:

$$\tau_{\lambda_2} = T(Sim(\tau_\alpha, \tau_{\gamma_2}), \tau_\beta) \leq min(Sim(\tau_\alpha, \tau_{\gamma_2}), \tau_\beta) \leq \tau_\beta \Longrightarrow \tau_{\lambda_2} \leq \tau_\beta$$

We deduce that $\tau_{\lambda_1} \leq \tau_{\lambda_2} \leq \tau_\beta$. This allows us to deduce that the more an observation is a weakening of the rule premise, the more the inference conclusion is a weakening of the rule conclusion.

3 Approximate Reasoning and Linguistic Modifiers

The concept of modifiers is often used in the framework of fuzzy logic. Their theoretical study has been pointed out for several years [9, 18]. They have been mostly used in expert and decision-making systems working with fuzzy linguistic variables. The use of modifiers is chiefly pointed out by [5], [6], [11], where it is shown how their use can be easy and can provide simplified inferences in a fuzzy logic system. Another approach concerns a symbolic view of linguistic modifiers [2, 3]. As we work on approximate reasoning in a symbolic context, we are interested in this last approach and we present, in section 3.1, the Generalized Symbolic Modifiers defined in [3].

A linguistic modifier is an operator which builds terms from a primary term. These modifiers are usually used in natural language. They correspond to adverbs such as *very*, *really*, *more or less*, etc. In fuzzy context, as well as in symbolic one modifiers are classified by means of their behavior [2, 5]. Two types of modifiers are distinguished. Firstly, the reinforcing modifiers reinforce the concept expressed by the term, therefore they cause a profit of precision (like *very*). Secondly, the weakening modifiers weaken the concept expressed by the term; therefore they cause a loss of precision (like *more*

[6] $\forall z$, if $x \leq y$ then $\mathcal{I}(z,x) \leq \mathcal{I}(z,y)$ and $T(z,y) \leq T(z,x)$.

or less). In the fuzzy sets theory, a linguistic modifier leads to a new fuzzy set $m(A)$ different from the fuzzy set A [18]. While in the multi-sets theory, a linguistic modifier generally preserves the same multi-set A but modifies the degree with which a variable X satisfies the concept A [2].

In the literature, El- Sayed and Pacholczyk [13] introduced new rules of GMP with free rules by using linguistic modifiers. In these rules, the observation multi-set is different from the premise multi-set, contrary to the rules of the GMP (7) and (8) where the observation and the premise correspond to the same multi-set. In this work, we consider that the observation is represented by the same multi-set as the premise, but with different truth-degrees. Moreover, we are interested in strong rules because they generalize free rules.

Linguistic modifiers may correspond to a form of similarity. For this reason, it is possible to use them in the GMP to evaluate the relation between premise and observation. Here, we are more interested in the mathematical aspect of the modifiers then in their linguistic aspect. Indeed, saying that $\mathcal{A}' = m(\mathcal{A})$ allows to determine the mathematical operations applied on $\mathcal{A}$ to lead to $\mathcal{A}'$. Moreover, their use will allow to have a coherent inference result, conform to the axioms (3). The inference diagram of the approximate reasoning with linguistic modifiers is as follows [9]:

$$\frac{\begin{array}{l}\text{If "}X \text{ is } \mathcal{A}\text{" then "}Y \text{ is } \mathcal{B}\text{"}\\ \text{"}X \text{ is } m(\mathcal{A})\text{"}\end{array}}{\text{"}Y \text{ is } m'(\mathcal{B})\text{"}} \tag{9}$$

To determine the inference conclusion $\mathcal{B}' = m'(\mathcal{B})$, it is enough to determine the modifier m'. This modifier is obtained from the modifier m and from the causality between the rule premise and conclusion. Our hypothesis in this work is that the obtained modifier m' is equal to m. It translates effectively the graduality between the predicates $\mathcal{A}$ and $\mathcal{B}$. Indeed, the modification observed on the rule premise will be applied with the same intensity to the rule conclusion. For example, given a rule $C \rightarrow D$ and an observation *very* C, the conclusion obtained is *very* D [12, 13, 14]. In this work, we consider the following GMP:

$$\frac{\begin{array}{l}\text{If "}X \text{ is } \mathcal{A}\text{" then "}Y \text{ is } \mathcal{B}\text{"}\\ \text{"}X \text{ is } m(\mathcal{A})\text{"}\end{array}}{\text{"}Y \text{ is } m(\mathcal{B})\text{"}} \tag{10}$$

3.1 Generalized Symbolic Modifiers

Linguistic symbolic modifiers were proposed by Akdag and al. in [2], they were generalized and formalized in [3] and named the Generalized Symbolic Modifiers. A multi-set can be seen as a symbolic scale. Thus, the data modification in multi-valued logic is done by a transformation of the scale and/or the degree of the multi-set. This leads to dilation or erosion of the scale.

A Generalized Symbolic Modifier (GSM) is a semantic triplet of parameters: *radius*, *nature* (i.e dilated, eroded or conserved) and *mode* (i.e reinforcing, weakening or centring). The radius is denoted ρ with $\rho \in \mathbb{N}^*$. The more ρ increases, the more the modifier is powerful [3].

Definition 1. *Let us consider a symbolic degree τ_i with $i \in \mathbb{N}$ in a scale $\mathcal{L}_M$ of a base $M \in \mathbb{N}^* \setminus \{1\}$, and $i < M$. A GSM m with a radius ρ is denoted m_ρ. The modifier m_ρ is a function which applies a linear transformation to the symbolic degree τ_i to obtain a new degree $\tau_{i'} \in \mathcal{L}_{M'}$ (where $\mathcal{L}_{M'}$ is the linear transformation of $\mathcal{L}_M$) according to a radius ρ such as:*

$$m_\rho : \mathcal{L}_M \rightarrow \mathcal{L}_{M'}$$
$$\tau_i \mapsto \tau_{i'}$$

The greatest degree in the base is denoted $MAX(\mathcal{L}_M) = M - 1$. The scale position in the base is denoted $p(\tau_i) = i$. A proportion is associated to each symbolic

Table 1. Definitions of weakening and reinforcing modifiers

MODE / NATURE	Weakening		Reinforcing	
Erosion	$\tau_{i'} = \tau_{max(0,i-\rho)}$ $\mathcal{L}_{M'} = \mathcal{L}_{max(1,M-\rho)}$	EW_ρ	$\tau_{i'} = \tau_i$ $\mathcal{L}_{M'} = \mathcal{L}_{max(i+1,M-\rho)}$	ER_ρ
			$\tau_{i'} = \tau_{min(i+\rho,M-\rho-1)}$ $\mathcal{L}_{M'} = \mathcal{L}_{max(1,M-\rho)}$	ER'_ρ
Dilation	$\tau_{i'} = \tau_i$ $\mathcal{L}_{M'} = \mathcal{L}_{M+\rho}$	DW_ρ	$\tau_{i'} = \tau_{i+\rho}$ $\mathcal{L}_{M'} = \mathcal{L}_{M+\rho}$	DR_ρ
	$\tau_{i'} = \tau_{max(0,i-\rho)}$ $\mathcal{L}_{M'} = \mathcal{L}_{M+\rho}$	DW'_ρ		
Conservation	$\tau_{i'} = \tau_{max(0,i-\rho)}$ $\mathcal{L}_{M'} = \mathcal{L}_M$	CW_ρ	$\tau_{i'} = \tau_{min(i+\rho,M-1)}$ $\mathcal{L}_{M'} = \mathcal{L}_M$	CR_ρ

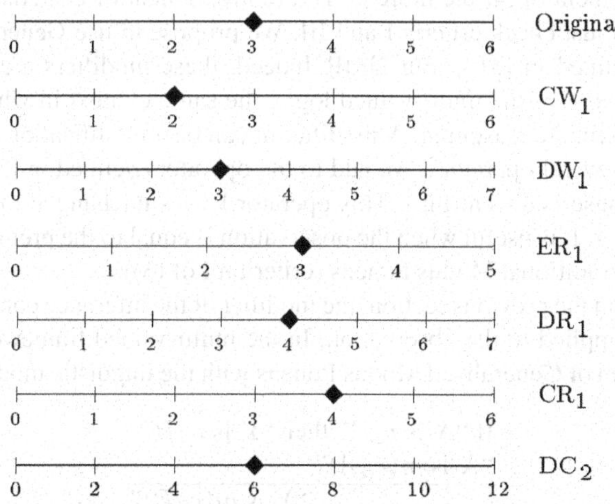

Fig. 2. Illustration of some symbolic linguistic modifiers

degree within a base denoted $Prop(\tau_i) = \frac{p(\tau_i)}{MAX(\mathcal{L}_M)}$ [3]. By analogy with fuzzy modifiers, the authors propose a classification of symbolic modifiers according to their behavior: weakening and reinforcing modifiers, and they add the family of central modifiers [2]. The first two families correspond to the modifiers which weaken or reinforce the initial value of $Prop(\tau_i)$. Thus, for the family of weakening modifiers we obtain $Prop(\tau_{i'}) < Prop(\tau_i)$, and for the family of reinforcing modifiers $Prop(\tau_{i'}) > Prop(\tau_i)$. The modifiers of the third family leave the rate $Prop(\tau_i)$ unchanged, but act as a zoom on the scale. Moreover, Generalized Symbolic Modifiers allow to erode, dilate or conserve the initial scale $\mathcal{L}_M$. The authors proposed three weakening modifiers, called EW, DW and CW for Eroded Weakening, Dilated Weakening and Conserved Weakening. In the same way, three reinforcing modifiers were defined: ER, DR and CR for Eroded Reinforcing, Dilated Reinforcing and Conserved Reinforcing. The definitions of the reinforcing and weakening modifiers are given in table 1, and an example of some modifiers is given in figure 2 [3]. The central modifiers proposed are EC (for Eroded Centring) and DC (for Dilated Centring). The definitions of EC and DC are[7]:

$$EC_\rho = \begin{cases} \tau_{i'} = \tau_{max(\lfloor \frac{i}{\rho} \rfloor, 1)} \\ \mathcal{L}_{M'} = \mathcal{L}_{max(\lfloor \frac{M}{\rho} \rfloor + 1, 3)} \end{cases} \qquad DC_\rho = \begin{cases} \tau_{i'} = \tau_{i\rho} \\ \mathcal{L}_{M'} = \mathcal{L}_{M\rho - \rho + 1} \end{cases}$$

4 An Approximate Reasoning of Type I

Axiomatic of approximate reasoning allows to distinguish two types of approximate reasoning. In this section, we introduce a symbolic GMP of type I based on linguistic modifiers. This integration of linguistic modifiers allows checking the axiomatic of approximate reasoning, contrary to approximate reasoning based on similarity [7, 16].

The proposed model of approximate reasoning has to check criterion II-1. This criterion enables to provide a gradual reasoning based on the meta-knowledge: "the more $\mathcal{A}'$ is a reinforcement of $\mathcal{A}$, the more $\mathcal{B}'$ is a reinforcement of $\mathcal{B}$". Also, this approximate reasoning must check criteria I and III. We propose to use Generalized Linguistic Modifiers defined in [3] in our GMP. Indeed, these modifiers are defined in the symbolic framework of the multi-valued logic, the same context in which we want to model our approximate reasoning. A modifier m can be a modification operator, or the composition of several operators. We add to the operators defined in [3] another operator CC (for Conserved Centring). This operator leaves unchanged the multi-set, i.e. $CC(v_\alpha A) = v_\alpha A$. It is useful when the observation is equal to the premise, thus, it will allows to check traditional Modus Ponens (criterion I of (3)).

As explained in the previous section, the modifier of the inference conclusion is equal to the modifier applied to the observation. In the multi-valued framework with strong rules, the diagram of Generalized Modus Ponens with the linguistic modifiers becomes:

$$\frac{\begin{array}{l}\text{If "}X \text{ is } v_\alpha A\text{" then "}Y \text{ is } v_\beta B\text{"} \\ \text{"}X \text{ is } m(v_\alpha A)\text{"}\end{array}}{\text{"}Y \text{ is } m(v_\beta B)\text{"}} \tag{11}$$

[7] $\lfloor . \rfloor$ is the floor function.

The fundamental stage to infer with this GMP is to determine the modifier m. To find a modifier which allows to pass from a couple degree/base $(\tau_\alpha, \mathcal{L}_M)$ to another couple $(\tau_{\alpha'}, \mathcal{L}_{M'})$, we can use the elementary operators who are CR, CW, ER and DW as well as the neutral operator CC. Given two couples degree/base $(\tau_\alpha, \mathcal{L}_M)$ and $(\tau_{\alpha'}, \mathcal{L}_{M'})$, the modifier m which transform the first couple to the second one can be obtained with the following algorithm:

Algorithm 1

- *If* $M = M'$ *and* $\alpha = \alpha'$ *then* $m = CC$.
- *If* $M = M'$ *and* $\alpha \neq \alpha'$, m *is equal to the operator* m_{ρ_1}, *defined below, which allows to modify the degree* τ_α.
- *If* $M \neq M'$ *and* $\alpha = \alpha'$, m *is equal to the operator* m_{ρ_2}, *defined below, which allows to modify the base* $\mathcal{L}_M$.
- *If* $M \neq M'$ *and* $\alpha \neq \alpha'$, m *is equal to the composition of the two operators:* $m_{\rho_1} \circ m_{\rho_2}$, *with* m_{ρ_1} *is the operator allowing to modify the degree and* m_{ρ_2} *is the one allowing to modify the base.*
- *Determination of the operator* m_{ρ_1} *acting on the degree: the radius of the operator is* $\rho_1 = |\alpha - \alpha'|$;
  - *If* $\alpha < \alpha'$ *then the operator is* CR_{ρ_1}.
  - *If* $\alpha > \alpha'$ *then the operator is* CW_{ρ_1}.
- *Determination of the operator* m_{ρ_2} *acting on the base: the radius of the operator is* $\rho_2 = |M - M'|$;
  - *If* $M < M'$ *then the operator is* DW_{ρ_2}.
  - *If* $M > M'$ *then the operator is* ER_{ρ_2}.

For example, to pass from $(\tau_6, \mathcal{L}_{20})$ to $(\tau_{15}, \mathcal{L}_{32})$, we apply CR_9 then DW_{12}, i.e. $DW_{12} \circ CR_9(\tau_6, \mathcal{L}_{20}) = (\tau_{15}, \mathcal{L}_{32})$.

These modifiers act on the truth-degree and on the base of the multi-set. However, in some cases, the degrees base is fixed by the expert at the system initialization time. So its modification is not wished. For that, it is recommended to use the modifiers which act only on the truth-degree of the multiset without changing its base, namely CR and CW. Moreover, we can use the other operators, but it is necessary to work with the proportion $Prop(\tau_i)$ so that the resulting multi-set has the same initial base.

Example 1. Given the list of truth-degrees $\mathcal{L}_7 = \{$not-at-all, very-little, little, moderately, enough, very, completely$\}$, and the rule:

"If a tomato is moderately red then it is enough ripe." $(\tau_\alpha = \tau_3, \tau_\beta = \tau_4)$

For the following fact: *"This tomato is little red."* $(\tau_{\alpha'} = \tau_2)$

the modifier m transforms "moderately" to "little", so we go from the truth-degree τ_3 to τ_2. Thus, $m = CW_1$ since the operator CW decreases truth-degrees. We apply this modifier to the rule conclusion.

$CW_1(enough) = moderately$. And so we deduce:

"This tomato is moderately ripe." $(\tau_{\beta'} = \tau_3)$

For the following fact: *"This tomato is enough red."*

the modifier which transform "moderately" to "enough" is CR_1. We apply this modifier to the rule conclusion: $CR_1(enough) = very$. We deduce:

"This tomato is very ripe."

We notice through this example that when the observation is a reinforcement of the premise (*enough* > *moderately*), the conclusion undergoes the same reinforcement intensity as the premise. Thus, in the suggested model, the inference consists to determine the linguistic modifier to pass from A to A' ($A' = m(A)$)), and then to apply it to the conclusion ($B' = m(B)$). If A is reinforced, B will be reinforced too, and if A is weakened B will be weakened. This model of approximate reasoning checks the type 1. Indeed, criterion I is checked by the modifier CC, criterion II-1 is checked by reinforcing modifiers, and finally criterion III is checked by weakening modifiers. The proposed GMP corresponds to an approximate reasoning of type I.

The results of our approximate reasoning based on linguistic modifiers cannot be obtained with approximate reasoning based on similarity [7, 16]. Indeed, when the observation is a reinforcement of the premise, our model takes into account this type of modification and consequently reinforces the conclusion, whereas approximate reasoning based on similarity applies a blind modification on the rule conclusion (without taking into account the direction of the modification), which may lead to contradict the axiomatic of approximate reasoning.

5 An Approximate Reasoning of Type II

In this part, we introduce a GMP of type II in a symbolic multi-valued logic. So, our aim is to provide a new model of approximate reasoning checking criterion II-2. In this case, it is not necessarily to use linguistic modifiers as gradual reasoning is not needed. A simple modification of Khoukhi model (8) can give a solution to our aim [16]. The resulted model also have to check criteria I and III.

We have demonstrate that Khoukhi model verifies both criteria I and III. The problem is that it does not check criterion II. So, in the Khoukhi model (8), the problem appears when the observation is a reinforcement of the rule premise, i.e. $\tau_\alpha < \tau_\gamma$. Thus, to ensure criterion II-2, the conclusion should be equal to the rule conclusion with the same truth-degree, i.e. $\tau_\lambda = \tau_\beta$:

If $\tau_\alpha < \tau_\gamma$ then $\tau_\lambda = \tau_\beta$
As in the Khoukhi model $\tau_\lambda = T(Sim(\tau_\alpha, \tau_\gamma), \tau_\beta)$, we want to impose:
$T(Sim(\tau_\alpha, \tau_\gamma), \tau_\beta) = \tau_\beta$

The neutral element of T-norm relation is τ_{M-1}. To have the equality $\tau_\lambda = \tau_\beta$, we impose that $Sim(\tau_\alpha, \tau_\gamma)$ verifies: $Sim(\tau_\alpha, \tau_\gamma) = \tau_{M-1}$. Thus, we propose to replace the similarity relation Sim by another relation temporarily named R. R must check the following properties:

1. If $\tau_\alpha < \tau_\gamma$ then $R(\tau_\alpha, \tau_\gamma) = \tau_{M-1}$,
2. If $\tau_\alpha \geq \tau_\gamma$ then $R(\tau_\alpha, \tau_\gamma) = Sim(\tau_\alpha, \tau_\gamma)$.

The first property ensures that when the observation is a reinforcement of the rule premise, i.e. the truth-degree of the multi-set A increases, the multi-set B will keep the same truth-degree of the rule conclusion. That allows to ensure criterion II-2 of (3). The second property ensures that when the observation is a weakening of the premise, i.e. the truth-degree of the multi-set A decreases, the multi-set B decreases with an intensity depending on $Sim(\tau_\alpha, \tau_\gamma)$. We return in this case to the Khoukhi model for ensuring criterion III. Moreover, this same property will enable us to guarantee the case of traditional Modus Ponens (criterion I). Indeed, when the observation is the rule premise ($\tau_\alpha = \tau_\gamma$), the result will be equal to the rule conclusion: $R(\tau_\alpha, \tau_\alpha) = Sim(\tau_\alpha, \tau_\alpha) = min\{\mathcal{I}(\tau_\alpha, \tau_\alpha), \mathcal{I}(\tau_\alpha, \tau_\alpha)\} = min\{\tau_{M-1}, \tau_{M-1}\} = \tau_{M-1}$, so we deduce that $\tau_\lambda = \tau_\beta$.

In order to verify the two targeted properties, we propose to use an implication operator $R(\tau_\alpha, \tau_\gamma) = \mathcal{I}(\tau_\alpha, \tau_\gamma) = \tau_\alpha \rightarrow \tau_\gamma$. Indeed, we know that every implication operator verifies the following property:

$$\text{if } \tau_\alpha < \tau_\gamma \text{ then } \tau_\alpha \rightarrow \tau_\gamma = \tau_{M-1}$$

The first property is then verified. Moreover, when $\tau_\alpha \geq \tau_\gamma$ we have $\mathcal{I}(\tau_\gamma, \tau_\alpha) = \tau_{M-1}$, so:

$$Sim(\tau_\alpha, \tau_\gamma) = min\{\mathcal{I}(\tau_\alpha, \tau_\gamma), \mathcal{I}(\tau_\gamma, \tau_\alpha)\} = \mathcal{I}(\tau_\alpha, \tau_\gamma)$$

The second property is also satisfied. Consequently, the implication operator $\tau_\alpha \rightarrow \tau_\gamma$ checks the two properties defined previously. We propose to use it instead of the similarity relation in the Generalized Modus Ponens:

$$\begin{array}{l} \text{If "}X \text{ is } v_\alpha A\text{" then "}Y \text{ is } v_\beta B\text{"} \\ \text{"}X \text{ is } v_\gamma A\text{"} \\ \hline \text{"}Y \text{ is } v_\lambda B\text{" with } \tau_\lambda = T(\tau_\alpha \rightarrow \tau_\gamma, \tau_\beta) \end{array} \quad (12)$$

Example 2. In this example, we consider the Lukasiewicz implication $\mathcal{I}_\mathcal{L}$ and the Lukasiewicz T-norm T_L(the most used in the multi-valued logic). Given the list of truth-degrees $\mathcal{L}_7 = \{$not-at-all, very-little, little, moderately, enough, very, completely$\}$, and the rule:

"If a tomato is moderately red then it is enough ripe." ($\tau_\alpha = \tau_3, \tau_\beta = \tau_4$)

For the following fact: *"This tomato is little red."* ($\tau_\gamma = \tau_2$)

the truth degree of the conclusion is $T_L(\tau_3 \rightarrow \tau_2, \tau_4) = T_L(\tau_5, \tau_4) = \tau_3$. The resulted degree τ_3 corresponds to the linguistic term "moderately". We deduce: *"This tomato is moderately ripe."*

For the following fact: *"This tomato is enough red."* ($\tau_{\gamma'} = \tau_4$)

the truth degree of the conclusion is:

$T_L(\tau_3 \rightarrow \tau_4, \tau_4) = T_L(\tau_6, \tau_4) = \tau_4$.

We obtain the conclusion: *"This tomato is enough ripe."*

According to the suggested model, when the observation is more precise than the rule premise ($\tau_\gamma > \tau_\alpha$), the inference conclusion is equal to the rule conclusion ($\tau_\lambda = \tau_\beta$). In addition, if the observation is equal or less precise than the premise ($\tau_\gamma \leq \tau_\alpha$), we find the result of Khoukhi model based on similarity. We thus deduce that our approach checks criteria II-2 and III, as well as criterion I of (3). The proposed GMP corresponds to an approximate reasoning of type II.

6 Conclusion

In this paper, we were interested in the approximate reasoning in a symbolic multi-valued framework. We presented a generalization of the approximate reasoning axiomatic, and show that the Khoukhi [16] approach of GMP in a multi-valued context does not check this axiomatic. More precisely, criteria II-1 and II-2 related to a reinforcement of the premise are not verified. Indeed, when the observation is a reinforcement of the rule premise, the inference conclusion may be a weakening of the rule conclusion.

We then proposed two models of approximate reasoning checking the axiomatic. The first proposed GMP is based on linguistic modifiers introduced in [3]. This GMP allows having a gradual reasoning based on the meta-knowledge: "the more the observation is a reinforcement of the premise, the more the inference conclusion is a reinforcement of the rule conclusion". The second rule is an amelioration of Khoukhi GMP [16], and it is based on implication operator.

In this work, we proposed a GMP based on linguistic modifiers that corresponds to the following inference diagram:

$$\frac{\begin{array}{l}\text{If "}X \text{ is } v_\alpha A\text{" then "}Y \text{ is } v_\beta B\text{"}\\ \text{"}X \text{ is } m(v_\alpha A)\text{"}\end{array}}{\text{"}Y \text{ is } m(v_\beta B)\text{"}}$$

An interesting perspective would be to consider the more general case where the linguistic modifiers applied on the rule conclusion is different from the one applied on the observation.

Moreover, as a future work it could be interesting to study the inference formalization for more complex strong rules, for example with multiple premises as in the following diagram:

$$\begin{array}{l}\text{If "}X_1 \text{ is } v_{\alpha_1} A_1\text{" and} \ldots \text{and "}X_n \text{ is } v_{\alpha_n} A_n\text{" then "}Y \text{ is } v_\beta B\text{"}\\ \text{"}X_1 \text{ is } v_{\delta_1} A_1\text{" and} \ldots \text{and "}X_n \text{ is } v_{\delta_1} A_n\text{"}\\ \hfill \text{"}Y \text{ is } v_\gamma B\text{"}\end{array}$$

References

1. Akdag, H., De Glas, M., Pacholczyk, D.: A qualitative theory of uncertainty. Fundamenta Informaticae 17, 333–362 (1992)
2. Akdag, H., Mellouli, N., Borgi, A.: A symbolic approach of linguistic modifiers. In: Information Processing and Management of Uncertainty in Knowledge- Based Systems, Madrid, pp. 1713–1719 (2000)
3. Akdag, H., Truck, I., Borgi, A., Mellouli, N.: Linguistic Modifers in a Symbolic Framework. I. J. of Uncertainty. Fuzziness and Knowledge-Based Systems 9(suppl.), 49–61 (2001)
4. Baldwin, J.F., Pilsworth, B.W.: Axiomatic approach to implication for approximate reasoning with fuzzy logic. Fuzzy sets and systems 3, 193–219 (1980)
5. Bouchon-Meunier, B.: Stability of linguistic modifiers compatible with a fuzzy logic. In: Yager, R.R., Saitta, L., Bouchon, B. (eds.) IPMU 1988. LNCS, vol. 313, pp. 63–70. Springer, Heidelberg (1988)
6. Bouchon-Meunier, B.: On the management of uncertainty in knowledgebased system. In: Holman, A.G., Kent, A., Williams, G.J. (eds.) Encyclopedia of Computer Science and Technology, Marcel Dekker (1989)
7. Bouchon-Meunier, B., Delechamp, J., Marsala, C., Rifqi, M.: Several forms of fuzzy analogical reasoning. In: FUZZ-IEEE 1997, Barcelona, pp. 45–50 (1997)
8. Bouchon, B., Jia, Y.: Gradual change of linguistic category by means of modifiers. In: Proc. 3rd International Conference IPMU, Paris, pp. 242–244 (1990)

9. Bouchon-Meunier, B., Kreinovich, V.: Fuzzy Modus Ponens as a Calculus of Logical Modifiers: Towards Zadeh's Vision of Implication Calculus. Information Sciences 116, 219–227 (1999)
10. Cornelis, C., Kerre, E.E.: Inclusion-Based Approximate Reasoning. In: Alexandrov, V.N., Dongarra, J., Juliano, B.A., Renner, R.S., Tan, C.J.K. (eds.) ICCS-ComputSci 2001. LNCS, vol. 2074, pp. 221–230. Springer, Heidelberg (2001)
11. Di Lascio, L., Gisolfi, A., Loia, V.: A New Model for Linguistic Modifiers. International Journal of Approximate Reasoning 15, 25–47 (1996)
12. Di Lascio, L., Gisolfi, A., Cortés, U.: Linguistic hedges and the generalized modus ponens. International Journal of Intelligent Systems 14, 981–993 (1999)
13. El-Sayed, M., Pacholczyk, D.: Towards a Symbolic Interpretation of Approximate Reasoning. Electr. Notes Theor. Comput. Sci. 82, 108–119 (2003)
14. Fukami, S., Mizumoto, M., Tanaka, K.: Some considerations of fuzzy conditional inference. Fuzzy sets and systems 4, 243–273 (1980)
15. Ginsberg, M.L.: Multivalued logics: a uniform approach to reasoning in artificial intelligence. Computational Intelligence 4, 265–316 (1988)
16. Khoukhi, F.: Approche logico-symbolique dans le traitement des connaissances incertaines et imprécises dans les systèmes à base de connaissances. PhD thesis, Université de Reims, France (1996)
17. Zadeh, L.: Fuzzy sets. Information and Control 8, 338–353 (1965)
18. Zadeh, L.: The concept of a linguistic variable and its application to approximate reasoning I-II-III. Information Sciences 8, 199–249; 8, 301–357; 9, 43–80 (1975)
19. Zadeh, L.: A theory of approximate reasoning. Machine Intelligence 9, 149–194 (1979)

Effect of Sensor Specific Crop Location on Wireless Sensor Network Performance

Hyun-joong Kang, Meong-hun Lee, and Hyun Yoe*

School of Information and Communications Engineering, Sunchon National Univ., 315 Maegok-dong, Sunchon-Si, Jeonnam 540-742, Korea
hjkang@mail.sunchon.ac.kr, {leemh777,yhyun}@sunchon.ac.kr

Summary. Because a sensor node is operated at a low power, a good communication can hardly be done when an obstacle exists in inter-node's communication point. In this study, a simulation environment was made using a shadowing model for the position of sensor node and the communication impact by crop in observing the crop's growth and a greenhouse environment. In such developed simulation environment, the simulation was conducted through changing ad-hoc routing protocols for the sensor node. We could validate that there was a difference in the performance of sensor node according to the installed position through the simulation result. Also, we could confirm that there were different characteristics respectively depending on the routing protocol. We consider that such result will be useful to provide strategies for designing the effective routing protocol for an agricultural environment such as the greenhouse in the future.

Keywords: WSN, greenhouse, routing, agriculture.

1 Introduction

In the last several decades, a size of a terminal has been gradually reduced through rapid development such as the increase of operating speed, miniaturization of Hardware size, and the increase in density of electronics, and its performance has been improved [1].

Through such miniaturization and the development of performance, a coin size's very small terminal appeared, and also the sensor node which measures and collects diversified environmental information using it has appeared [2].

The sensor node measures the physical phenomena and transmits it to the collection node through configuring inter-sensor node's self-operated networks. The study that automates the variety of equipment using the sensor node has been executed, and the research field is ranged to almost all fields including a road, harbors, building, military affairs, agriculture and home [3].

Especially in the agriculture, there is the study to create the optimum environment where the crop can grow through periodically collecting information related to the crop such as the temperature, humidity and intensity of illumination with the sensor node [4]. In the agricultural environment like the greenhouse where the large number of crops grows, the values including the temperature, humidity and intensity of illumination are

* Corresponding author.

R. Lee and H.-K. Kim (Eds.): Computer and Information Science, SCI 131, pp. 219–229, 2008.
springerlink.com

different in the lower part close to the surface of land, middle part and the top part of the greenhouse respectively [5]. When the physical environment information is measured from the different position using the sensor node in such greenhouse, it is likely that the sensor node fails in the transmission and reception of data due to the communication jamming caused by obstacle like crop [6].

This study is intended to virtually simulate the obstruction of propagation by such crop through the application of shadowing model. Also, the study is intended to apply diversified ad-hoc routing protocols and to search the effective routing protocol by comparing a transfer efficiency and energy consumption rate. In addition, the study is aimed at being useful for researching the suitable routing method for the crop in the greenhouse through such comparison.

This paper is made up as follows. Diversified ad-hoc routing protocols to apply to the simulation are firstly explained. Then, the simulation environment applying the shadowing model is described and then results derived from the application diversified routing protocols are compared. Finally, the proper routing method is proposed through preceded comparison.

2 Related Researches

This chapter is intended to state the proper ah-hoc based routing protocol suitable for the sensor node, the sensor network and the sensor network to be used for the simulation.

2.1 Sensor Network

2.1.1 Sensor Node

There are many parts on the sensor node: the processor, the radio communication unit, the battery and the various sensor modules. Groups of this sensor node form the Sensor network. Sensor node monitors temperature, humidity and many other environmental conditions in the field and transmits to the sink node.

2.1.2 Sensor Network

As all the sensors are not able to transmit data to the sink node directly, some send data through a few nodes [7]. In other words, the node is both sensing and acting as a router which forwards the data.

In a normal network, the route which goes through the least number of hops is chosen. However, the Sensor network usually takes the amount of battery as the criterion.

There are many places where Sensor network is implemented such as in the military, home, traffic, medical, science and agriculture areas all of which collect environmental information to sense physical phenomenon.

2.2 Routing Protocols

2.2.1 DSR [8]

DSR basically uses an algorithm of Source Routing [9]. By using the Source Routing in DSR, an origin mobile terminal adds the route gained by executing the path discovery

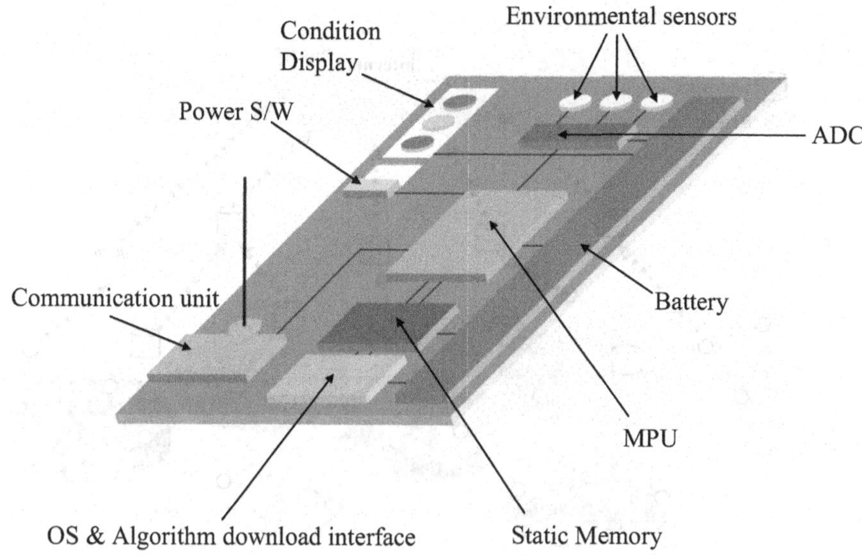

Fig. 1. A simple sensor node example: ① MPU: low clock rate, suitable for simple program or algorithm ② Battery: support for few months only, can be extended by external energy like solar panel ③ Static memory: no dynamic allocation allowed, just support data read from it ④ Communication unit: consumes energy more than any other modules ⑤ OS & Algorithm download interface: download updated or edited version of small size OS & Algorithm from Desktop computer ⑥ Environmental sensors: measure the environmental condition, but the measured value is analogue. So the ADC unit converts this analogue value into digital.

procedures when the path is necessary in the data packet.s header and then transmits it. DSR keeps the route cache in stead of route table. When the path is discovered, the operation of route cache is stored in the route cache. As the numeral of path stored in the route cache is 'Path-Inf, Path-FIFO-32, Path-Gen-64 and others', there are many algorithms. When the new path information is obtained at higher limited path number, firstly-discovered path information is deleted and the new path information is stored for the update of route cache path.

2.2.2 DSDV [10]

Destination Sequenced Distance Vector (DSDV) is based on the Bellman-Ford routing [11] system that is used in the Wired Network, and prevents an occurrence of routing loop by topology change through using the destination sequence number. Each node maintains the route information for other all nodes in the routing table. The update of routing table is done in the form of full dump and incremental dump. Full dump as the method which broadcasts all routing information of the node to another node is used when there is more routing information to update. On the other hand, the incremental dump broadcasts only newly-modified routing information when the routing information is modified. The routing information of each node has the form of [destination address, metric, sequence number, next hop]. Among them, [destination

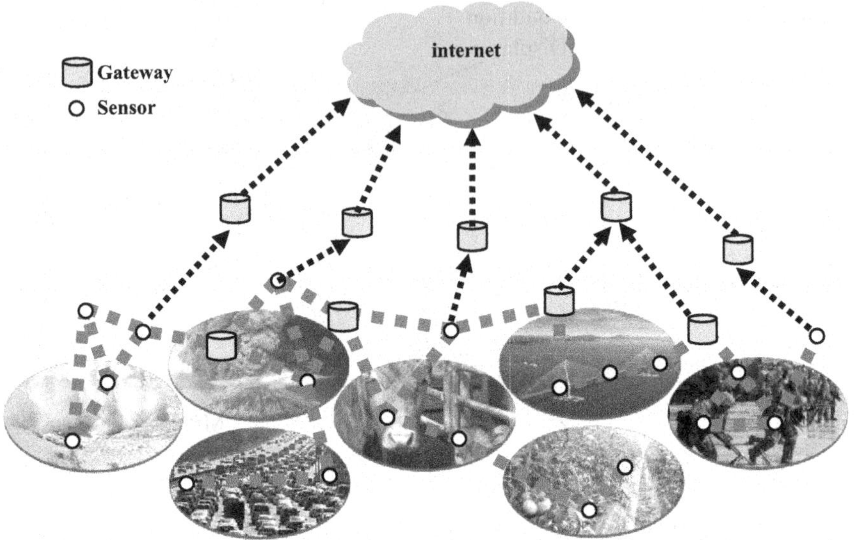

Fig. 2. Sensor Network applications

address, metric, sequence number] information is broadcasted to the neighbor node through full/incremental dump. A sequence number is increased when the new routing information is generated or updated.

2.2.3 AODV [12]

AODV stands for Ad hoc On-demand Distance Vector, and is the protocol proposed for solving the problem of Dynamic Source Routing (DSR) protocol in the ad-hoc network. This routing protocol as very light routing protocol for the ad-hoc network is defined in RFC 3561 of IETF. This is the routing protocol available for ensuring the routing while using smaller memory. In particular, this routing protocol is the on-demand routing protocol to be used only when it delivers data. Accordingly, because it is not used when it doesn't deliver data, there are small overheads by routing. The routing mode is used to search the routing path when the wireless node in ad-hoc network requires the routing path to deliver data. Accordingly, when it has been used to deliver data, the routing path can be used as it is without searching again. But, if not, the message for searching the routing path called 'RREQ' floods the network to search the routing path.

3 Simulation Model

3.1 Motivation

In the place like the greenhouse, the crop is closely cultivated by considering productivity. When the crop is planted in the soil at the beginning, the size of crop is small and there is the low area of the crop. Accordingly, there are many proper places to install the sensor. That is, at the beginning, the sensor is installed in various positions such as a lower part, middle part and top part of the greenhouse, and then the physical phenomena information of greenhouse is measured. But, as time goes by, the crop grows and the sensors installed in the greenhouse show errors in the communication gradually. Because of scattered re-transmitted caused by such communication error, the sensor node consumes more energies. Especially, such communication error seriously occurs in the lower part of the sensor node where the crop grows closely. Moreover, because the sensor node installed in the lower part is used for the purpose of measuring the soil's humidity and temperature, information transmitted by the sensor node of the lower part is highly important.

3.2 Agriculture Applications

Because the sensor nodes scattered in the greenhouse can hardly receive the supply of powers, how the limited resources can be used efficiently becomes the important issue. Hence, recently, the method that efficiently operates the energy through improving the sensor node's protocol in saving the energy has been studied a lot [13, 14, 15]. But, as we know, the study that efficiently operates the energy of sensor node according to the application is still insufficient.

Fig. 3. The example of crop cultivated in the greenhouse (Tomato): As shown in the above figure, when the crop grows, the crop grows compactly so that it is difficult to secure a visual field.

In the agricultural environment, the application can be classified into two sensing methods; the method that measures the physical phenomena with the constant or rapid cycle and the method that measures it with the cycle of several tens of minutes.

1) Sensing with constant or rapid cycle
For the crop cultivated in the place that an instant observation such as a flood, snow and frost is required [16], the sensor node needs the constant observation in such environment. Fortunately, such physical phenomena don't require considerably more sensor nodes in measuring. It is because the range is very small even though the phenomenon locally occurs.

2) Sensing with the cycle of several tens of minutes
In the place like the greenhouse, there is the bigger change than the change of outdoor environment. It is because an indoor air is not connected with the outside. Because of the air in the greenhouse that exists in an independent space, the temperature sharply goes up in the daytime, but on the contrary, the temperature sharply goes down at night. Such change pattern steadily occurs in the cycle of tens of minutes. Accordingly, the sensor node that measures such environmental information has longer sensing cycle than other nodes. It can be operated for several months with the existing technology without the external power supply [17].

In executing such application-based study, the impact by crop will become one of the factors to determine the sensing cycle and routing method, when the greenhouse is sensed.

We made the wireless network topology like the above figure using NS-2 Simulator [18]. As shown in the figure, we formed the crop that was the obstacle to inter-node link communication as the simulation environment through modifying the shadowing model [19].

3.3 Shadowing Model of NS-2 Simulator [18, 20]

3.3.1 Shadowing Model

To introduce random events, the shadowing model utilizes a random variable X. The shadowing model requires a reference distance d0 to calculate the average received FS signal strength Pr,FS(d0). The path loss exponent depends on the simulated environment and is constant throughout simulations. Values vary between two (free space) and six (indoor, non-line-of-sight). X is normal distributed with an aver-age of zero and a standard deviation σ (called shadow deviation). It is non-variable and reasonable values vary between three (factory, LOS) and twelve (outside of buildings). Values for β and σ were empirically determined.

3.4 Simulated Shadowing Model

In the above Fig 4, several sensor nodes are hindered by crops planted in the greenhouse for the communication. In NS-2, we tried to make such effect through the shadowing model. For the value to be used in the shadowing model, we mainly used the value that was actually measured through the propagation test in the actual environment, and we

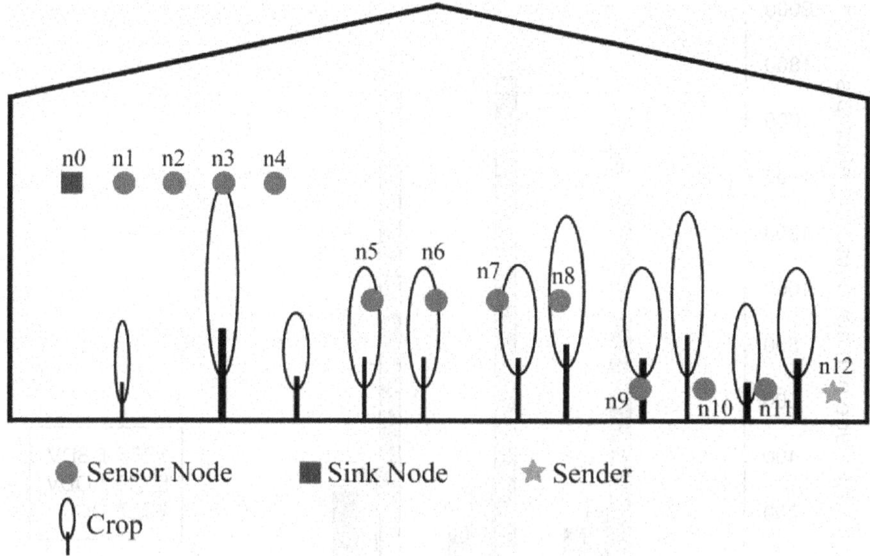

Fig. 4. Greenhouse model of simulation

applied the value using the numerical value that we tested indoors among such standard values.

4 Simulation

4.1 Simulation Parameters Used

The studied scenario is composed of 13 nodes, and 4 nodes are located in the lower part, middle part and top part respectively. There is the sink node as the collection node at the end of the top part. For the routing protocol, we used three ah-hoc routing protocols that are generally used in NS-2 Simulation. Through the application of shadowing model, we virtually implemented the lower part and middle part, and the lower part. Then, we simulated by properly adjusting the shadowing model parameters.

For the shadowing model, we used the value reported by actual measurement as the standard. It was based on the shadowed urban area value and outdoor value defined in ns-2. We supposed that only one sensor node(Sender) of the lower end generated the data, and generated the UDP packet once per 5 seconds constantly. The generated packet.s size was 256 bytes, and the simulation time was 2000s. The energy model was used to indicate how much the sensor node used the energy. The initial energy was 10000Joules and there were tx= 0.660joules, and rx=0.395joules.

4.2 Simulation with Routing Protocols

Fig. 5 indicates how many packet routes each node attempts when the sensor node transmits data to the top part from the lower part.

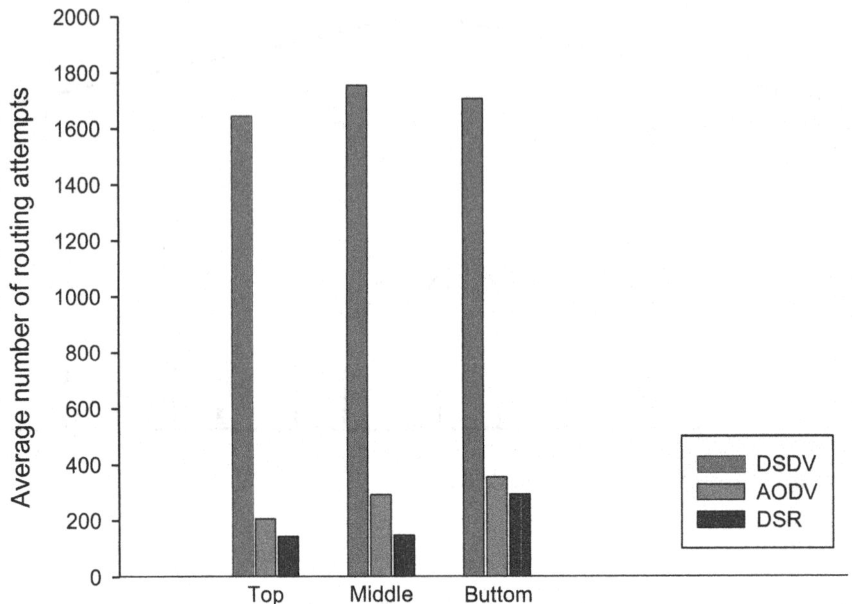

Fig. 5. Average number of routing attempts

As shown in the figure, we can confirm that the attempts of routing forward was decreased from the lower part that the crops are more distributed to the top part that the crops are less distributed in all routing protocols.

Fig. 6 indicates how many data were delivered to the sink node (destination node) as the inter-arrival is increased to 5 sec. from 1 sec. The reason that DSR is higher than other protocols as shown in the figure is that DSR indicates the stable transmission rate through reliability-based communication.

After considering Fig 5 and 6, the reason why the overall efficiency of DSDV is lower than other protocols even though DSDV generates more routing packets seems to be because all nodes in the network exchange the routing table that "All deliverable destination" and "hop count at each destination" are recorded in order for DSDV to keep all routes, but the change of topology occurs a bit because of no movement of nodes according to our scenario. Also, because AODV and DSR is the on-demand routing protocol, the periodic overhead is relatively small. Accordingly, the attempts of routing forward seem to be small.

Fig. 7 indicates how more energy the nodes of the lower part, middle part and top part consumed. Notwithstanding the different routing method, the energy consumption was bigger in the lower part. It is because the routing seemed to be more attempted due to a tendency of errors.

After considering the figures, the following contents must seem to be considered to the routing protocol and the sensor node for the crop. The routing protocol must be designed considering the crop's growth. As we know, the existing routing methods that have the shadowing effect by obstacles have been studied, but the scenario that

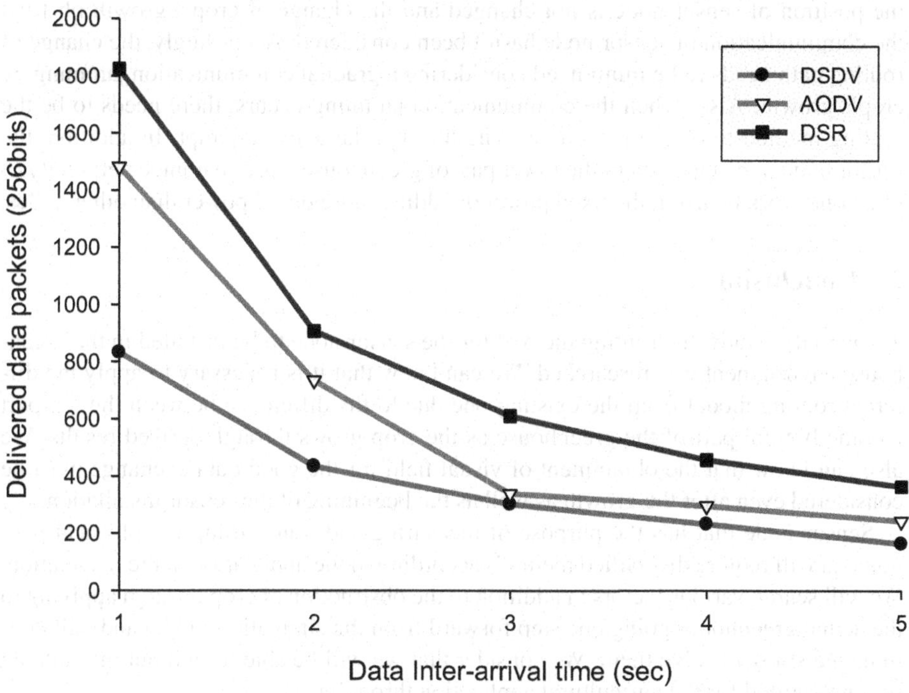

Fig. 6. Delivered data packets (pkt/256bits)

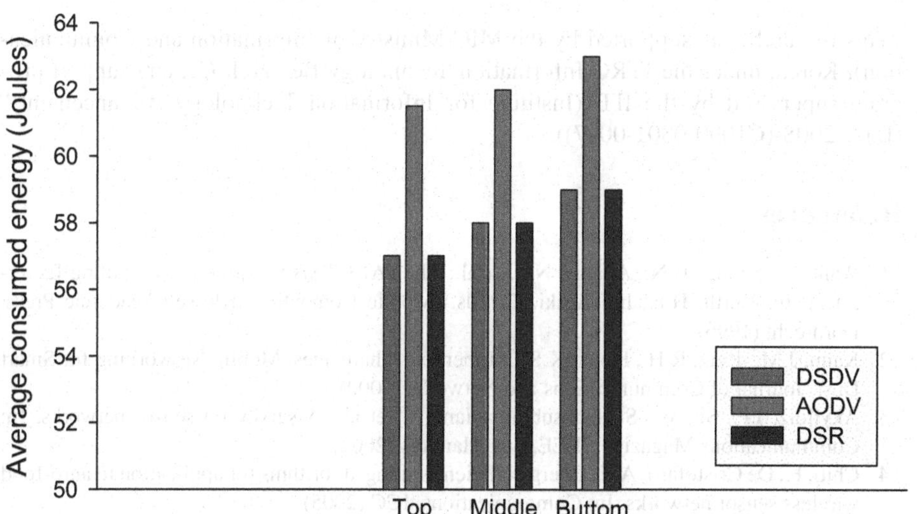

Fig. 7. Average consumed energy (Joules)

the position of sensor node is not changed and the change of crop's growth obstruct the communication of sensor node hasn't been considered. Accordingly, the change of routing path needs to be minimized considering a gradual communication jamming by crop's growth. Also, when the communication jamming occurs, there needs to be the routing method to obtain the stable route through the active attempt. In addition, the sensor node to be installed in the lower part of greenhouse must guarantee perpetuity of sensor network by using the fixed power or adding more initial power than other nodes.

5 Conclusion

Through this study, the routing method for the sensor node to be installed in the greenhouse environment was researched. We can know that it is necessary to apply the different routing model from the existing one due to the difference between the top part and the bottom part of the greenhouse as the crop grows through derived results. We also can know that the obtainment of visual field for the good data exchange must be considered even after the growth as well as the beginning of the sensor installation.

Sensor node that has the purpose of measuring and transmitting the physical phenomena will require diversified methods according to the application in the installation. We will search various factors in addition to the obstruction of crop through applying to the actual greenhouse going one step forward from the simulation model and will continue the study to solve them. We consider that we will be able to find out the suitable routing method for the agricultural application through it.

Acknowledgement

"This research was supported by the MIC(Ministry of Information and Communication), Korea, under the ITRC(Information Technology Research Center) support program supervised by the IITA(Institute for Information Technology Advancement)" (IITA-2008-(C1090-0801-0047)).

References

1. Want, R., Schilit, B.N., Adams, N.I., et al.: The PARCTAB Ubiquitous Computing Experiment. In: Korth, H.F., Imielinski, T. (eds.) Mobile Computing. Kluwer Academic Press, Dordrecht (1996)
2. Kahn, J.M., Katz, R.H., Pister, K.S.J.: Emerging Challenges: Mobile Networking for Smart Dust. Journal of Communications and Networks (2000)
3. Akyildiz, I.F., Su, W., Sankarasubramaniam, Y., et al.: A survey on sensor networks. In: Communications Magazine. IEEE, Los Alamitos (2002)
4. Chiti, F., De Cristofaro, A.F.: Energy efficient routing algorithms for application to agro-food wireless sensor networks. In: Communications, ICC (2005)
5. Jeong, S.W., Chung, J.H., Limet, K.H., et al.: A study on Indoor Temperature Distribution of Cultivated Paprika at greenhouse in Summer. In: SAREK, pp. 435–440 (1998)
6. DeCruyenaere, J.P., Falconer, D.: A shadowing model for prediction of coverage in fixed terrestrialwireless systems. In: Vehicular Technology Conference, VTC 1999-Fall, IEEE VTS 50th (1999)

7. Shah, R.C., Rabaey, J.M.: Energy Aware Routing for Low Energy Ad Hoc Sensor Networks. IEEE WCNC 2002 1, 350–355 (2002)
8. Johnson, D., Maltz, D.D.: Dynamic source routing in ad hoc wireless networks. In: Imielinski, T., Korth, H. (eds.) Mobile Computing, pp. 153–181. Kluwer Academic Publishers, The Netherlands (1996)
9. Dixon, R.C., Pitt, D.A.: Addressing, bridging, and source routing. IEEE Network 2(1), 25–32 (1988)
10. Perkins, C., Bhagwat, P.: Highly dynamic destination-sequenced distance-vector (dsdv) routing for mobile computers. In: Proceedings of ACM SIGCOMM (1994)
11. Cheng, C., Riley, R., Kumar, S.P.R., Garcia-Luna-Aceves, J.J.: A loop-free Bellman-Ford routing protocol without bouncing effect. In: ACM SIGCOMM 1989, pp. 224–237 (September 1989)
12. Perkins, C., Royer, E.: Ad hoc On demand Distance Vector Routing. In: Proc. of 2nd IEEE Workshop on Mobile Computing Systems and Applications (February 1999)
13. Schurgers, C., Srivastava, M.B.: Energy efficient routing in wireless sensor networks. MILCOM (2001)
14. Heinzelman, W.R., Chandrakasan, A.P., Balakrishnan, H.: Energy-Efficient Communication Protocol for Wireless Microsensor Networks. In: Proc. 33rd Hawaii Int'l. Conf. Sys. Sci. (January 2000)
15. Shah, R.C., Rabaey, J.M.: Energy Aware Routing for Low Energy Ad Hoc Sensor Networks. In: Proc. IEEE Wireless Commun. and Network Conf., vol. 1, pp. 350–355 (March 2002)
16. Burrell, J., Brooke, T., Beckwith, R.: Vineyard computing: sensor networks in agricultural production. In: Pervasive Computing, IEEE, Los Alamitos (2004)
17. MICA2 Datasheet: http://www.xbow.com/Products/Product_pdf_files/Wireless_pdf/MICA2_Datasheet.pdf
18. Fall, K., Varadhan, K.: The ns Manual (formerly ns Notes and Documentation). The VINT Project, A Collaboration between researches at UC Berkeley, LBL, USC/ISI and Xerox PARC
19. Habaebi, M.H., Abduljali, E., Ragaibi, T., et al.: Effect of sensor specific body location on wireless network routing performance. Electronics Letters (2008)
20. Gruber, I., Knauf, O., Li, H.: Performance of Ad Hoc Routing Protocols in Urban Environments. In: Proceedings of European Wireless (2004)

An Industrial Study Using UML Design Metrics for Web Applications

Emad Ghosheh[1], Sue Black[2], and Jihad Qaddour[3]

[1] AT & T, One Bell Center
Saint Louis, MO, USA
eg2534@att.com

[2] Department of Information and Software Systems,
Harrow School of Computer Science
University of Westminster
London HA1 3TP, UK
s.e.black@westminster.ac.uk

[3] School of Information Technology
Illinois State University,
Normal, IL 61790-5150, USA
jqaddou@ilstu.edu

Summary. Many web applications have evolved from simple HTML pages to complex applications that are difficult to maintain. In order to control the maintenance of web applications quantitative metrics and models for predicting web applications maintainability must be used. This paper introduces new design metrics for measuring the maintainability of web applications from class diagrams. The metrics are based on Web Application Extension (WAE) for UML and measure the design attributes of size, complexity, and coupling. The paper describes an experiment carried out using a CVS repository from a US telecoms web application. A relationship is established between the metrics and maintenance effort measured by the number of lines of code changed.

Keywords: Web applications, metrics, maintainability.

1 Introduction

A high percentage of World Wide Web applications incorporate important business assets and offer a convenient way for businesses to promote their services through the Internet. Many of these web applications have evolved from simple HTML pages to complex applications that are difficult to maintain. This can be explained by the Laws of Software Evolution [1]. Two Laws [1] that affect the evolution of web applications are:

- First Law-Continuing change: A program used in the real world must change or eventually it will become less useful in the changing world.
- Second Law-Growing complexity: As a program evolves it becomes more complex and extra resources are needed to preserve and simplify its structure.

In order to control the maintenance of web applications, quantitative metrics for predicting web applications maintainability must be used. Web applications are different

R. Lee and H.-K. Kim (Eds.): Computer and Information Science, SCI 131, pp. 231–241, 2008.
springerlink.com

from traditional software systems because they have special features such as hypertext structure, dynamic code generation and heterogeneity that cannot be captured by traditional and object-oriented metrics. Hence metrics for traditional systems cannot be applied to web applications. Many web application metrics have been proposed [2,3,4,5], but it is critical to validate them to give confidence that they are meaningful and useful.

This paper describes an industrial experiment that shows the relationship between the metrics and the maintenance effort, measured by the number of lines of code changed. The remainder of this paper is organized as follows: Section II provides a review on related work. Section III describes the industrial case study carried out at a telecommunication company. Section IV discusses results and analysis. Finally, section V provides a conclusion.

2 Related Work

2.1 General

Modeling is a technique used to represent complex systems at different levels of abstraction, and helps in managing complexity. UML [6] is an object-oriented language [7] that can be used to model object-oriented systems. Web applications are not inherently object-oriented, therefore, it is difficult to use UML to model web applications, but UML has now been enhanced with extensions to capture the various elements of web applications. Conallen proposed an extension of UML for web applications [7], Fig 1 shows the elements of the Conallen web application model.

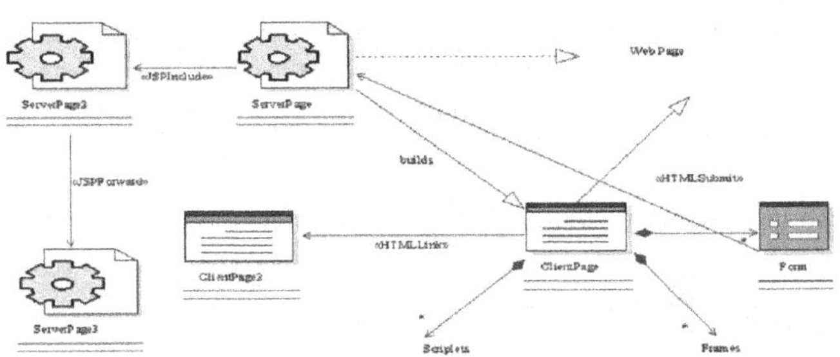

Fig. 1. Web Applications Model

The important elements of Conallen's model are as follows: [7]:

- *Web Page*: A web page is the primary element of a web application. It is modeled with two separate stereotyped classes, the client page and the server page. The client page contains client side scripts and user interface formatting. The server page contains server methods and page scoped variables.
- *Relationships*: The model defines the following relations between different components: builds, redirects, links, submit, includes, and forwards. The builds relationship is a directional relationship from the server page to the client page. It shows

the HTML output coming from the server page. The redirects relationship is a directional relationship that requests a resource from another resource. The links relationship is an association between client pages and server or client pages. It models the anchor element in HTML. The links relationship can have parameters which are modeled as attributes in the relationship. The submit relationship is a relationship between the form and the server page that processes it. The include relationship is a directional association between a server page and another client or server page. The forward relationship is a directional relationship between a server page and a client or server page. This presents delegating the server request to another page.

- *Forms*: Forms are defined to separate the form processing from the client page. The form element contains field elements. Forms are contained in client pages. Each form submits to a different action page.
- *Components*: Components run on the client or server page. ActiveX controls and Applets are examples of components.
- *Scriplet*: A scriplet contains references to components and controls that are re-used by client pages.
- *Framesets*: A frameset divides the user interface into multiple views each containing one web page. Frames can contain more than one client page, but they must contain at least one client page.
- *XML*: An XML element is a hierarchical data representation that can be passed back and forth between client and server pages.

2.2 UML Design Metrics for Web Applications

Most of the studies related to maintainability metrics have looked at structured and object-oriented systems. Little work has been done in this regard on web applications. Maintainability can be defined as:

the ease with which a software system or component can be modified to correct faults, improve performance or other attributes, or adapt to a changed environment [8, 9].

Most studies use source code metrics for measuring maintainability. Some studies in the object-oriented domain measure maintainability using understandability and modifiability [10, 11, 12]. Some of these studies used UML diagrams for measuring maintainability metrics. These models can be used in the web domain. However, the UML diagrams must be adapted to the web domain by using UML extensions and stereotypes.

There are a couple of studies in the web domain. Web Application Maintainability Model (WAMM) [4] is one which used source code metrics and the maintainability was measured using the Maintainability Index. Mendes [13] *et al* used design and authoring effort as the dependent variables while the independent variables were based on source code metrics. In [14] design metrics were introduced based on W2000 which is a UML like language. In the study the dependent variables were variations of design effort. The independent variables were measured from the presentation, navigational and information models. For more details on web application maintainability approaches see [4, 13, 14] and for the authors previous research upon which this experiment is based please refer to [2, 15].

This study defines metrics for the following design attributes: size, complexity, and coupling. The metrics are based on Web Application Extension (WAE) for UML and measure attributes of class diagrams. Table 1 provides a description of the metrics. The metrics use the different components of WAE as units of measurement. The metrics are categorized into the following categories: Size Metrics(NServerP, NClientP, NWebP, NFormP), Structural Complexity metrics(NLinkR, NSubmitR, NBuildsR, NForwardR, NIncludeR), Coupling Metrics(WebControlCoupling). In the next section the metrics are used in an empirical case study, to show how they can be applied and used in the real world.

3 Empirical Case Study

The proposal of new metrics is not helpful if their practical use is not proved through empirical studies. This study is an explorative case study conducted at a telecommunication company in the US. The data is collected from a CVS repository.

3.1 Case Study Context

The web application used is from the telecommunication Operational Support System (OSS) domain. It is a provisioning application which is used to provision and activate the wireless service in the network. We refer to the web application as ProvisionApp. ProvisionApp has around 10,000 users of which 2,500 are concurrent. It is a critical application that is used by customer care advocates to resolve provisioning issues for wireless subscribers. ProvisionApp is built using the latest web technologies and frameworks such Struts [16], and EJBs [17] and uses Oracle [18] for the database. Table 2 shows the characteristics of ProvisionApp.

The dependent variable for this study is maintenance effort measured by the number of lines of code changed. Maintenance effort has been used in some studies to validate object-oriented metrics [19, 20]. The independent variables are the metrics defined in Table 1.

Table 1. Web Application Design Metrics

Metric Type	Description
Size	Total number of server pages (NServerP)
	Total number of client pages (NClientP)
	Total number of web pages (NWebP)=(NServerP + NClientP)
	Total number of form pages (NFormP)
Structural Complexity	Total number of link relationships (NLinkR)
	Total number of submit relationships (NSubmitR)
	Total number of builds relationships (NbuildsR)
	Total number of forward relationships(NForwardR)
	Total number of include relationships(NIncludeR)
Control Coupling	Number of relationships over number of web pages: WebControlCoupling = (NLinkR + NSubmitR + NbuildsR + NForwardR + NIncludeR)/ NWebP)

Table 2. Characteristics of ProvisionApp

Characteristic	**ProvisionApp**
Application Domain	Telecom
Age	4
Language	Java
Web Server	WebLogic 7.0
Application Server	WebLogic 7.0
Framework	Struts 1.1
Database	Oracle
Configuration Management Tool	CVS
Design Tool	Rational Rose

3.2 Hypothesis

This study is trying to answer the following question: Is there a relationship between the metrics identified in Table 1(NServerP, NClientP, NWebP, NFormP, NLinkR, NSubmitR, NBuildsR, NForwardR, NIncludeR, WebControlCoupling) and maintenance effort?

The following hypotheses are investigated:

- H1: the higher the size metrics of the class diagram, the higher the maintenance effort
- H2: the higher the structural complexity metrics of the class diagram, the higher the maintenance effort
- H3: the higher the coupling metrics of the class diagram, the higher the maintenance effort

3.3 Data Collection

In this experiment an IBM tool: Rational XDE [21] is used to reverse engineer the ProvisionApp web application. Rational XDE combines visual modeling with a Java forward and reverse engineering tool. The user can import the source code into Rational XDE and generate class diagrams from the imported code. The ProvisionApp code was

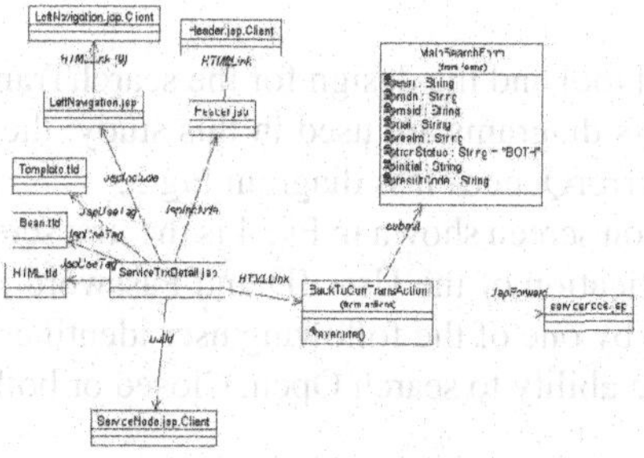

Fig. 2. Search Transaction Design

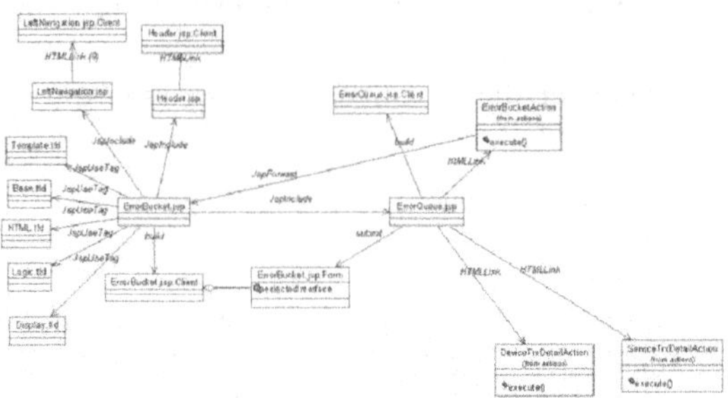

Fig. 3. Error Queue Design

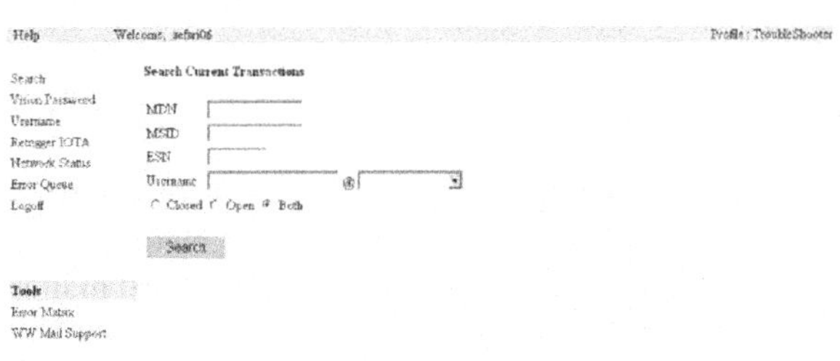

Fig. 4. Search Transaction Screen

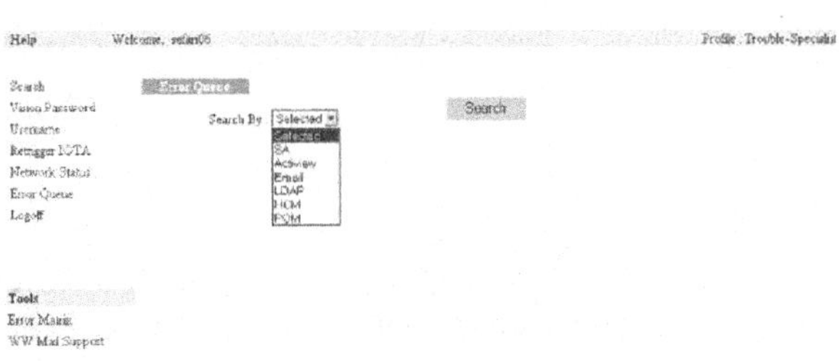

Fig. 5. Error Queue Screen

imported to the Rational tool and the design for the searchTransaction, and errorQueue was generated. Two class diagrams are used in this study: the searchTransaction class diagram Fig 2, and the errorQueue class diagram Fig 3.

The Search Transaction screen shown in Fig 4 is the first screen presented to the User upon successful authentication of the User ID and Password. It provides the ability to Search for a transaction by one of the following user identifiers: MDN, MSID, ESN or NAI. It also provides the ability to search Open, Closed or both transactions.

The Error Queue screen shown in Fig 5 has the following abilities:

- Provides a list of Error Transactions by System/Network Element
- Provides the ability to sort Erred transactions by Error or Create Date/Time
- Provides ability to get 'more info' about the Transaction

The metrics were computed from the class diagrams and recorded in Table 3. The maintenance effort was computed by adding the total number of lines added or deleted from a class diagram during a maintenance task.

4 Analysis and Results

4.1 Analysis

The main objective in this study is to explore the relationship between the metrics identified in Table 1 and maintenance effort. Accordingly, we focused on the relationship between these and the amount of maintenance effort which was computed by adding the total number of lines added or deleted from a class diagram during a maintenance task. Analysis of variance (ANOVA) was conducted to determine if the mean Lines of Code for the high and low metric values were significantly different, to be able to accept hypothesis H1, H2 and H3. The unit of measurement used is Lines of Code changed. 20 Data points were collected for the Search Transaction and Error Queue diagrams shown in Figures 6 and 8. The first step of the analysis procedure is to check the normality of the data. If the data is substantially normal, then mean can be used to summarize the data. The Shapiro-Wilk W test is used because it is one of the most reliable test for non-normality for small to medium sized samples [22]. The null hypothesis for this test

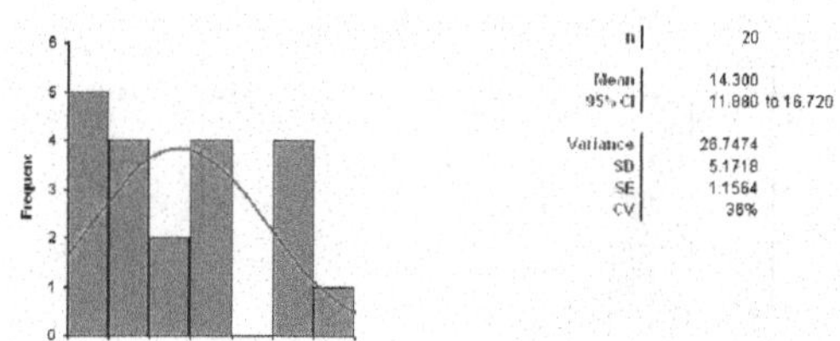

Fig. 6. Search Transactions Histogram

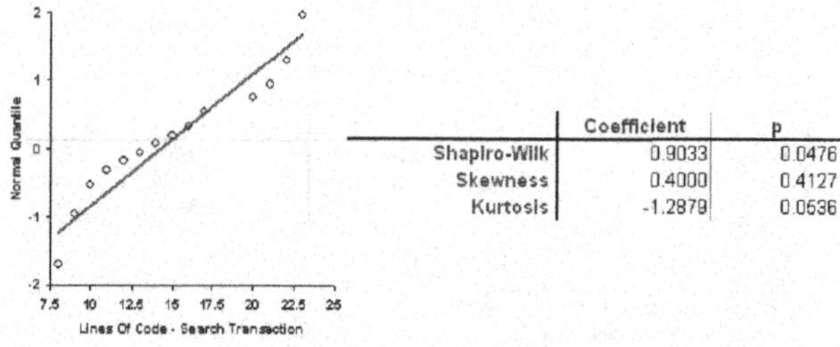

	Coefficient	p
Shapiro-Wilk	0.9033	0.0476
Skewness	0.4000	0.4127
Kurtosis	-1.2879	0.0536

Fig. 7. Search Transactions Line Plot

is that the data is normally distributed. The chosen alpha level is 0.02. If the p-value is less than 0.02, then the null hypothesis that the data are normally distributed is rejected. If the p-value is greater than 0.02, then the null hypothesis is not rejected.

4.2 Results

Figures 6 and 8 show the data for Search Transaction and Error Queue diagrams plotted against a cumulative frequency distribution. The heights of the columns for both diagrams are roughly shaped on the superimposed blue line which means the data is normally distributed. Comparing against the standard S-shaped cumulative frequency distribution is sometimes difficult, so Fig 7 and Fig 9 show the data for both diagrams transformed so it can be compared against a straight-line. Both the Search Transaction and Error Queue diagrams follow the blue straight-line which means the data is normally distributed. Figure 7 shows the p-value of 0.0476 for the Search Transaction diagram which means the null hypothesis is not rejected and the data is normally distributed. Figure 9 shows the p-value of 0.6772 for the Error Queue diagram which means the null hypothesis is not rejected and the data is normally distributed. Since, the data is normally distributed the mean can be used to summarize the data. In Table 3 the mean is used to summarize Lines of Code. Table 3 shows the mean Lines of Code for the Search Transaction and Error Queue class diagrams. The Error Queue class diagram has higher size metrics, structural complexity metrics, and higher coupling metrics than the Search Transaction class diagram. It is expected to have greater maintenance effort. When looking at the ANOVA results for the Lines of Code, one can notice that the mean Lines of Code for the Error Queue diagram is 19.85 Lines of Code which is higher than the mean Lines of Code for the Search Transaction diagram around 14.3

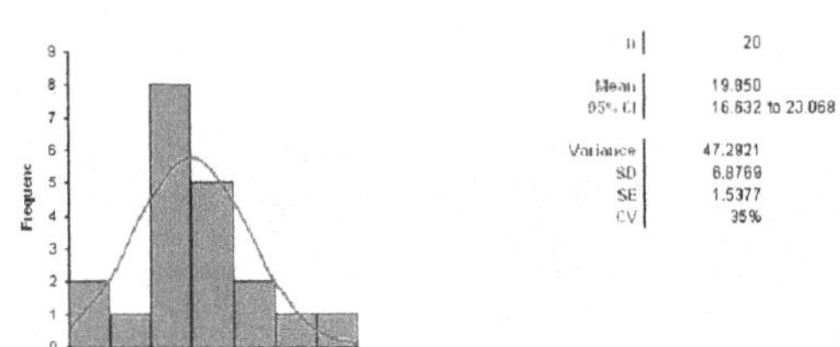

Fig. 8. Error Queue Histogram

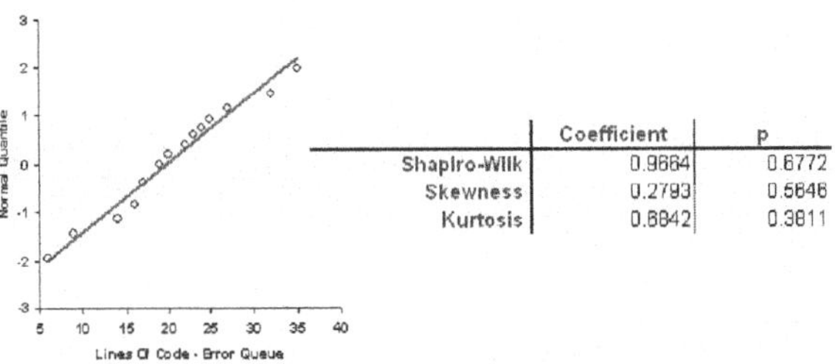

	Coefficient	p
Shapiro-Wilk	0.9664	0.6772
Skewness	0.2793	0.5646
Kurtosis	0.6842	0.3811

Fig. 9. Error Queue Line

Table 3. Results & Analysis

Metric Type	Metric Name	Error Queue	Search Transaction
Size	NServerP	4	4
Size	NClientP	4	3
Size	NWebP	8	7
Size	NFormP	1	1
Structural Complexity	NLinkR	13	11
Structural Complexity	NSubmitR	1	1
Structural Complexity	NBuildsR	4	3
Structural Complexity	NForwardR	1	1
Structural Complexity	NIncludeR	3	2
Structural Complexity	NUseTagR	5	3
Control Coupling	WebControlCoupling	3.38	3
Lines of Code	Mean Lines of Code	19.85	14.3

Table 4. ANOVA Analysis

Name	F	P
Search Transaction vs Error Queue Lines of Code	8.32	.0064

LOC. One can notice that the differences are statistically significant (P value of 0.0064) as shown in Table 4. Thus, one can say that a class diagram with higher size metrics, higher structural complexity metrics, and higher coupling metrics requires more Lines of Code. Thus we can accept hypothesis H1, H2, and H3. As a conclusion, this study can serve as a basis for future studies, and can provide a first indication of the use of the newly introduced metrics.

5 Conclusions and Future Work

This study has introduced metrics for measuring the maintainability of web applications from class diagrams. The metrics are based on Web Application Extension (WAE) for UML. In this study an industrial experiment at a Telecommunication Company is carried out in order to show the relationship between the metrics and the maintenance effort measured by the number of lines of code changed. Analysis of variance (ANOVA) was conducted and showed that the mean Lines of Code has a relationship with the metrics. The exploratory experiment showed that we can accept hypothesis H1, H2, and H3. Thus, higher size metrics, higher structural complexity metrics, and higher coupling metrics result in higher maintenance effort.

The results just give a first indication of the usefulness of the UML design metrics. In the future we will provide statistical analysis to determine the strength of the relationship between the individual metrics and the measured maintenance effort.

References

1. Lehman, M., Ramil, J., Wernick, P., Perry, D., Turski, W.: Metrics and laws of software evolution the nineties view. In: Proceedings of the 4th International Software Metrics Symposium, pp. 20–32. IEEE Computer Society Press, Los Alamitos (1997)
2. Ghosheh, E., Qaddour, J., Kuofie, M., Black, S.: A comparative analysis of maintainability approaches for web applications. In: Proceedings of the 4th ACS/IEEE International Conference on Computer Systems and Applications, p. 247. IEEE Computer Society Press, Los Alamitos (2006)
3. Mendes, E., Mosley, N., Counsell, S.: Early web size measures and effort prediction for web costimation. In: Proceedings of the 9th International Software Metrics Symposium, pp. 18–39. IEEE Computer Society Press, Los Alamitos (2003)
4. DiLucca, G., Fasolino, A., Tramontana, P., Visaggio, C.: Towards the definition of a maintainability model for web applications. In: Proceeding of the 8th European Conference on Software Maintenance and Reengineering, pp. 279–287. IEEE Computer Society Press, Los Alamitos (2004)
5. Ruhe, M., Jeffery, R., Wieczorek, S.: Using web objects for estimating software development effort for web applications. In: Proceedings of the 9th International Software Metrics Symposium, pp. 30–37. IEEE Computer Society Press, Los Alamitos (2003)
6. Fowler, M.: UML Distilled: A Brief Guide to the Standard Object Modeling Language, 2nd edn. Addison-Wesley, Reading (2005)
7. Conallen, J.: Building Web Applications with UML, 2nd edn. Addison-Wesley, Reading (2003)
8. Bhatt, P., Shroff, G., Misra, A.: Dynamics of software maintenance. ACM SIGSOFT Software Engineering Notes 29(4), 1–5 (2004)
9. Oman, P., Hagemeister, J.: Construction of testing polynomials predicting software maintainability. Journal of Software Systems 27(3), 251–266 (1994)
10. Briand, L., Bunse, C., Daly, J.: A controlled experiment for evaluating quality guidelines on the maintainability of object-oriented designs. IEEE Transactions on Software Engineering 27(06), 513–530 (2001)
11. Mario, M., Manso, E., Cantone, G.: Building UML class diagram maintainability prediction models based on early metrics. In: Proceedings of the 9th International Software Metrics Symposium, pp. 263–278. IEEE Computer Society Press, Los Alamitos (2003)
12. Kiewkanya, M., Jindasawat, N., Muenchaisri, P.: A methodology for constructing maintainability model of object-oriented design. In: Proceedings of the 4th International Conference on Quality Software, pp. 206–213. IEEE Computer Society Press, Los Alamitos (2004)
13. Mendes, E., Mosley, N., Counsell, S.: Web metrics - estimating design and authoring effort. IEEE Multimedia 08(01), 50–57 (2001)
14. Baresi, L., Morasca, S., Paolini, P.: Estimating the design effort of web applications. In: Proceedings of the 9th International Software Metrics Symposium, pp. 62–72. IEEE Computer Society Press, Los Alamitos (2003)
15. Ghosheh, E., Black, S., Qaddour, J.: An introduction of new UML design metrics for web applications. International Journal of Computer & Information Science 8(4), 600–609 (2007)
16. Goodwill, J.: Mastering Jakarta Struts, 1st edn. Wiley, Chichester (2002)
17. Roman, E., Ambler, S., Jewell, T., Marinescu, F.: Mastering Enterprise JavaBeans, 2nd edn. Wiley, Chichester (2001)
18. Loney, K., Koch, G.: Oracle9i: The Complete Reference, 1st edn. McGraw-Hill Osborne Media, New York (2002)

19. Li, W., Henry, S.: Object-oriented metrics that predict maintainability. Journal of Software Systems 23(2), 111–122 (1993)
20. Li, W., Henry, S., Kafura, D., Schulman, R.: Measuring object-oriented design. Journal of Object-Oriented Programming 8(4), 48–55 (1995)
21. Boggs, W., Boggs, M.: Mastering Rational XDE, 1st edn. Sybex (2003)
22. Conover, W.: Practical Nonparametric Statistics, 3rd edn. Wiley, Chichester (1999)

An Ontology-Based Model for Supporting Inter-process Artifact Sharing in Component-Based Software Development

Hye-Kyeong Jo and In-Young Ko

School of Engineering, Information and Communications University
119 Munji-Ro, Yuseong-Gu, Daejeon, Republic of Korea
{hgcho,iko}@icu.ac.kr

Summary. To improve software reuse, it is essential to allow developers to effectively find and reuse existing components that meet the requirements of software development. Also, relevant artifacts need to be reused across the entire software development life cycle. In this paper, we propose an ontology-based model by which developers can semantically describe and find artifacts based on the common aspects and properties of component-based software development processes. Using our model, developers, who work on different projects that are developed under different software processes, can share artifacts that satisfy similar functional or non-functional requirements. We have applied our models to RUP and ADDMe, the most popular software development processes in the Korean military domain. This paper also proposes the artifact classification mechanism based on the model and an example of classifying artifacts of RUP and ADDMe. We developed a prototype system to show the effectiveness of the approach.

1 Introduction

Nowadays, because of the advancement of component based software frameworks and standards, we have more and more components that are candidates to be reused for multiple projects. To improve software reuse, it is essential to allow developers to effectively find and reuse existing artifacts that meet the requirements of software development. Not only software components but also relevant artifacts need to be reused across the entire software development life cycle. In addition, artifacts that are developed by using different software development processes need to be effectively shared.

Frakes defines that software reuse is the use of existing software or software knowledge to build new software [1]. In this definition, software and software knowledge include components and related artifacts such as documents and models. There have been many efforts to improve reusability in developing software. One of the most popular efforts includes constructing and using software components by using Component-Based software Development (CBD). CBD96 [2], Catalysis [3], and UML Components [4] are well-known CBD processes. They provide models and processes to help developers build and reuse components, and to allow the composition of applications from existing components.

R. Lee and H.-K. Kim (Eds.): Computer and Information Science, SCI 131, pp. 243–254, 2008.
springerlink.com

For developers to effectively find and reuse components and related artifacts that are developed under different CBD processes, we need to have common description models that explicitly represent the generic properties of artifacts. Using these models, we can automate the process of comparing and identifying similar types of artifacts developed for different applications. This will improve reusability of the artifacts across different software development processes and different projects.

We propose an ontology-based artifact description model that is for representing essential properties of common artifacts. We have developed the artifacts classification mechanism for CBD processes. These processes are commonly used in the Korean defense software domain. By using the description model and the artifact classification mechanism, artifacts can be semantically described and searched based on a certain set of requirements. A prototype tool has been developed for developers to effectively find and reuse artifacts developed in different processes and projects.

Our work contribution includes a reuse framework that enables developers to find appropriate components and related artifacts that meet the early requirements of a software development project. The reuse process can be systematically guided and supported in the overall life cycle of software development.

This paper is organized as follows. Section 2 describes an ontology-based artifact description model. Section 3 explains the structure of RUP and ADDMe and the artifact classification mechanism. Section 4 presents the method to semantically describe artifacts. Section 5 explains a prototype implementation of our approach. Section 6 describes the related works about component search and classification, and CBD processes. Conclusions and future works are discussed in Section 7.

2 The Ontology-Based Artifact Description Model

This section describes an ontology-based artifact description model that includes common concepts and properties of various artifacts that are the keys to finding and reusing artifacts.

Figure 1 shows the ontology-based artifact description model. In Figure 1, the concept *artifact* has properties such as *writtenBy, kind, name, version, id, synonym, creationDate,* and so on. The property *kind* indicates the type of artifact such as text, diagram, or code. The property *synonym* is for keeping terms that are equivalent to the artifact name. In addition, the concept *project* provides us with project information such as the used software process and the developed system. The concept *person* and *organization* help us understand who has developed the current project and what organization has employed him. The concept *process* gives information about which step, which activity, and which task the current artifact belongs to. This description model will be evolved over time. More information will be represented at the model in the future.

The model enables developers to register and retrieve artifacts based on their semantic information. It provides artifact search information in a threedimensional perspective for developers; ***project*** where search artifacts were generated in (①), ***process*** used to create the artifacts (②), ***person*** to make them (③). It is used as a base model for the sharing and reuse of artifacts in different projects and processes.

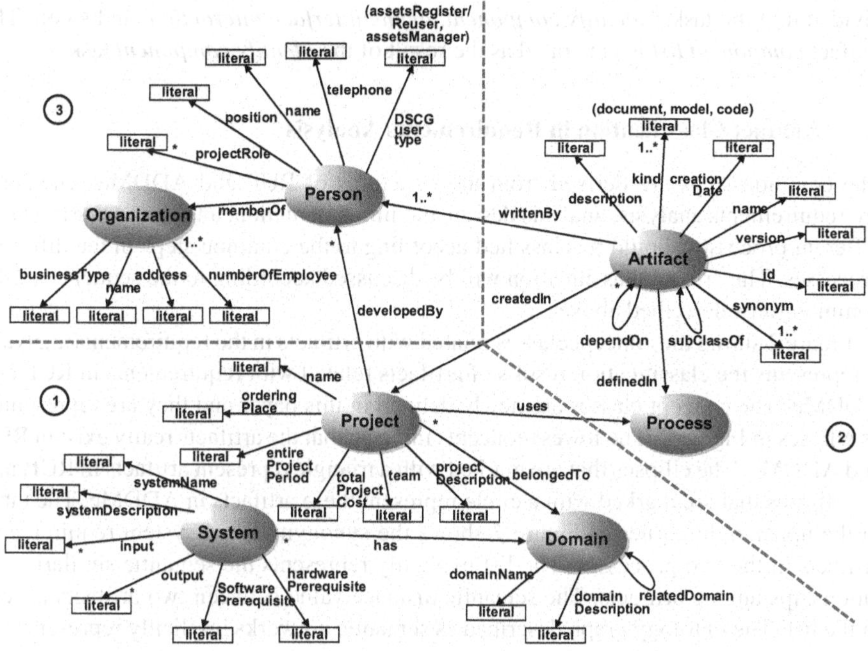

Fig. 1. The artifact description model

3 The Artifact Classification Mechanism

This section introduces a mechanism to classify artifacts and to discover the semantic similarity relationships among artifacts. RUP and ADDMe are used as examples to illustrate the artifact classification mechanism. They are software processes used to develop components in the Korean defense domain. The structures of RUP and ADDMe will be first described in order to explain the artifact classification mechanism.

3.1 The Structures of RUP and ADDMe

RUP and ADDMe are inspected one after the other, and then the structures of them are described. RUP has both phases and workflows as core elements. The phases of RUP include inception, elaboration, construction, and transition. The workflows of RUP include business modeling, requirements elicitation, analysis and design, implementation, testing, and deployment. A workflow consists of workers, artifacts, and activities. The activity is a basic action for a developer to produce artifacts in a workflow.

ADDMe is composed of 4 steps, 12 activities, and 37 tasks. The steps of ADDMe include analysis, design, implementation and testing, and delivery. Each step may have some activities. An activity may be further discomposed into tasks. Artifacts are developed by completion of a specific task. For instance, the activity, *outline design*, is divided into the tasks: *identify component, define interface interaction,* and so on. The artifact *component list* is generated as the result of the *identify component* task.

3.2 Artifact Classification in Requirements Analysis

The common steps are derived from the structures of RUP and ADDMe. The steps are requirements analysis, analysis design and implementation, and test. All artifacts in different processes should be classified according to the common steps in the different processes. The artifact classification will be discussed according to the sequence of the common steps described above.

Figure 2 shows the concept classes related with artifacts in the requirements analysis. It represents the classification results of artifacts related with requirements in RUP and ADDMe. The concept classes are newly defined in this paper and they are represented as ellipses in Figure 2. The lowest concepts indicate that the artifacts really exist in RUP and ADDMe. The ellipses that are marked with a triangle represent artifacts in RUP, and the ellipses that are marked with a circle represent these artifacts in ADDMe. The table in the upper right corner of Figure 2 shows the synonym list of system requirements artifacts in the two processes. The list explicitly represents the semantic similarity relationships among artifacts. The semantic distance value between two artifacts is zero in the list. The ontology graph described as semantic networks implicitly represents the semantic similarity of the artifacts.

As an example of the listing of synonyms, *dictionary* in ADDMe is a synonym for *glossary* in RUP. The *use-case* specification artifact in ADDMe is a synonym for the *use-case model* artifact and the *use-case story board* artifact in RUP. It means that the *use-case specification* artifact has the contents of the two artifacts in RUP. If two artifacts hold closely similar information to each other, they become the synonym for each other in our approach.

The synonym list and the ontology graph are used to identify the semantic relationships between any artifacts in different software processes. They help developers search and reuse artifacts produced by the different processes with the process currently used in the developing system. All the properties of concepts in Figure 1 are inherited by all artifact concepts in Figure 2.

As you can see in Figure 2, many artifacts that exist in ADDMe are similar to those found in RUP. The reason is because most development teams go through the process of business requirements analysis, system requirements analysis, architecture design, component design, and implementation. So it is reasonable to have the semantic similarity relationships among artifacts in different software processes.

The concept classes with no marks were newly generated in here. They are the concepts such as *business requirements analysis artifact, system requirements analysis artifact*, and so on. They are not real-world artifacts. They are used as criteria for the artifacts classification when we want to classify artifacts in a new software process.

Two artifact classification ontologies were also developed for the analysis design and implementation step, and the test step. They represent a semantic relationship among artifacts in a software process and different processes. Due to a shortage of space, a description of the two ontologies is omitted herein.

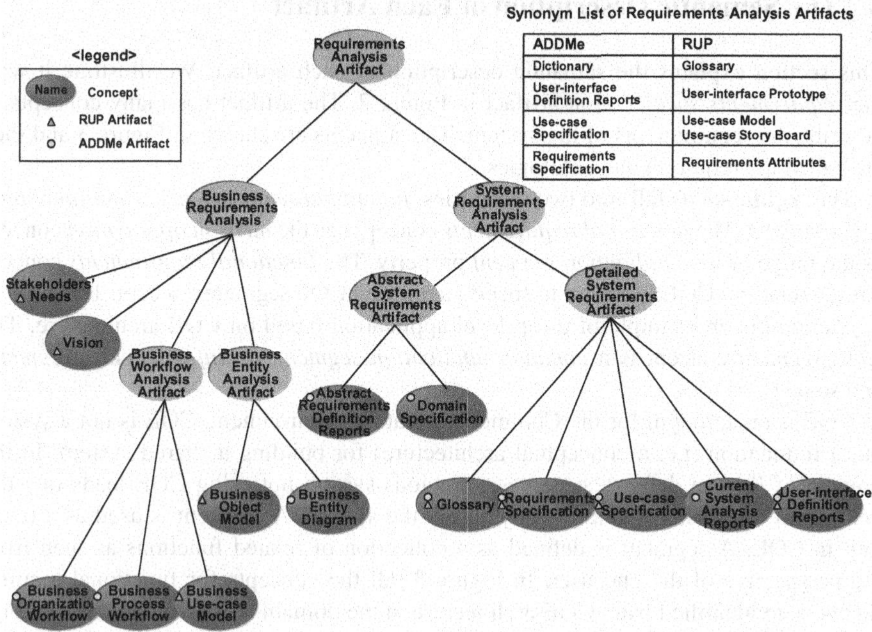

ADDMe	RUP
Dictionary	Glossary
User-interface Definition Reports	User-interface Prototype
Use-case Specification	Use-case Model Use-case Story Board
Requirements Specification	Requirements Attributes

Fig. 2. The artifact classification ontology for the requirements analysis step

3.3 The Artifact Classification Mechanism

The artifact classification mechanism is implemented in the following sequence:

1. Define the common steps from phases in different software development processes.
2. Generate new ontology concept classes used by criteria for artifacts classification after we analyze the semantic relationship among artifacts in different processes.
3. Classify the artifacts in each software development process.
4. Define synonyms for artifacts in different processes.
5. Extract the properties of concepts from real-world artifacts.
6. Define that each artifact is associated with any artifacts that exist in other processes.
7. Complete ontology graphs for artifacts in different processes.

The classification mechanism helps us to build the artifact classification ontologies for a new CBD process. Whenever we classify artifacts in a new software development process, the concepts are propagated and the ontology graphs are extended. It is more practicable to calculate the semantic distance between artifacts and to distinguish the semantic similarity relationship between artifacts by using the ontologies.

The concepts and properties can also be used to automatically generate templates for artifact descriptions. The automatically generated templates will help developers easily register artifacts in the repository and the templates will guide developers for documentation on artifacts.

4 The Semantic Description of Each Artifact

This section explains the semantic description of each artifact. We illustrate it with the *requirements specification* artifact in Figure 2. The artifact has many concepts to describe the requirements of the system. The concepts are shown in Figure 3 and they are called the requirements ontologies.

The *requirements* fall into two categories: *functional requirements* or *nonfunctional requirements*. The *functional requirements* concept has the *mission application* concept as the range of the *application segment* property. The *functional requirements* concept has the relation 'COESegment' to specify a relevant COE segment concept. The *mission application* is an example of a top-level application based on COE architecture. The *COE segment* concept is a *common application segment, an infrastructure segment*, and so on.

COE is an acronym for the Common Operating Environment. COE is not a system but a foundation (i.e. a conceptual architecture) for building a shared system. In the domain of Korean defense systems, a previous system not using COE leads to redevelopment of the same functionality across the system. A segment is used as a reuse unit in COE. A segment is defined as a collection of related functions as seen from the perspective of the end user. In Figure 3, all the concepts for functional requirements were identified from COE architecture in the domain of Korean defense. In Figure 3, all the concepts for non-functional requirements were defined by IEEE standard

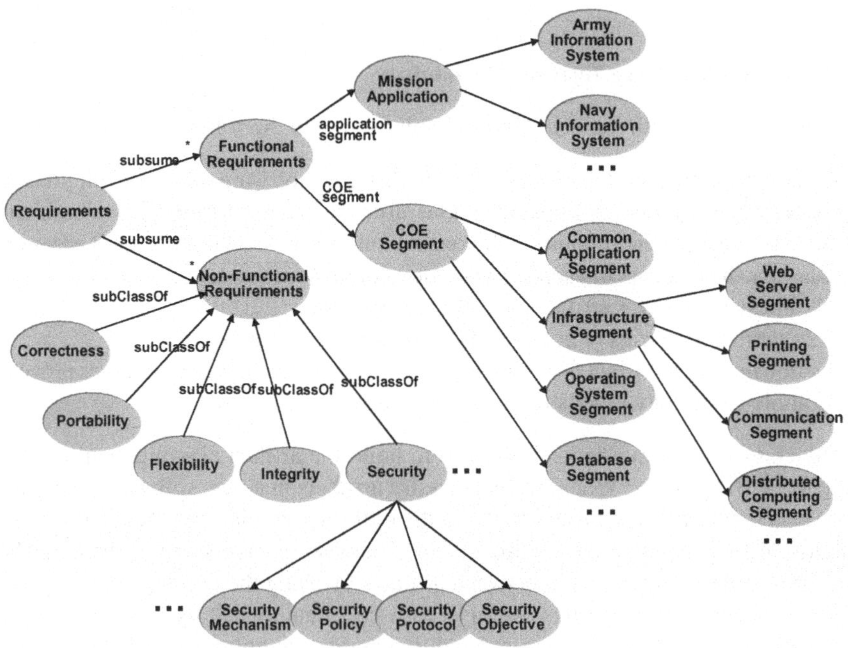

Fig. 3. The requirements ontology

1061-1998 [5]. The concepts related to security were developed by the USA Navy security ontology [6].

Figure 3 shows that an artifact can be semantically described by an ontology in our model. It helps developers easily search artifacts to meet their requirements across different projects. The ontology description is used to give efficient search information to a description of real-world artifact.

Gruber defines that ontology as an explicit specification of a conceptualization [7]. The ontology has been used in various research areas and its use has been expanded.

5 Implementation

Our approach mentioned above is applied to a real-world artifact and then the result is described below. The *requirements specification* artifact has all the properties of the *artifact* concept in Figure 1 such as *id, name, kind, description, version*, and so on. Figure 4 shows the contents of the artifact written in XML. Real-world artifacts were written in XML/XSL/XSD for the implementation of a prototype. The XML description of an artifact helps developers semantically describe a realworld artifact in details. It also helps attach proper semantic tags to the artifact by using the ontologies, as shown in Figure 4. The *requirements specification* artifact was specified by the set of requirements for developing the system. The type of each requirement is indicated with the

```
<Artifact>
  <Kind> Document </Kind>
  <Name> Requirements Specification </Name>
  <Description> Defines and Specifies Requirements </Description>
  <Version> 1.0 </Version>
  <CreationDate> 30/10/2007 </CreationDate>
  <WrittenBy> Gil-Dong Hong </WrittenBy>
  <Id> 1R15a </Id>
  <Project> Military Situation Reporting System </Project>
  <Process> ADDMe </Process>
     ...
</Artifact>
```

```
<RequirementsTable>
  <Requirement>
    <ReqId> RN_S1 </ReqId>
    <Type> Functional.CommonApplicationSegement.AlertsGenerationHandling </Type>
    <Description> This system shall provide alarm function
                  when sensors perceive a fire is happened </Description>
    <Precondition> N/A </Precondition>
    <Priority> High </Priority>
  </Requirement>
  <Requirement>
    <ReqId> RN_S2 </ReqId>
    <Type> NonFunctional.Security </Type>
    <Description> This system shall communicate with IP packets
                  which are encrypted as MD5 format </Description>
    <Precondition> N/A </Precondition>
    <Priority> High </Priority>
  </Requirement>
     ...
</RequirementsTable>
```

Fig. 4. An example of XML-based artifact specification

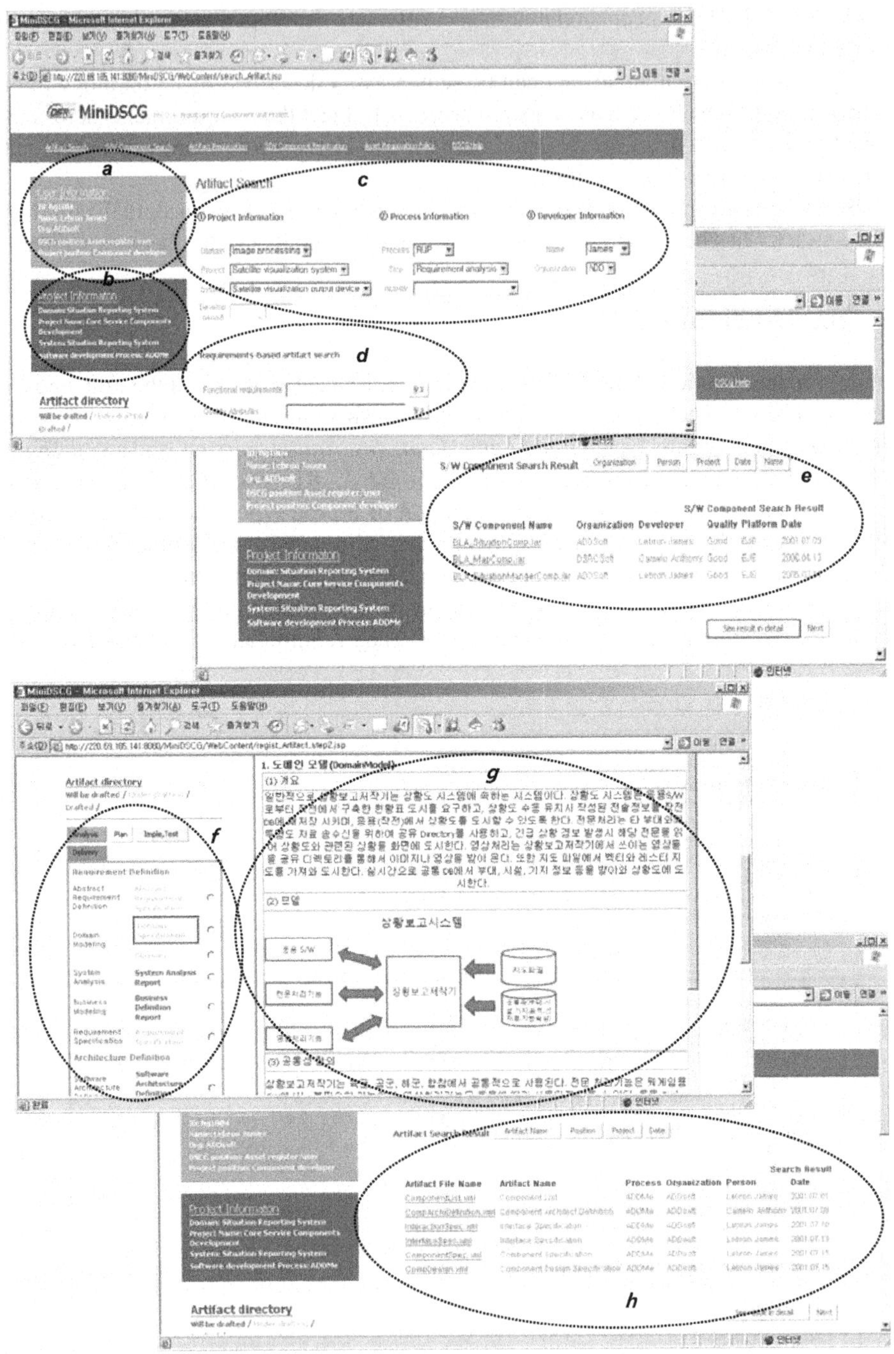

Fig. 5. A prototype implementation

requirements ontology. For example, the RN_S2 requirement is a non-functional requirement for security.

Figure 5 shows the prototype for the implementation of our approach. When a developer builds an artifact in ADDMe, he or she would want to compose the artifact by reusing related artifacts in RUP from the repository. The reused artifacts would be those developed for other similar projects and by other developers.

In Figure 5, the ellipse a drawn with a dashed line represents the login user information. Because a person can participate in several projects, the ellipse b indicates the current login project which a user wants to produce artifacts for. The ellipse ***c*** shows the inputs of search data provided in a three-dimensional perspective; ***project, process,*** and ***person*** are shown in Figure 1. The ellipse ***d*** contains forms to input search data based on the ontology of *requirements*. The ellipse ***e*** presents the search results for software components. The results can be sorted by developer, file name, and/or development platform such as EJB and COM/COM+. The ellipse ***f*** represents steps, activities, tasks, and artifacts in a software development process used for developing the current project. Developers can know the status of each artifact based on the font color of the artifact name text; an artifact which is under drafting, drafted, and will be drafted. The ellipse ***g*** shows the instance data to the *domain specification* artifact, which had been written in XML/XSL/XSD. The instance data is for the Korean military situation reporting system. The ellipse ***h*** presents the search results for artifacts. It contains the xml files corresponding to search artifacts. The files had already been built by other developers and stored in the repository for artifacts and software components.

6 Related Works

Section 6 describes related works such as component search and classification as well as the component based software development process. Because software components and related artifacts are reused by accessing a repository, this section addresses approaches for component search and classification in a repository of software components. When we proceed with a software development process, software components and related artifacts are produced. So some CBD processes will also be discussed.

6.1 Component Search and Classification

We need methods to classify and search components in a repository. Ostertage roughly divided the methods into free text keywords, facets index, and semantic net [8]. Mili classified them into simple keywords and string matching, facetbased classification and search, signature matching, and behavior matching [9]. The above free text keywords method is similar to the simple keywords and string matching method. The methods search components using text keywords but they do not have semantic information associated with the keywords. The facets index method is similar to the facet-based classification and search method. The method means that keywords are searched by experts in programs and documents and then they are sorted by a faceted classification scheme.

Also, it is labor intensive, because the method should search the keywords and then sort them by the facets. Signature matching is a method to search components by using the signatures of a component. The method cant specify the nonfunctional aspects of a component. Behavior matching means to search components by comparing the actual output to the expected output. If a component functions complexly and the component malfunctions, it is difficult to apply the matching method to searching for a component.

The semantic net method is used to overcome these problems; it is also called a semantic search. Sugumaran and Storey referred to information semantically represented with ontologies and then searched a suitable component to meet requirements [10]. Khemakhem and Drira developed a search engine called SEC [11], and they suggested the discovery ontology and integration ontology. The discovery ontology specifies the functional and non-functional features. The integration ontology describes the problem-solving methods used to specify the structural features of components. The two ontologies were not fully mentioned in [11] and they were not implemented into SEC. Yao and Etzkorn proposed the design of a component repository, which is a semantic-based approach to the use of natural language processing, conceptual graph, web service, and domain ontology [12]. The approach receives input representing user-defined queries in a natural language, which converts the queries into concept graphs and then the graphs are again converted into the corresponding semantic representation. The above mentioned researches are focused on the ontology-based storage and retrieval of software components. They did not consider the semantic description of other artifacts except for software components. The approaches also did not cover the overall life cycle of software development.

6.2 Component-Based Software Development Process

CBD is for developing a software system with components. RUP [13], CBD 96, Catalysis, and UML component are the most prominent CBD processes. RUP is widely used in the various domains. Strictly speaking, it is not a CBD process but a framework for the generation of a software process. RUP is a use-case driven process, meaning that most of the activities in phases got started with use-cases. RUP does not have the explicit artifacts to identify the component and to describe component interfaces. RUP is focused on just classes and objects, not components.

ADDMe was developed and standardized in the national defense domain in Korean. ADDMe was affected by RUP and other CBD processes [14]. ADDMe development was finished up in 2006. ADDMe is so explicit that it would be easy to apply it to real-world projects. ADDMe is conformed to ISO/IEC 12207 and MIL-STD-498. ISO/IEC 12207 is an international standard for software life cycle processes. The purpose of MIL-STD-498 is to define the artifacts that developers should create while building software systems in the defense domain.

7 Conclusions

To provide the basis for sharing and reusing artifacts across different projects and processes, we have developed the artifact description model, the artifact classification mechanism and ontologies. It has been shown that an artifact can be described in XML based on our research. Our approach allows developers to systematically describe and search artifacts for performing a specific CBD process. The approach also enables users to find related artifacts based on the semantic relationship between artifacts that were developed in different projects. It is accomplished through the artifact classification ontology. We also developed a prototype to demonstrate the applicability of our approach.

We are currently extending the artifacts classification for other major CBD processes. We are also building a distributed artifact repository system for the various developers in the Korean defense domain. We believe that our approach benefits developers by allowing the reuse of software artifacts transparently throughout a software development life cycle. Our approach also supports requirement traceability that enables developers to consistently find and manage relevant artifacts based on their requirements in each step of software development process. We also envision that effective artifacts management supported by our approach will enhance the development of software components for reuse, and the development of individual applications with reusing software components and other artifacts.

Acknowledgments

This work was partially supported by Defense Acquisition Program Administration and Agency for Defense Development under the contract (2008-SW-12-IJ-01). We would like to thank Mr. Sungjin Park at Korean Agency for Defense Development for his support for accessing ADDMerelated references. We would also like to thank Mr. Junki Lee and Mr. Beomjun Jeon for their help on implementing the prototype.

References

[1] William, B.F., Kyo, K.: Software reuse research: status and future. IEEE Trans. Softw. Eng. 31(7) (2005)
[2] Sterling Software, The COOL:Gen component standard 3.0 (1999)
[3] Desmond, F.D., Alan, C.W.: Objects, components and frameworks with UML. Addison-Wesley, Reading (1998)
[4] John, C., John, D.: UML components: a simple process for specifying component-based software. Addison-Wesley, Reading (2000)
[5] Pierre, B., Robert, D., Alain, A., et al.: The guide to the software engineering body of knowledge. IEEE Softw. 16, 35–44 (1999)
[6] Anya, K., Jim, L., Myong, H.K.: Security ontology for annotating resources. In: OTM Conferences, pp. 1483–1499 (2005)
[7] Thomas, R.G.: Toward principles for the design of ontologies used for knowledge sharing. Int. J. Hum.-Comput. Stud. 43, 907–928 (1995)
[8] Eduardo, O., James, H., Ruben, P.D., et al.: Computing similarity in a reuse library system: an AI-based approach. ACM Trans. Softw. Eng. Methodol. 1(3), 205–228 (1992)
[9] Ali, M., Rym, M., Roland, M.: A survey of software reuse libraries. Ann. Softw. Eng. 5(2), 349–414 (1998)
[10] Vijayan, S., Veda, C.S.: A semantic-based approach to component retrieval. ACM SIGMIS Database 34(3), 8–24 (2003)
[11] Sofien, K., Khalil, D., Mohamed, J.: SEC: a Search Engine for Component based software development. In: Biham, E., Youssef, A.M. (eds.) SAC 2006. LNCS, vol. 4356, pp. 1745–1750. Springer, Heidelberg (2007)

[12] Haining, Y., Letha, E.: Towards a semantic-based approach for software reusable component classification and retrieval. In: ACMSE 2004, pp. 110–115 (2004)
[13] Ivar, J., Grady, B., James, R.: The unified software development process. Addison-Wesley, Reading (1999)
[14] Agency for Defense Development, ADDMe manuals, Korea (2006)

Empirical Investigation of Metrics for Fault Prediction on Object-Oriented Software

Bindu Goel and Yogesh Singh

University School of Information Technology, Guru Gobind Singh Indraprastha University, Kashmere gate, Delhi (India)
bindu_delus@yahoo.com, ys66@rediffmail.com

Summary. The importance of software-quality classification models which can predict the modules to be faulty, or not, based on certain software product metrics has increased. Such predictions can be used to target improvement efforts to those modules that need it the most. The application of metrics to build models can assist to focus quality improvement efforts to modules that are likely to be faulty during operations, thereby cost-effectively utilizing the software quality testing and enhancement resources. In the present study we have investigated the relationship between OO metrics and the detection of the faults in the objectoriented software. Fault prediction models are made and validated using regression methods for detecting faulty classes and discover the number of faults in each class. The univariate and multivariate logistic regression models are made by taking the dependent variable as the presence of fault or not. While linear regression models are built using the number of faults as dependent variable. The results of the two models are compared and an investigation on the metrics is presented.

Keywords: fault prone classes, prediction models, software engineering, OO metrics, logistic regression, linear regression.

1 Introduction

An early detection of faulty software modules is of importance both for reduction of development cost and assurance of software quality. Like if it is not fixed in the early phases, the cost for fixing it after delivery of the software is between 5 and 100 times higher [1]. The measures collected at the early phases of development have proved to be useful indicators of fault predictions. These early identification or prediction of faulty modules can help in regulating and executing the quality improvement activities. The quality indicators are generally evaluated indirectly by collecting data from earlier projects. The underlying theory of software quality prediction is that a module currently under development is fault prone if a module with the similar product or process metrics in an earlier project (or release) developed in the same environment was fault prone [2]. Measuring structural design properties of a software system, such as coupling, cohesion, or complexity, is a promising approach towards early quality assessments. Quality models are built to quantitatively describe how these internal structural properties relate to relevant external system qualities such as reliability or maintainability.

Fault Prediction has been conducted so far by many researchers. A number of case studies have provided empirical evidence that by using regression analysis techniques;

R. Lee and H.-K. Kim (Eds.): Computer and Information Science, SCI 131, pp. 255–265, 2008.
springerlink.com

highly accurate prediction models for class fault-proneness can be built from existing OO design measures [3, 4]. Munson and Khoshgoftaar used software complexity metrics and the logistic regression analysis to detect fault-prone modules [5]. Basili et al. also used logistic regression for detection of fault-proneness using object-oriented metrics [6]. Fenton et al. proposed a Bayesian Belief Network based approach to calculate the fault-proneness [7].Some recent studies [8-14] report the use of object-oriented metrics to predict fault-proneness and number of faults by applying various statistical methods and neural network techniques. The accuracy of the prediction models in these studies is usually quantified by some measure of goodness of fit of the model predictions to the actual fault data. Most of them used some kind of software metrics, such as program complexity, size of modules, object-oriented metrics, and so on, and constructed mathematical models to calculate fault-proneness [15, 16].

The purpose of building such models is to apply them to other systems (e.g., different systems developed in the same development environment), in order to focus verification and validation efforts on fault-prone parts of those systems. In this paper, we build a fault-proneness prediction model based on regression techniques and count the number of defects in each class. The model is validated by the mentioned evaluation parameters. Also an investigation of the results by the two models is presented.

The paper is organized as follows: Next Section gives the data set used and summarizes the metrics studied for the model. Followed by section that presents the data analysis. Next Section gives the research methodology followed in this paper and the results of the study. The paper is concluded by conclusions.

2 Description of Data and Metric Suite

In this section we present the data set used for the study. Also a brief introduction to the metric suite is also given.

2.1 Dataset

The present study makes use of public domain data set KC1 posted on-line at the NASA Metrics Data Program web site [17]. The data in KC1 was collected from a storage management system for receiving/processing ground data, which was implemented in the C++ programming language. It consisted of 145 classes that comprise of 2107 methods, and a total of 40K lines of code. It consisted of 60 faulty classes and the rest of the classes without any fault.

2.2 Metric Suite and Fault Proneness

Table 1 shows the different types of predictor software metrics (independent variables) used in our study. It is common to investigate large numbers of measures, as they tend to be exploratory for studies. The Static OO measures [18] collected during the development process are taken for the study. We have taken all these measures as independent variables. They include various measures of class size, inheritance, coupling and cohesion. They are all at class levels. The traditional static complexity and size metrics

Table 1. Metric suite at class level used for the study

Metric	Definition
Coupling_between_objects(CBO)	The number of distinct non-inheritance-related classes on which a class depends.
Depth_of_inheritance(DIT)	The level for a class. Depth indicates at what level a class is located within its class hierarchy.
Lack_of_cohesion(LCOM)	It counts number of null pairs of methods that do not have common attributes.
Fan_in(FANIN)	This is a count of calls by higher modules.
Response_for_class(RFC)	A count of methods implemented within a class plus the number of methods accessible to an object class due to inheritance.
Weighted_methods_per_class(WMPC)	A count of methods implemented within a class (rather than all methods accessible within the class hierarchy).
Number_of_children(NOC)	It is the number of classes derived from a specified class.
Cyclomatic_Complexity(CC)	It is a measure of the complexity of a modules decision structure. It is the number of linearly independent paths.
Source_lines_of_code(SLOC)	Source lines of code that contain only code and white space.
Num_of_methods_per_class(NMC)	This is the count of methods per class

include well known metrics, such as McCabe's Cyclomatic complexity and executable lines of code [19]. These metrics are posted at the site for method level but they are converted to class level and their average values are used for the study. Another traditional metric Fan-in given by Henry and Kafura [20] is also used. It is used to measure inter class coupling by counting the number of calls passed to the class. Also we have computed a metric called number of methods per class. It counts the number of modules in a class. The underlying fact for this is the more the number of methods per class more is it fault prone. The dependent variable used is Defect level metric which refers to whether the module is fault prone or fault-free. So our dependent variable or the response variable is the defect level whose response is true if the class contains one or more defects and false if does not contain a fault. We have converted the variable in numeric giving it the value1 if it contains fault and 0 otherwise. For linear regression modeling the dependent variable is the actual number of faults in the class. A relationship between the metrics and fault proneness can be drawn as follows:

- The size related metrics gives the length or the size of the class. The larger the size of the class more likely it is to be fault prone.

- Inheritance related metrics (dit, noc) this measures an amount of potential reuse of the class. The more reuse a class might have, the more complex it may be, and the more fault prone the classes may be.
- Coupling metrics measures the degree of inter dependence among the components of a software system. High coupling makes a system more complex; highly interrelated. So more the couplings more fault prone it may be.
- The complexity metrics of a class specifies the complexity of a modules decision structure. More complex the class more prone it is to faults.
- The cohesion metrics specifies class cohesion and classes with lower cohesion than its peers is considered to be more fault prone.

2.3 Defect Analysis

In this section we calculated the number of defects in each class. The defect is given at the module level in the MDP dataset. We have mapped that to class level to indicate the number of defects in each class. There are a total number of 669 defects associated with 60 faulty classes. There are 85 classes in the system without any defect. Table 2 presents the distribution of defects in each class of the data.

Table 2. Distribution of defects in the data

No. of classes	No. of associated Defects in classes	% of classes
85	0	58.62
3	1	2.07
5	2	3.45
8	3	5.52
8	4	5.52
2	5	1.38
5	6	3.45
3	7	2.07
3	8	2.07
2	9	1.38
2	11	1.38
2	13	1.38
2	14	1.38
1	16	0.69
2	17	1.38
2	19	1.38
2	20	1.38
1	22	0.69
2	23	1.38
1	24	0.69
1	26	0.69
1	31	0.69
1	42	0.69
1	101	0.69
145	669	100.00

3 Research Methodology and Analysis Results

In this section, we will present the analyses we performed to discover the relationships between the different metrics and the number of defects found in each class. We have employed egression analysis methods, which are widely used to predict an unknown variable based on one or more known variables. We chose logistic regression [21] to model the relationship between the metrics and fault-prediction. They classified the classes as faulty or non-faulty. We applied linear regression [22] as well to predict the number of defects. We have made univariate and multivariate models using both the methods. These techniques have been applied by other noted researchers for similar kind of predictions [16,23].

3.1 Data Analysis

Within the case study, the distribution max, mean, standard deviation and variance of each independent variable is examined. Low variance measures do not differentiate classes very well and therefore are likely to be less useful. Presenting and analyzing the distribution of measures is important for the comparison of different metrics used. All the measures with more than six non-zero data points are considered for further analysis. The Table 3 presents the descriptive statistics as shown below. There are 145 classes for which values for all the metrics are available. The NOC and FANIN have low variance and are unlikely to be significant in the results. But all the measures are included for the modeling as different dimensions are captured by each metrics.

Table 3. Descriptive statistics of the data used

	CBO	DIT	LCOM	FANIN	RFC	WMPC	NOC	CC	SLOC	NMC
Mean	8.32	2.00	68.72	.63	34.38	17.42	.21	2.46	11.83	14.53
Median	8.00	2.00	84.00	1.00	28.00	12.00	.00	1.86	8.67	10.00
Std. Dev	6.38	1.26	36.89	.69	36.20	17.45	.69	1.73	12.12	15.26
Variance	40.66	1.58	1360.77	.48	1310.65	304.47	.49	2.99	146.88	232.99

3.2 Logistic Regression Analysis

Logistic Regression (LR) [21] is widely applied to the data where the dependent variable (DV) is dichotomous i.e. it is either present or absent. It is used to predict the likelihood for an event tooccur, e.g., fault detection. LR unlike linear regression (where the goal is to determine some form of functional dependency like linear between the explanatory variables and dependent variable) does not assume any strict functional form to link explanatory variables and the probability function. LR is based on maximum likelihood and assumes that all observations are independent. Outlier analysis is done to find data points that are over influential and removing them is essential. To identify multivariate outlier we calculate for each data point the Mahalanobis Jackknife distance. Details on outlier analysis can be found in [24].

Multivariate Logistic regression (MLR) model is defined by the following equation (if it contains only one independent variable (IV) , then we have a univariate model):

$$\frac{\pi(X_1, X_2, \ldots, X_N)}{1-\pi} = \frac{Prob(event)}{Prob(noevent)} = e^{B_0+B_1X_1+\ldots\ldots\ldots+B_NX_N}$$

Where Xis are the independent variables and is the probability of occurrence of a fault. MLR is done in order to build an accurate prediction model for the dependent variable. It looks at the rela tionships between IVs and the DV, but considers the former in combination, as covariates in a multivariate model, in order to better explain the variance of the DV and ultimately obtain accurate predictions. To measure the prediction accuracy, different modeling techniques have specific measures of goodness of Fit of the model. In the present study following statistics are used to illustrate and evaluate the experimental results obtained:

- B_is are the estimated regression coefficients of the LR equation. They show the extent of the impact of each explanatory variable on the estimated probability, and, therefore, the importance of each explanatory variable. The larger the absolute value of the coefficient, the stronger the impact (positive or negative, according to the sign of the coefficient) of the explanatory variable on the probability of a fault being detected in a class.
- p, the statistical significance of the LR coefficients, provides an insight into the accuracy of the coefficients estimates. A significance threshold of $\alpha = 0.05$ (i.e., 5% probability) has often been used to determine whether a variable is a significant predictor.
- R^2, called r-square coefficient is the goodness of fit, not to be confused with least-square regression R^2. Both are built upon very different formulae, even though they both range between 0 and 1. The higher R^2, the higher the effect of the model explanatory variables, the more accurate the model. However this value is rarely high for LR. We have used Cox and Snell's R-Square in the present study which is an imitation to the interpretation of multiple R-Square for binary LR.

It is known that the examined metrics are not totally independent and may capture redundant information. Also the validation studies described here are exploratory in nature, that is, we do not have a strong theory that tells us which variables should be included in the prediction model and which not. In this state of affairs, a stepwise selection process can be used, where prediction models are built in a stepwise manner, where each step consists of one variable entering or leaving the model. The general backward

Table 4. Univariate Logistic Regression model

Covariates	CBO	DIT	LCOM	FANIN	RFC	WMPC	NOC	CC	SLOC	NMC
Coefficient	.207	.054	.000	.351	.010	.025	-.432	.400	.076	.076
Constant	-2.167	-.456	-.343	-.575	-.679	-.789	-.269	-1.336	-1.256	-1.384
P(sig)	.000	.687	.987	.153	.060	.030	.184	.001	.000	.000
R^2	.263	.001	.000	.014	.027	.040	.016	.093	.147	.140

elimination procedure starts with a model that includes all independent variables. Variables are selected one at a time to be deleted from the model, until a stopping criterion is fulfilled. MLR model is also tested for multicollinearity. If X1,...,Xn are the covariates of the model. Then a principal component analysis on these variables gives Imax to be the largest eigenvalue, Imin the smallest eigenvalue of the principal components. The conditional number is then defined as $\lambda = \sqrt{I_{max}/I_{min}}$. A large conditional number (i.e., discrepancy between minimum and maximum eigen value) indicates the presence of multicollinearity and should be under 30 for acceptable limits [25]. The model is made using all the 145 classes for the training. The model consists of various metrics as shown in univariate and multivariate models shown in Table 4. and 5. The metrics DIT, LCOM, FANIN, and NOC are shown as insignificant in the univariate model. NOC, CC and FANIN are not there in the multivariate model. A classifier at threshold =0.5 is used to classify the classes as presented in Table 5. The test for multicollinearity for the model is done by computing the conditional number is $\sqrt{(2.521/2.321)} = 1.04$ which is well below 30. Also no influential outlier was found for the model. The accuracy of the model is 79.31%, recall is 71.67% and the precision is 76.79%.

The model is also evaluated for accuracy in terms of correctness and completeness, the two standard measures used for classifications. Correctness is the number of classes correctly classified as fault prone, divided by the total number of classes classified as fault-prone. The larger the correctness, the lesser the effort spent on detecting fault-free classes and hence improvement in the efficiency. In the present case the correctness is 76.79%.The other parameter completeness [12] is defined as the number of faults in

Table 5. Multivariate Logistic Regression model

Covariates	CBO	DIT	LCOM	RFC	WMPC	SLOC	NMC	Const.
Coefficient	.23	-1.3	-.02	.04	-.24	.07	.26	-.92
P(sig)	.001	.002	.028	.091	.079	.021	.061	.150
-2 Log likelihood	118.792							
R^2	.416							

Table 6. Classification Results

LR Model		predicted NFP Po<=0.5	predicted FP Po>0.5	Total
actual	NFP	72	13	85
	FP	17 (101 faults)	43 (568 faults)	60 (669 faults)
Total		89	56	145

classes classified as fault-prone, divided by the total number of faults in the system. The larger the completeness values, more the number of faults in faulty predicted classes detected. The completeness in the present study is 568 out of 669 i.e. 84.90%.

3.3 Linear Regression Analysis

Logistic regression is best used for classification problems. But in the present study there are many classes with varying number of defects. So this situation can be only modeled using linear regression [26]. In linear regression, the explanatory variables are the standardized metrics while the dependent variable is the number of defects in a class.

The accuracy of the model can be measured by the goodness of fit of the model and statistical significance of the parameters. Two important test statistics are used: the multiple coefficients of de termination R2 and the t-value. R^2 is the variance percent in the dependent variable that is accounted for by the regression model. e standardized coefficients or betas are an attempt to make the regression coefficients more comparable when often the independent variables are measures in different units.

The univariate and multivariate model results are as shown in the table 7 and 8 respectively. The metrics DIT, LCOM, FANIN, and NOC appears again as insignificant predictors for modeling number of defects. In the multivariate model two of the earlier selected metrics are present i.e. DIT and NMC. While CC makes a surprise entry. The R^2 value is .422 which means 58% variance is not explained by the model. The rest of the variance which is not captured can be attributed to external environmental and psychological factors which are not accounted for by the given metric set. There could be several other possible reasons to it.

Table 7. Univariate Linear Regression model

Covariates	CBO	DIT	LCOM	FANIN	RFC	WMPC	NOC	CC	SLOC	NMC
Coefficient	.623	.320	.037	1.573	.082	.288	-1.635	2.057	.288	.417
Constant	-.565	3.973	2.062	3.613	1.807	-.409	4.963	-.438	1.203	-1.440
P(sig)	.000	.658	.131	.228	.001	.000	.208	.000	.000	.000
R^2	.134	.001	.016	.010	.074	.215	.011	.107	.104	.343

Table 8. Multivariate Linear Regression model

Covariates	DIT	CC	NMC	Const.
Coefficient	-1.987	2.151	.387	-2.316
t	-2.961	4.340	8.297	-1.599
Beta	-.230	.342	.544	---
R^2	.422			
Adjusted R^2	.410			

4 Conclusions

The paper attempts to present the model with which metrics calculated from the source code of object-oriented systems can be used predict the fault proneness and the number of defects in a class. The main observations on the metrics are as following:

1. NOC cannot be used at all for fault-prediction. This is consistently not useful.
2. CBO is a good predictor of fault-prediction. It is a better metric to classify than predict the number of defects.
3. DIT is a poor individual predictor but appears in both the multivariate model. This proves that depth in inheritance is responsible for faults.
4. LCOM is also a poor predictor of fault.
5. The SLOC metric did fairly well and due to its ease in collection can be a useful predictor.
6. Another size metric NMC has performed fairly well and can be collected easily.
7. Both the models predict the similar individual metrics like DIT, LCOM, FANIN, and NOC as poor predictors.
8. CC is unreliable as it could not model the presence of faults but can be effective in calculating number of faults.
9. RFC appears as better predictor in second model than first one.
10. Both the models conclude similarly and hence proves that metrics can be significantly useful in modelling fault-prediction.

However the results can be optimistic as model is tested on the same datasets. The model could further be generalized over the software systems from different environments. We have made the model using measures collected from the early analysis and design artifacts and are based on the measurement of the source code. They might not be indicative of the final developed system. So the use of predictive models based on early artifacts and their capability to predict the quality of the final system still remains to be investigated. Also there should be datasets from similar environments so that model from existing system measures can be applied to other systems being developed.

Also like any empirical study, our conclusions are biased according to what data we used to generate them. For example, the sample used here comes from NASA and NASA works in a particularly unique market niche. Nevertheless, it can be contested that results from NASA are relevant to the general software engineering industry. The applicability of the statistical techniques (logistic and linear regression) is also established to predict the fault prediction models. Both have proved to be viable methodologies. The application of regression technique in predicting software fault prediction can be compared to other viable methodologies. And the results of the two can be compared. A future direction in the study could be to map the fault proneness of the classes with effort and time required to correct them during development as well as maintenance phase. Also Study can be conducted to use these metrics to predict different maintenance metrics.

References

[1] Boehm, B., Basili, V.: Software Defect Reduction Top 10 Lists. IEEE Computer 34(1), 135–137 (2001)

[2] Khoshgoftaar, T.M., Allen, E.B.R., Munikoti, F.D., Goel, R.N., Nandi, A.: Predicting fault-prone modules with case-based reasoning. In: ISSRE 1997, the Eighth International Symposium on Software Engineering, pp. 27–35. IEEE Computer Society, Los Alamitos (1997)

[3] Briand, L., Wüst, J., Daly, J., Porter, V.: Exploring the Relationship between Design Measures and Software Quality in Object-Oriented Systems. Journal of Systems and Software 51, 245–273 (2000); Also Technical Report ISERN-98-07

[4] Briand, L., Wüst, H., Lounis, S.: Investigating Quality Factors in Object-Oriented Designs: an Industrial Case Study. In: Proceedings of the 21st International Conference on Software Engineering, ICSE 1999, Los Angeles, USA, pp. 345–354 (1999)

[5] Munson, J.C., Khoshgoftaar, T.M.: The detection of fault-prone programs. IEEE Trans. on Software Engineering 18(5), 423–433 (1992)

[6] Basili, V.R., Briand, L.C., Melo, W.L.: A validation of object oriented metrics as quality ind icators. IEEE Trans. On Software Engineering 22(10), 751–761 (1996)

[7] Fenton, N.E., Neil, M.: A critique of software defect prediction models. IEEE Trans. on Software Engineering 25(5), 675–689 (1999)

[8] Thwin, M.M.T., Quah, T.-S.: Application of neural network for predicting software development faults using object-oriented design metrics. In: Proceedings of the 9th International Conference on Neural Information Processing, pp. 2312–2316 (2002)

[9] Kamiya, T., Kusumoto, S., Inoue, K.: Prediction of fault-proneness at early phase in object-oriented development. In: Proceedings of the Second IEEE International Symposium on Object-Oriented Real-Time Distributed Computing, pp. 253–258 (1999)

[10] Emam, L., Wüst, J., Daly, J.W.: The prediction of faulty classes using object- oriented design metrics. Journal of Systems and Software 56, 63–75 (2001)

[11] Briand, L., Wüst, J., Daly, J.W.: Exploring the relationships between design measures and software quality in object-oriented systems. Journal of Systems and Software 51, 245–273 (2000)

[12] Briand, L.C., Wüst, J., Daly, J.W.: Assessing the applicability of fault-proneness models across object-oriented software projects. IEEE Transactions on Software engineering 28(7), 706–720 (2002)

[13] Glasberg, D., Emam, K.E.: Validating object-oriented design metrics on a commercial java application. Technical Report NRC/ERB-1080 (2000)

[14] Mao, Y., Sahraoui, H.A., Lounis, H.: Reusability hypothesis verification using machine learning techniques: a case study. In: Proceedings of the 13th IEEE International Conference on Automated Software Engineering, pp. 84–93 (1998)

[15] Briand, L., Wuest, J.: The Impact of Design Properties on Development Cost in Object-Oriented Systems. International Software Engineering Research Network (1999)

[16] Yu, P., Systa, T., Muller, H.: Predicting Fault-Proneness using OO Metrics: An Industrial Case Study. In: Sixth European Conference on Software Maintenance and Reengineering, Budapest, Hungary (2002)

[17] Metrics Data Program, NASA IV&V Facility, `http://mdp.ivv.nasa.gov/`

[18] Chidamber, S.R., Kemerer, C.F.: A Metrics Suite for Object Oriented Design. IEEE Transactions on Software Engineering 20(6), 476–493 (1994)

[19] McCabe, T.J.: A complexity measure. IEEE Transactions on Software Engineering SE-2(4), 308–320 (1976)

[20] Henry, S., Kafura, D.: Software structure metrics based on information flow. IEEE Transactions on Software Engineering SE-7(5), 510–518 (1981)

[21] Hosmer, D., Lemeshow, S.: Applied Logistic Regression. Wiley-Interscience, Chichester (1989)
[22] Neter, J., Wasserman, W., Kutner, M.H.: Applied Linear Statistical Models, 3rd edn. Richard D.Irwin (1990)
[23] Basili, V.R., Briand, L.C., Melo, W.L.: A validation of object oriented metrics as quality indicators. IEEE Trans. On Software Engineering 22(10), 751–761 (1996)
[24] Barnett, V., Price, T.: Outliers in Statistical Data. John Wiley & Sons, Chichester (1995)
[25] Belsley, D., Kuh, E., Welsch, R.: Regression Diagnostics: Identifying Influential Data and Sources of Collinearity. John Wiley & Sons, Chichester (1980)
[26] Gyimothy, T., Ference, R., Siket, I.: Empirical Validation of Object –Oriented Metrics on open source software for fault prediction. IEEE Transactions on Software Engineering 31(10) (October 2005)

Two $O(n^2)$Time Fault–Tolerant Parallel Algorithm for Inter NoC Communication in NiP

Nitin*, Chauhan Durg Singh**, and Sehgal Vivek Kumar***

Jaypee University of Information Technology (JUIT), Waknaghat, via Kandaghat, P.O. Dumehar Bani, District Solan–173215, Himachal Pradesh, India
{delnitin,ds.chauhan,sehgal.vivek}@juit.ac.in

Summary. Networks–in–Package (NiPs) is a solution for inter Networks–on–Chip (NoCs) communication, where power dissipation, performance, and size of processing package are the major factors. NiP is a group of NoCs mounted on a single package. In this paper, we proposed a new method, which merges the two communications, local communication between the Intellectual Property (IP) cores in a NoC, and global communication between NoCs in NiP. In addition to this, we proposed two $O(n^2)$time Faulttolerant Parallel Algorithms for setting of Inter NoC Communication in NiP. The proposed algorithm allows different NoCs in NiP to communicate in parallel using either fault-tolerant Hexa Multistage Interconnection Network (HXN) or fault tolerant Penta Multistage Interconnection Network (PNN).

Keywords: Networks–in–package, Networks–on–chip, Intellectual property, Multi-stage Interconnection network, Hexa network, Penta network, Fault–tolerance, Parallel algorithm, Interconnectonchip.

1 Introduction

1.1 Networks–in–Package

The NiPs designs provide integrated solutions to challenging design problems in the field of multimedia and real time embedded applications. The main characteristics of NiP platforms are as:

1. Networking between chip-to-chip in a single package.
2. Lower development cost than NoC approach.
3. Lower power consumption.
4. High performance.
5. Small area.

* Corresponding Author, Member IEEE and Member ACM, Senior Lecturer, Department of Computer Science Engineering and Information Technology. E-mail: delnitin@gmail.com

** Member UGC India, Professor and Vice Chancellor. E-mail: pdschauhan@gmail.com

*** Member IEEE and Member ACM, Senior Lecturer, Department of Electronics and Communication Engineering. E-mail: vivekseh@gmail.com

R. Lee and H.-K. Kim (Eds.): Computer and Information Science, SCI 131, pp. 267–282, 2008.
springerlink.com

Along with these characteristics, there are various trends and challenges in NiP such as:

1. Communication versus computation.
2. Deep submicron effects
3. Power
4. Global synchronization
5. Heterogeneity of functions.

On-chip wire delays have become more critical than gate delays and recently synchronization problems between Intellectual Properties (IPs) are more apparent. This trend only worsens as the clock frequencies increases and the feature sizes decreases [1]. However, the design process of NiP-based systems borrows some of its aspects from the parallel computing domain; it is driven by a significantly different set of constraints. Large, complex multiprocessor-based System-on-Chip (SoC) [2, 3] platforms are already well into existence, and, according to common expectations and technology roadmaps, the emergence of billion-transistor chips is just around the corner. The complexity of such systems calls for a serious revisiting of several on chip communication issues.

We focus on an emerging paradigm that effectively addresses and presumably overcomes the many onchip interconnection and communication challenges that already exist in todays chips or will likely occur in future chips. This new paradigm is commonly known as the network–on–chip paradigm. The NoC paradigm is one, if not the only one, fit for the integration of an exceedingly large number of computational, logic, and storage blocks in a single chip (otherwise known as a SoC). Notwithstanding this school of thought, the adoption and deployment of NoCs face important issues relating to design and test methodologies and automation tools. In many cases, these issues remain unresolved.

1.2 Networks–on–Chip

Networks–on–Chip is an emerging paradigm for communications within large VLSI systems implemented on a single silicon chip. In a NoC system, modules such as processor cores, memories and specialized IP blocks exchange data using a network as a "public transportation" sub–system for the information traffic. A NoC is constructed from multiple point–to–point data links interconnected by switches (also known as routers), such that messages can be relayed from any source module to any destination module over several links, by making routing decisions at the switches. A NoC is similar to a modern telecommunications network, using digital bit-packet switching over multiplexed links. Although packet–switching is sometimes claimed as a necessity for NoCs, there are several NoC proposals utilizing circuit–switching techniques. This definition based on routers is usually interpreted so that a single shared bus, a single crossbar switch or a point–to–point network are not NoCs but practically all other topologies are. This is somewhat confusing since all above mentioned are networks (they enable communication between two or more devices) but they are not considered as network–on–chips. Note that some articles erroneously use NoC as a synonym for mesh topology although NoC paradigm does not dictate the topology. Likewise, the regularity of topology is sometimes considered as a requirement, which is obviously, not the case in research concentrating on "application–specific NoC topology synthesis".

The wires in the links of the NoC are shared by many signals. A high level of parallelism is achieved, because all links in the NoC can operate simultaneously on different data packets. Therefore, as the complexity of integrated systems keeps growing, a NoC provides enhanced performance (such as throughput) and scalability in comparison with previous communication architectures Of course; the algorithms must be designed in such a way that they offer large parallelism and can hence utilize the potential of NoC.

The adoption of NoC architecture is driven by several forces: from a physical design viewpoint, in nanometer CMOS technology interconnects dominate both performance and dynamic power dissipation, as signal propagation in wires across the chip requires multiple clock cycles. NoC links can reduce the complexity of designing wires for predictable speed, power, noise, reliability, etc., thanks to their regular, well controlled structure. From a system design viewpoint, with the advent of multi–core processor systems, a network is a natural architectural choice. A NoC can provide separation between computation and communication; support modularity and IP reuse via standard interfaces, handle synchronization issues, serve as a platform for system test, and, hence, increase engineering productivity.

Although NoCs can borrow concepts and techniques from the well-established domain of computer networking, it is impractical to blindly reuse features of "classical" computer networks and symmetric multiprocessors. In particular, NoC switches should be small, energy–efficient, and fast. Neglecting these aspects along with proper, quantitative comparison was typical for early NoC research but nowadays they are considered in more detail. The routing algorithms should be implemented by simple logic, and the number of data buffers should be minimal. Network topology and properties may be application–specific. NoCs need to support quality of service, namely achieve the various requirements in terms of throughput, end–to–end delays and deadlines. To date, several prototype NoCs have been designed and analyzed in both industry and academia but only few have been implemented on silicon. However, many challenging research problems remain to be solved at all levels, from the physical link level through the network level, and all the way up to the system architecture and application software.

Most NoCs are used in embedded systems, which interact with their environment under more or less hard time constraints. The communication in such systems has a strong influence on the global timing behavior. Methods are needed to analyze the timing, as average throughput as well as worst–case response time [2]. However, from a VLSI design perspective, the energy dissipation profile of the interconnect architectures is of prime importance as the latter can represent a significant portion of the overall energy budget. The silicon area overhead due to the interconnect fabric is important too. The common characteristic of these kinds of architectures is such that the processor/storage cores communicate with each other through high–performance links and intelligent switches and such that the communication design is represented at a high abstraction level [4]. The different NoC topologies are already used in [5] and these topologies give different communication structure in NoC.

In this paper, we propose a new method, which merges the two communications, local communication (between the IP cores in a NoC), and global communication between NoCs in NiP. In addition to this, we propose fault-tolerant parallel algorithm for setting up inter NoC communication in NiP. The proposed algorithm allows different NoCs in NiP to communicate in parallel using either fault-tolerant HXN Multi–stage Interconnection Network (MIN) or fault tolerant PNN MIN. Section 2 describes the general NiP architecture including the fault tolerant parallel algorithm designed to provide parallel communication among different NoCs followed by the conclusions and references.

2 General NiP Architecture: A Communication Platform for NoCs

In NiP, any NoC can be used as source or destination with any configuration (homogeneous or heterogeneous). The flow of data from source IP to destination IP is in the form of packets. Since the source IP and destination IP are not on single chip, the communication between them is made through an on–package interconnect which is highly fault tolerant. The data flows through different OSI layers in local communication zone and global communication. The different NoC chips on a package can communicate with each other through an Interconnect–on–Chip (IoC) on same package.

The general architecture of NiP resembles with the Open System Interconnection (OSI) Model. The Physical layer refers to all that concerns the electric details of wires, the circuits and techniques to drive information (drivers, repeaters, layout), while the Data Link level ensures a reliable transfer regardless of any unreliability in the physical layer and deals with medium access (sharing/contention). At the Network level there are issues related to the topology and the consequent routing scheme, while the Transport layer manages the end-to-end services and the packet segmentation/reassembly. Upper levels can be viewed merged up to the Application as a sort of adaptation layer that implements services in hardware or through part of an operating systems and exposes the NoC infrastructure according to a proper programming model, e.g. the Message Passing (MP) paradigm.

NiP is a specific approach, which provides the common interface through which all NoC chips can communicate with each other more efficiently and robustly. It contains three different types of building blocks, appropriately interconnected to each other, and a patented network topology that promises to deliver the best price/performance trade–off in future NiP applications:

1. The network interface (NW IF), on chip NW IF connect the IP core with the on-chip network and the off chip NW IF connect the NoC chip with the Interconnected Network–on–Chip in order to communicate with other NoC chip on the same package.
2. The Router, which is responsible for the flow of data traffic across the network and for the Quality of Service (QoS) offered by the network.
3. The physical link, which is responsible for the actual propagation of the signals across the network and to/from the external IPs and subsystems.

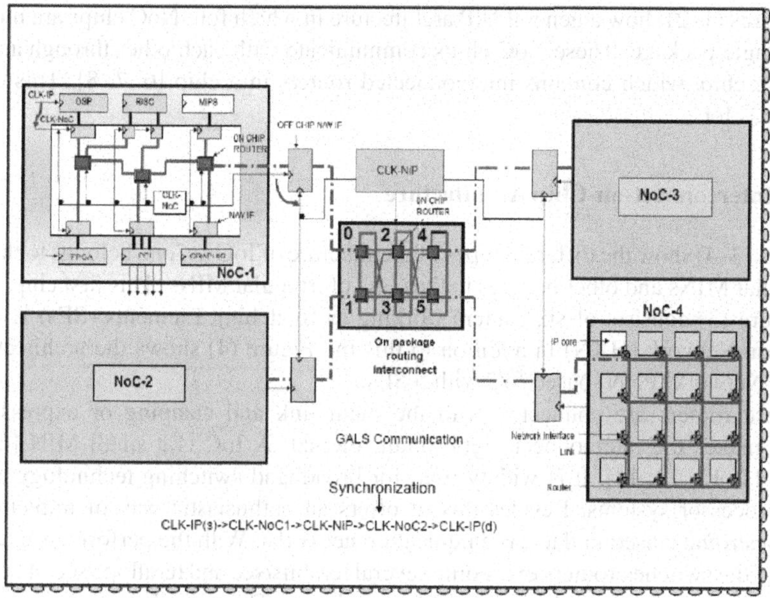

Fig. 1. Parallel Communication among various NoCs using HXN MIN

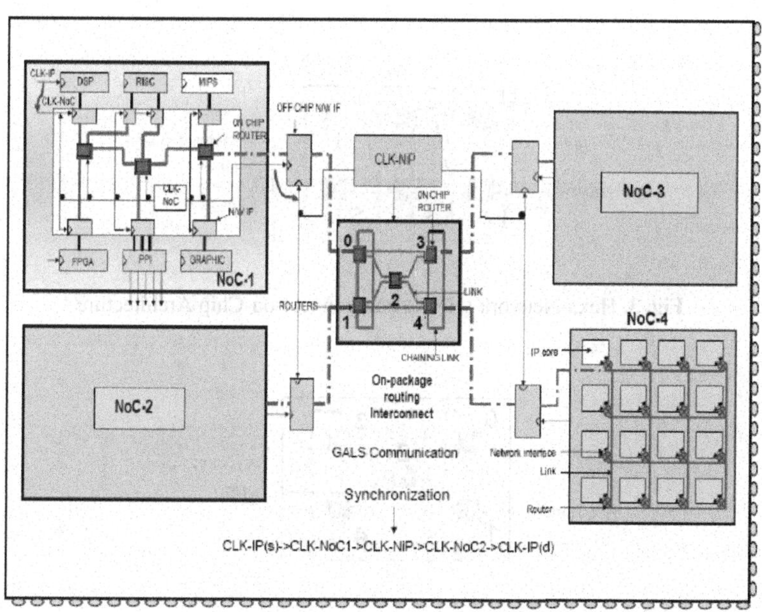

Fig. 2. Parallel Communication among various NoCs using PNN MIN

Figures (1–2) show a general NiP architecture in which four NoC chips are mounted on a single package. These NoC chips communicate with each other through an intermediate chip, which contains interconnected routers in a chip [6, 7, 8]. This chip is known as IoC.

2.1 Interconnect-on-Chip Architecture

Figures (3–4) show the different types of architecture of IoC i.e. one belongs to the class of regular MINs and other belongs to the class of irregular MINs. This first chip, shown in Figure (3) consists of six routers working as Switching Elements (SEs) is known as Hexa–Network (HXN) in addition to this the Figure (4) shows the architecture of Penta–Network (PNN) based IoC with 5 SEs.

These routers are connected with the main link and chaining or express links, which makes the interconnect highly fault tolerant. A IoC is a small MIN [6, 7, 8, 9, 10, 11, 12, 13, 14]. It is widely used for broadband switching technology and for multiprocessor systems. Besides this, it offers an enthusiastic way of implementing switches/routers used in data communication networks. With the performance requirement of the switches/routers exceeding several terabits/sec and teraflops/sec, it becomes imperative to make them dynamic and fault–tolerant. The typical modern day application of the MIN includes fault–tolerant packet switches, designing multicast, broadcast router fabrics while SoC and NoC are hottest now days.

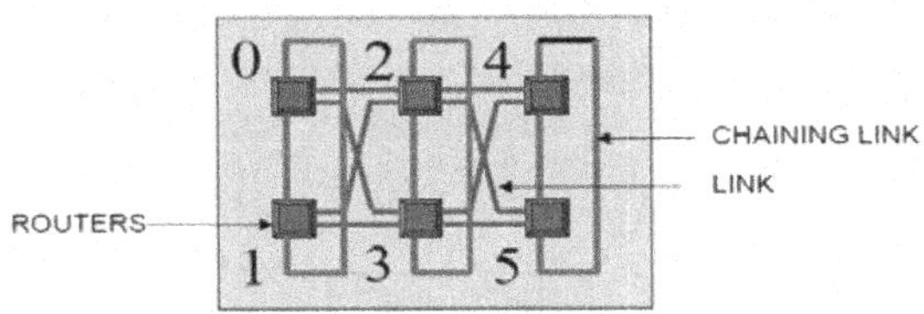

Fig. 3. Hexa-Network (HXN) Interconnect-on-Chip Architecture

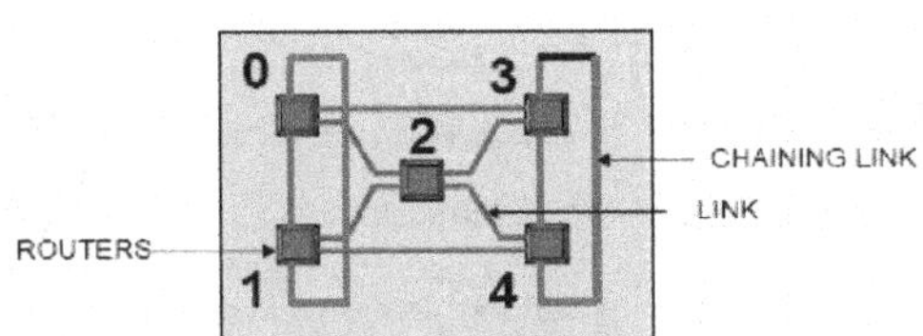

Fig. 4. Penta-Network (PNN) Interconnect-on-Chip Architecture

2.2 Fault-Tolerant Parallel Algorithm for Dynamic Communication among NoCs in NiP Using HXN MIN

```
ALGORITHM_FAULT-TOLERANT_HEXA_NETWORK
_________________________________________________________

Input: n, stores Number of parallel processing NoC's.
Source, a NoC type array that stores the Source NoC
Numbers a part of NoC structure.
Destination, a NoC type array that stores the Destination
NoC Number a part of NoC structure.
Payload, a part of NoC structure that holds the data
generated as Source NoC's.
Output: Payload.
_________________________________________________________
BEGIN
FOR (I=0 to 6) DO /*Initialization of elements of
                    Switching Element structure*/
 Info = NULL
 Info1 = NULL
 Source = -999
 Destination = -999
END FOR
FOR (I = 0 to 4) DO
 Payload = 0 /*Initializing payload of all the 4 NoC's to
              0*/
 Number = I /*Numbering of all the 4 NoC's from 0 to 3*/
END FOR
 Get the Number of parallel communicating NoC's, n
FOR (l = 0 to n) DO
 Get the Source and Destination NoC's S[l], D[l]
 Get the respective payloads
END FOR
FOR(x = 0 to 6) DO
BEGIN
/*Stage1: Transferring the" Payload" values from the
NoC's to the respective "Info" values of Switching
Element structure*/
FOR (y = 0 to n-1) DO
 Transfer the payloads of NoC's to the Info of next
immediate Switching Elements
 Transfer the Source NoC Number to the Source of
Switching Element
 Transfer the Destination Number to the Switching Element
END FOR
/*Stage2: If the communicating NoC's are on the same side
checking the source and respective destination Number of
```

```
the Switching Element and transferring it to the next
Switching Element having empty "Info" value*/
FOR (t = 0 to 2) DO
Check Source and respective Destination pair DO
 Check Info of chain linked Switching Element DO
  IF Info = NULL
   Transfer Info into this Switching Element
   Transfer the Source and Destination Number to this
   Switching Element
/*If "Info" of first linked Switching Element is not
empty then transfer the packet to the "Info" of second
linked Switching Element*/
ELSE
 Check Info of first linked Switching Element DO
  IF Info = NULL
   Transfer Info into Switching Element
   Transfer the Source and Destination Number to the
Switching Element
ELSE
 Check Info of second linked Switching Element DO
  IF Info = NULL
   Transfer Info into Switching Element
   Transfer the Source and Destination Number to the
Switching Element
/*Stage2: If communicating NoC's are on the opposite side
checking the source and respective destination no. of the
Switching Element and transferring it to Switching
Element having empty "Info" value*/
Check Source and respective Destination pair DO
Check Info of first linked Switching Element DO
Transfer Info into this Switching Element
 IF Info = NULL
  Transfer the Source and Destination Number to this
  Switching Element
/*If the "Info" of first linked Switching Element is not
empty then transfer of packet to 'Info" of second linked
Switching Element*/
  ELSE
   Check Info of second linked Switching Element DO
   Transfer Info into Switching Element
    IF Info = NULL
     Transfer the Source and Destination Number to the
     Switching Element
/*If the "Info" of first and second linked Switching
Elements are not empty then transfer "Info" to third
linked Switching Element*/
  ELSE
   Check Info of chain linked Switching Element DO
```

```
   Transfer Info into Switching Element
    IF Info = NULL
     Transfer the Source and Destination Number to the
     Switching Element
   /*Stage3*/
    Check Source and respective Destination pair DO
    Check Info of first linked Switching Element DO
     IF Info = NULL DO
      Transfer Info into this Switching Element
      Transfer the Source and Destination Number to this
      Switching Element
/*If the "Info" of first linked Switching Element is not
empty then transfer of packet to "Info" of second linked
Switching Element*/
  ELSE
   Check Info of second linked Switching Element DO
    IF Info = NULL DO
     Transfer Info into Switching Element
     Transfer the Source and Destination Number to the
     Switching Element
/*If the "Info" of first and second linked Switching
Elements are not empty then transfer "Info" to third
linked Switching Element*/
  ELSE
   Check Info of chain linked Switching Element DO
    IF Info = NULL DO
     Transfer Info into Switching Element
     Transfer the Source and Destination Number to the
     Switching Element
/*Stage4: If the linked NoC is the destination i.e. NoC-0
for SE[0], NoC-1 for SE[1] NoC-3 for SE[4], NoC-4 for
SE[5] then transfer the "Info" of the Switching Element
to their respective linked NoC*/
  IF Destination of SE[0]=0 DO
   Transfer Info to payload of NoC-0
/*If the linked NoC is not the destination then transfer
the "Info" of Switching Element to the next empty
Switching Element. If not then destroy the packet*/
    ELSE
     Transfer Info to first linked Switching Element
  END IF
    IF Destination of SE[1]=1 DO
     Transfer Info to payload of NoC-1
      ELSE
       Transfer Info to first linked Switching Element
    END IF
    IF Destination of SE[3]=2 DO
     Transfer Info to payload of NoC-2
```

```
        ELSE
          Transfer Info to first linked Switching Element
      END IF
      IF Destination of SE[4]=3
        Transfer Info to payload of NoC-3
        ELSE
          Transfer Info to first linked Switching Element
      END IF
  /*Stage5: Transfer the Info of the Switching Element that
  could not be transferred in the previous stage) to the
  destination NoC*/
  Transfer Info of Switching Element to the payload of the
  Destination NoC.
      END FOR
    END BEGIN
   END FOR
  END BEGIN
```

Complexity: The time Complexity of the above-mentioned algorithm (i.e. ALGORITHM_FAULT-TOLERANT_HEXA_NETWORK) K) is $O(n^2)$.

2.3 Fault–Tolerant Parallel Algorithm for Dynamic Communication Among NoCs in NiP Using PNN MIN

```
ALGORITHM_FAULT-TOLERANT_PENTA_NETWORK
_____________________________________________________________

Input: n, stores Number of parallel processing NoC's.
Source, a NoC type array that stores the Source NoC
Numbers a part of NoC structure.
Destination, a NoC type array that stores the Destination
NoC Number a part of NoC structure.
Payload, a part of NoC structure that holds the data
generated as Source NoC's.
Output: Payload.
_____________________________________________________________
BEGIN
FOR (I = 0 to 4) DO /*Initialization of Elements of
                      Switching Element structure*/
 Info = NULL
 Info1 = NULL
 Source = -999
 Destination = -999
END FOR
FOR (I = 0 to 3) DO
 Payload = 0 /*Initializing payload of all the 4 NOC's to
```

```
             0*/
 Number = I /*Numbering of all the 4 NOC's from 0 to 3*/
END FOR
 Get the Number of parallel communicating NoC's, n
FOR (l = 0 to n) DO
 Get the Source and Destination NoC's S[l], D[l]
 Get the respective payloads
END FOR
FOR(x = 0 to n) DO
BEGIN
/*Stage1: Transferring the "Payload" values from the
NoC's to the respective "Info" values of Switching
Element structure*/
FOR (y = 0 to n-1) DO
 Transfer the payloads of NoC's to the Info of next
 immediate Switching Elements
 Transfer the Source NoC Number to the Source of
 Switching Element
 Transfer the Destination Number to the Switching Element
END FOR
/*Stage2: If the communicating NoC's are on the same side
checking the source and respective destination Number of
the Switching Element and transferring it to the next
Switching Element having empty "Info" value*/
FOR(t = 0 to 2)DO
Check Source and respective Destination pair DO
 Check Info of first linked Switching Element DO
  IF Info = NULL
   Transfer Info into this Switching Element
   Transfer the Source and Destination Number to this
   Switching Element
  END IF
/*If "Info" of first linked Switching Element is not
empty then transfer the packet to the "Info" of Second
linked Switching Element*/
ELSE
 Check Info of Second linked Switching Element DO
  IF Info = NULL
   Transfer Info into Switching Element
   Transfer the Source and Destination Number to the
   Switching Element
  END IF
/*Stage2: If communicating NoC's are on the opposite side
and Checking the source and respective destination Number
of the Switching Element and transferring it to Switching
Element having empty "Info" value*/
Check Source and respective Destination pair DO
Check Info of first linked Switching Element DO
```

```
 IF Info = NULL
  Transfer Info into this Switching Element
  Transfer the Source and Destination Number to this
  Switching Element
 END IF
/*If the "Info" of first linked Switching Element is not
empty then transfer of packet to "Info" of Second linked
Switching Element*/
ELSE
 Check Info of Second linked Switching Element DO
  IF Info = NULL
   Transfer Info into Switching Element
   Transfer the Source and Destination Number to the
   Switching Element
  END IF
/*If the "Info" of first and Second linked Switching
Elements are not empty then transfer "Info" to third
linked Switching Element*/
ELSE
 Check Info of third linked Switching Element DO
  IF Info = NULL
   Transfer Info into Switching Element
   Transfer the Source and Destination Number to the
   Switching Element
  END IF
/*Stage3: If the linked NoC is the destination i.e. NoC-0
for SE[0], NoC-1 for SE[1] NoC-3 for SE[2], NoC-4 for
SE[3] then transfer the "Info" of the Switching Element
to their respective linked NoC*/
IF Destination of SE[0]=0
   Transfer Info to payload of NoC-0
/*If the linked NoC is not the destination then transfer
the "Info" of Switching Element(SE) to the next empty
Switching Element. If not then destroy the packet*/
ELSE
 Transfer Info to first linked Switching Element
END IF
   IF Destination of SE[1]=1
    Transfer Info to payload of NoC-1
     ELSE
      Transfer Info to first linked Switching Element
   END IF
   IF Destination of SE[3]=2
    Transfer Info to payload of NoC-2
     ELSE
      Transfer Info to first linked Switching Element
   END IF
   IF Destination of SE[4]=3
```

```
    Transfer Info to payload of NoC-3
     ELSE
      Transfer Info to first linked Switching Element
   END IF
/*Stage4: Transfer the Info of the Switching Element
(that could not be transferred in the previous stage) to
the destination NoC*/
Transfer Info of Switching Element to the payload of the
Destination NoC.
    END FOR
   END BEGIN
  END FOR
END BEGIN
```

Complexity: The time Complexity of the above-mentioned algorithm (i.e. ALGORITHM_FAULT-TOLERANT_PENTA_NETWORK) is $O(n^2)$.

2.4 Mean Time to Repair–MTTR for Interconnect–on–Chip

In fault–tolerant MINs, it is always expected that the detection of a fault in Switching Elements initiate the repair of the fault, to protect the SEs from the occurrence of a second, extremely harmful, fault. Only the conservative approximation of the MTTF of single fault tolerant MIN assumes the repair of the faults. Let the constant failure rate of individual switches be λ and the constant repair rate is μ.

Now suppose we have a MIN with M switches and N as network size. For a single fault-tolerant network, the Markov chain model is shown in Figure (5). A Markov chain describes at successive times the states of a system. At these times, the system may have changed from the state it was in the moment before to another or stayed in the same state. The changes of state are called transitions. The Markov property means the system is memory less, i.e. it does not "remember" the states it was in before, just "knows" its present state, and hence bases its "decision" to which future state it will transit purely on the present, not considering the past. Here the Markov chain model is represented with three conservative states: State A represents the nofault state; State B represents the single-fault state, while State C is the two-fault state. The IoC network can tolerate more than one fault. Now it is assumed that if the MIN reaches State C, it has failed. Since the schemes presented in this paper can tolerate more than one faulty switch in many cases, this model should give a lower bound for the mean time to failure

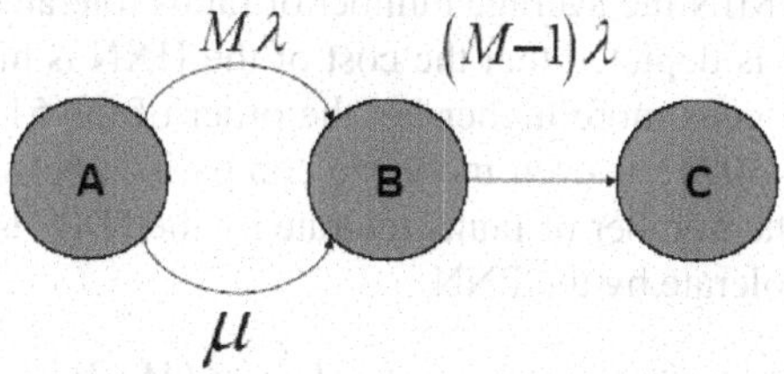

Fig. 5. A Markov Reliability Model for a MIN with repair

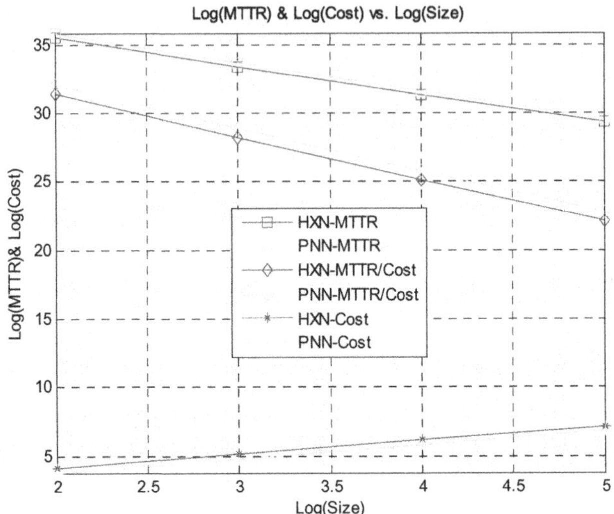

Fig. 6. MTTF under Repair of PNN and HXN along with the MTTF Lower bounds

Table 1. Values of MTTF under Repair of Comparative Networks

MINs Attributes	MTTR and Cost vs. Size			
	N=4	N=8	N=16	N=32
HXN–MTTR	50004500000	11365795450	2718451087	665418883
PNN–MTTR	62505000000	13891250000	3290625000	801850961.6
HXN–MTTR/Cost	2778027778	315716540.3	37756265.1	4620964.465
PNN–MTTR/Cost	4464642857	496116071.4	58761160.71	7159383.586
HXN–Cost	18	36	72	144
PNN–Cost	14	28	56	112

(MTTF) of the system [10, 11, 12]. See Equation (1) used for the calculation of MTTF with repair for the HXN and PNN MINs.

Let $\lambda = 10^{-6}$ / *hour* and (which are typical values used for the calculation). In Figure (6) the MTTR for HXN and PNN is examine for sizes from 4 x 4 to 32 x 32 and values are tabulated in Table (1). From graph, it seems that with increase in the size of the MIN, the MTTR improvement factor is actually decreasing. This is due to the conservative assumption that exactly one fault is successfully tolerated. In reality, with increasing size of the MIN the average number of faults tolerated increases. Moreover, from the Figure (6) it is depicted that the cost of the HXN is higher in comparison to the PNN. As the size goes more higher i.e. the order of the 512 x 512, 1024 x 1024, 2048 x 2048, 4096 x 4096, or even more the comparison between the cost increases more. Nevertheless, the number of faults tolerate by the HXN is higher in comparison to the number faults tolerate by the PNN.

$$\Phi_{MTTF\ with\ repair} = \frac{1}{(M-1)\lambda} + \frac{(M-1)\lambda + \mu}{(M-1)N\lambda^2} \quad (1)$$

3 Conclusions

In this chapter, we propose a new method, which merges two communications, local communication between the IP cores in a NoC, and global communication between NoCs in NiP. In addition to this, we propose two $O(n_2)$time faulttolerant Parallel Algorithms for setting of Inter NoC Communication in NiP. The proposed algorithm allows different NoCs in NiP to communicate in parallel and can tolerate faults. All communicating modules communicate together using either fault–tolerant Hexa multi–stage interconnection network or fault tolerant Penta multi–stage interconnection network. Both the HXN and PNN MINs are single fault-tolerant. The MTTR comparison of the MINs suggested that the HXN has the highest cost in comparison to the PNN MIN but the MTTR values of the HXN are low in comparison to the PNN, which signifies that the online repairing of the HXM MIN if fast as compared to the PNN MIN.

4 Future Scope

The work presented here can be applied on any kind of on-chip interconnects topology in IoC. In addition to this, we can find out the controllability and absorbability for each NoC as well as IoC. Moreover, one can design a condensed compartmental network for stochastic communication in NiP. The method for stochastic modeling is very useful to calculate the latency only if; we use the inflow and outflow in a NoC in NiP architecture. Furthermore, we can use this work to design a state observer, which can predict the total expected delay. This state observer can help us to synchronize the global communication by setting the frequency of NiP clock as function of stochastic communication.

Acknowledgments. The authors would like to thank the editor and the anonymous reviewers for their excellent comments and review reports. Independently me, eternally grateful to my Mom, Dad, Brother, Wife and Sweet Daughter for all of the love and support they have given me over the years. Without their nourishment, I never would have had the chance to succeed. This research work is dedicated to my Shri Shri 1008 Swami Shri Paramhans Dayal Sacchidanand Maharaja Ji and the loving memories of my departed Maternal Grandfather and Grandmother, who continues to guide me in spirit.

References

1. Kangmin, L., Se-Joong, L., Donghyun, K., Kwanho, K., Gawon, K., Joungho, K., Hoi-Jun, Y.: Networks-on-chip and Networks-in-Package for High-Performance SoC Platforms. In: 1st IEEE Asian-Solid State Circuits Conference, pp. 485–488 (2005)
2. Siebenborn, A., Bringmann, O., Rosenstiel, W.: Communication Analysis for Network-on-Chip design. In: Proceedings of the International Conference on Parallel Computing in Electrical Engineering
3. Benini, L., De Micheli, G.: Networks on Chips: A new SoC Paradigm. IEEE Computer, 70–78 (2002)

4. Murali, S., Micheli, G.: SUNMAP: A Tool for Automatic Topology Selection and Generation for NoCs. In: IEEE DAC, San Diego, California, USA, pp. 914–919 (2004)
5. Bjerregaard, T., Mahadevan, S.: A Survey of Research and Practices of Network-on-Chip. ACM Computing Surveys 38(1), 1–51 (2006)
6. Lawrie, D.H.: Access and alignment of data in an array processor. IEEE Transactions on Computers C-24, 1145–1155 (1975)
7. Patel, J.H.: Performance of processor-memory interconnection for multiprocessors. IEEE Transactions on Computers 30, 771–780 (1981)
8. Feng, T.Y.: A survey of Interconnection Networks. Computer 14, 12–27 (1981)
9. Shen, J.P.: Fault-tolerance Analysis of Several Interconnection Networks. In: Proceedings of International Conference on Parallel Processing, pp. 102–112
10. Duato, J., Yalamanchili, S., Ni, L.: Interconnection Networks: An Engineering Approach. IEEE Press, Los Alamitos (1997)
11. Dally, W., Towles, B.: Principles and Practices of Interconnection Networks. Morgan Kaufmann, San Francisco (2004)
12. Nitin: Component level reliability analysis of fault-tolerant hybrid MINs. WSEAS Transactions on Computers 9(5), 1851–1859 (2006)
13. Nitin, S.V.K., Bansal, P.K.: On MTTF Analysis of a Fault-tolerant Hybrid MINs. WSEAS Transactions on Computer Research 2(2), 130–138 (2007)
14. Nitin, S.V.K., Sharma, N., Krishna, K., Bhatia, A.: Path–length and routing–tag algorithm for hybrid irregular multi-stage interconnection networks, pp. 652–657. IEEE Computer Society Press, Los Alamitos (2007)

Author Index